9th International Scientific Conference on Building Defects (Building Defects 2017)

MATEC Web of Conferences Volume 146 (2018)

Ceske Budejovice, Czech Republic
23 - 24 November 2017

Editor:

Ingrid Juhasova Senitkova

ISBN: 978-1-5108-5613-4

Printed from e-media with permission by:

Curran Associates, Inc.
57 Morehouse Lane
Red Hook, NY 12571

Some format issues inherent in the e-media version may also appear in this print version.

Printed by Curran Associates, Inc. (2018)

For additional information, please contact EDP Sciences – Web of Conferences at the address below.

EDP Sciences – Web of Conferences
17, Avenue du Hoggar
Parc d'Activité de Courtabœuf
BP 112
F-91944 Les Ulis Cedex A
France

Phone: +33 (0) 1 69 18 75 75
Fax: +33 (0) 1 69 28 84 91

contact-edps@webofconferences.org

Additional copies of this publication are available from:

Curran Associates, Inc.
57 Morehouse Lane
Red Hook, NY 12571 USA
Phone: 845-758-0400
Fax: 845-758-2633
Email: curran@proceedings.com
Web: www.proceedings.com

TABLE OF CONTENTS

Preface

We are pleased to welcome you to the conference proceedings of the ninth International Conference on Building Defects (Building Defects 2017) held on 23rd – 24th November 2017 at the Institute of Technology and Business (ITB) in České Budějovice, Czech Republic. Conference is organized on behalf of His Magnificence prof. Marek Vochozka, MSc., MBA., Ph.D., rector of the Institute of Technology and Business in České Budějovice and under the honorary patronage of the Czech Chamber of Certified Engineers and Technicians (CKAIT).

The current conference is the ninth continuous of Building Defects conferences, which solves especially the causes of defects during the building processes and afterwards defects of the building themselves as well as the defect of building structures. The conference is the premiere chance for the presentation of new advances and research results in the fields of civil engineering. The special section of the 9th International Scientific Conference Building Defects is focused on Building Information Modeling (BIM) in design, construction and operations of buildings. The main aim of the conference is to the present the scientific research results and its practical solutions in the field of civil engineering, taking into account the principles of sustainable development. Contemporary civil engineering issues and its future development. The progressive technologies and built environment disciplines are presented through the development of new and innovative BIM tools and techniques. It is our pleasure to announce the participation of expert speakers from various universities and research institutes in this two-day event.

A lot of thanks are given to all authors of submitted papers for their participation. They contributed a great deal of effort and creativity to produce this work. We are happy they choose the 9th International Conference on Building Defects to present to their excellent work. Thanks also to all editors and reviewers, who donated substantial time from their busy schedules to carefully read and evaluate the submitted scientific papers.

The conference proceedings contains 44 papers, which are divided into 3 sections: Building Information Modeling (BIM), Mechanical & Materials Engineering and Building Design & Technology. Each contributed paper was rigorously reviewed by 2 anonymous reviewers who were drawn from the International Scientific Committee.

At last but not at least, we would like to appreciate to all supporting organizations for their support and encouragement and got making the event a success. In addition, I would like to thank also to Organizing Committee for their great preparing work.

We are looking forward to the anniversary 10th International Conference Building Defects 2018 that will be held next year on November at the same location. We hope that it will be interesting and enjoying at least as all of its ninth predecessors.

prof. Ingrid Juhásová Šenitková, MSc., CSc.
Conference Guarantee
ITB in České Budějovice

edp sciences

MATEC Web of Conferences

Statement of Peer review

In submitting conference proceedings to MATEC Web of Conferences, I certify to the Publisher that I adhere to the **Policy on Publishing Integrity** of the journal in order to safeguard good scientific practice in publishing.

1. All articles have been subjected to peer review administered by the proceedings editors.
2. Reviews have been conducted by expert referees, who have been requested to provide unbiased and constructive comments aimed, whenever possible, at improving the work.
3. Proceedings editors have taken all reasonable steps to ensure the quality of the materials they publish and their decision to accept or reject a paper for publication has been based only on the merits of the work and the relevance to the journal.

Title, date and place of the conference

9th International Scientific Conference Building Defects (Building Defects 2017)
23rd - 24th November 2017

The Institute of Technology and Business in České Budějovice
České Budějovice, Czech Republic

Proceedings editor(s):

prof. Ingrid JUHÁSOVÁ ŠENITKOVÁ, MSc., CSc.

Date and editor's signature

Implementation of BIM-technologies in the educational program of the architectural university

Alexandr Krivonogov [1,*], *Galina* Zakharova[2], *Sergey* Kruglikov[3] , and *Sergey* Plotnikov[3]

[1] URFU, Institute of Civil Engineering, 620002 Mira str., 19, Ekaterinburg, Russia
[2] USUAA, Department of Applied Informatics, 620075, K. Libknekhta str., 23, Ekaterinburg, Russia
[3] URFU, Graduate School of Economics and Management, 620002 Mira str., 19, Ekaterinburg, Russia

Abstract. BIM-technologies are the dynamically developing trend in a world construction practice. In conditions of a lack of qualified specialists and lack of appropriate educational standards, we offer our practice-oriented methodology for teaching BIM (Building Information Modeling), which is based on the use of new information technologies. Also some projects implemented in BIM and GreenBIM technologies at the Applied Informatics Department are presented.

1 Introduction

At present, the construction industry is developing very rapidly. In the context of the need for designing in a short time, increasing volumes of project documentation, calculating performance characteristics, interaction of a large number of participants in the life cycle of a construction object, the integrated information model of a building becomes an integral attribute. The BIM-model as an intellectual source of huge amount of interrelated information about the construction object and BIM-technology as a basis for managing this information [1], as well as a platform for interrelations of all specialists in design, construction, operation, reconstruction and demolition processes are confidently entering into the practice of innovative companies in the world of design and construction industry.

Taking into account the experience of such countries as Great Britain, the USA, Singapore, Australia, etc., Russia has become involved in the process of introducing BIM-technologies into design and construction practice. BIM-technologies are supported at the state level: in 2014 the Ministry of Construction of the Russian Federation issued the decree "On the approval of the plan for the gradual introduction of information modeling technologies in the field of industrial and civil construction". In 2017, the "roadmap" was approved for the implementation of information modeling technologies for buildings at all stages of the life cycle.

The approved document reflects such an important aspect as standardization and provides the development of BIM national standards for the design, construction, operation and demolition phases of buildings, as well as the alignment of normative and technical

* Corresponding author: kai5407@gmail.com

documents and estimated standards used in construction in accordance with the classifier of construction resources.

In conditions of the large deficit of qualified personnel, who possess new skills and is able to implement BIM-technologies, the government has an understanding of the need to train specialists to use the information modeling in the field of industrial and civil construction. We need educational standards and new educational programs that meet new requirements.

The system of education, being conservative (or better to say resistant to change), cannot react quickly to such challenges, systematically BIM-technologies in universities are practically not taught. Nevertheless, it seems to us important to find ways to introduce innovative technologies in the educational process for the formation of specialists with new corresponding skills.

In this year, 2017, we have organized and are launching a new educational program in the Urals Federal University "Artificial Intelligence in the Capital Management", where we will teach the newest technologies, including those related to construction projects. Today, investments in new technologies in construction are incredibly small; it is here that the opportunities for a huge economic effect are laid.

Further, taking into account our experience since 2009, when we firstly turned to the BIM, we will refer to the methodology of teaching new technologies and some degree projects carried out under our direction by students of the direction "Applied Informatics in the Architecture" at the Ural State Architectural and Art University.

2 BIM as a new technology platform

2.1 New information technologies in the construction

Modern information technologies supported by the electronic industry, software, telecommunications, make up the basis of the current techno-economic paradigm. Its scientific basis is achievements in the field of artificial intelligence and interdisciplinary approaches.

New technologies have an impact on all sectors of the economy. World technological trends with great speed are included in all branches of human activity, including in architectural and construction practice. The technology of building information modeling is associated with these innovations and develops in this direction. Let us briefly characterize these innovative technologies.

Cloud technologies. With their application the new concept of Connected BIM was appeared, when all processes, people and technologies are interconnected in the information environment. This technology allows you to manage quickly and efficiently the entire life cycle of the object.

Technologies "smart house" and "smart city". They are also based on cloud services and on the new technology *Internet of things (IoT)* when all objects (things) we deal with are connected. IoT allows you to organize an independent interconnection of various objects by means of sensors and wireless networks and monitor and control all parameters, for example, in a room to maintain a comfortable environment and to solve many other tasks.

Virtual and augmented reality (VR/AR). 3D visualization of construction objects is carried out in interactive graphics in the real time with immersing a user in a virtual environment with the help of special equipment – glasses or helmets. You can view a building from the outside and inside, have an additional information in the form of the infographic and interact with the environment.

3D scanning of the terrain, the combination of data with the GIS system and the formation of a 3D object model in reference to the terrain. The use of *unmanned aerial vehicles* to obtain current video information about the current state of construction.

3D printing of buildings, small architectural forms and interior objects. There are examples of printed houses in China, Europe, also in Russia.

Robots programmed for the construction of various structures, for example, "growing" bridges.

Artificial Intelligence. It has unlimited possibilities for solving optimization problems based on machine learning and neural network models and algorithms.

2.2 BIM evolution

According to Autodesk, one of the leaders in the software market of BIM, we currently live in the Era of connection (Connected BIM) – these are related BIM processes of information transmission throughout the life cycle of the object. The predecessors are the Era of documentation (AutoCad) and the Era of optimization (Building Information Modeling).

There are also 3 levels of BIM: from the zero level 2D-design, to the first level BIM – 3D-models and 2D-drawings. The second level additionally involves the exchange of information in different formats. The third level is an integrated information model of an object in which one change is made for all participants. The latter level is not implemented in any country in the world – for example, in the UK, for the time being implement BIM technology level 2.

The integrated building information model contains a comprehensive description of the facility, ensures the coordinated work of all groups of participants in the development and implementation of the project. The technology allows you to visualize and calculate the future object with the selected materials, have several options with a cost estimate, optimize the constructive part, clean up collisions of engineering equipment and architectural solutions, extract information for compiling a cost estimate for construction, a model for further operation of the facility, which ultimately raises efficiency of the project at all its stages.

Thus, we see that the information model is a multidimensional concept. If it is structured, as for example, in the work on recommendations for the implementation of BIM projects (Pennsylvania University) [2], then a lot of BIM scenarios or as they write BIM Uses are highlighted, these are components of BIM technologies tied to the planning, design, construction and operation stages. There are 25 scenarios at this resource, some of them are: modeling of existing conditions, cost estimation, design, construction analysis, light analysis, energy analysis, mechanical analysis, LEED compliance assessment, compliance, service planning, building system analysis, planning in case of emergency.

To date, this is the mostly complete and up-to-date view on the BIM.

3 BIM in the educational program

3.1 Practical-oriented methods of the teaching

The Department of the Applied Informatics delivers interdisciplinary specialists in the field "The Applied Informatics" with the profile "The Applied Informatics in the Architecture". We teach students IT-technologies, including relevant sections in such disciplines as "Computer-aided design (CAD)", "The Architectural design", "The Urban ecology", "Environmental factors in the architecture", and many others. We have also developed the system that allows us to comprehensively teach the principles of environmental design and building information modeling through the 6 following practice-oriented methods. Some additional information is published in [3, 4].

1. The introduction to the CAD course of modern software tools in the field of BIM and Green BIM such as Autodesk Revit, Graphisoft Archicad. In the Navisworks program, students are introduced to the principles of 4D and 5D-modeling. Autodesk 3DS Max and AutoCAD packages are also used in the learning process. Students are able to import 3D models into the Unity 3D and Unreal Engine software environment to organize interactive project management in the real time for VR. To work with the relief and design of infrastructure objects, AutoCAD Civil 3D is used. The development of project concepts is carried out using the InfraWorks 360 platform. Data bases such as Autodesk's Vault system and CSoft's TDMS system for the cataloging and integration of construction objects and for collective project development were mastered.

2. Discussions of the best world practices in the field of environmental design by the examples of winners of Pritzker and other awards. For interactive lessons students prepare for discussions the topics including the best world examples of «green» architecture [5].

3. Excursions to the objects certified according to the «green» standards in our city and studying the features of the facilities in other regions. We have visited the cottage village Ekodolia near Ekaterinburg to study the experience of building the house of class A+ [6].

4. The cooperation with organizations implementing building automation systems "smart home", the certification systems in according with «green» standards, with experts on energy-efficient housing construction. We regularly invite experts from leading enterprises in the field of energy-efficient housing construction to meet with students, such as the Institute of the Passive House (Moscow), the companies "Teplokrepost" and "D-systems" (Ekaterinburg), the "Bureau of Engineering" (St.-Petersburg) and many others.

5. The organization of conferences and roundtables with the participation of students and our graduates, the participation in scientific conferences on the introduction of advanced technologies in the urban environment. We also organize the participation of students in scientific and practical seminars organized by companies that produce new materials and technologies ("Knauf", "Izover", "Teplit"), in major international exhibitions and conferences in Ekaterinburg ("Innoprom", "Forum 100+", "City Spot").

6. The participation of students in international and Russian competitions. During several years, students of the department take part in the international competition of the Saint-Gobain company for the design of multi-comfort buildings in different countries. In the Autodesk international competition our student Alexander Ivlev in 2012 received the main prize in the nomination "Give the shape to the future" with the project of an eco-house controlled by the automatic system "Smart House".

3.2 Examples of diploma projects using Green BIM technologies

This section includes graduation projects using BIM and Green BIM. These are works related to real objects – industrial, public and residential buildings.

Fig. 1 shows a phased design of innovation Techno Park for manufacture of modular products. It was developed an effective parameterized digital Revit model, parameters determine the behaviour of each element. The result of this technology is significantly increased the design quality and production time of the project.

In the work on the Fig. 2 possibilities and advantages of using BIM technologies in the reconstruction of architectural objects are studied using the example of the residential group in Madrid. This project was presented at the international competition of Saint-Gobain company. BIM is important both for the restoration of historical architecture objects and for planning the reconstruction of objects.

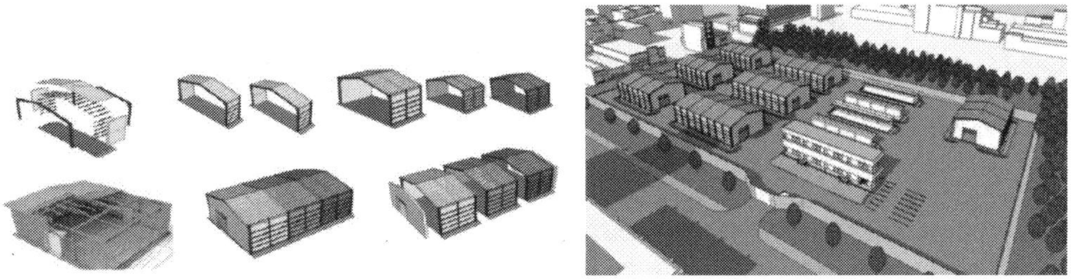

Fig. 1. Stages of designing of innovative Techno Park.

BIM model has all the necessary information. The use of BIM technologies in the reconstruction of buildings is a fairly new and unexplored direction. BIM also allows for designing a reconstruction taking into account new standards for environmental and energy-saving requirements.

Fig 2. The reconstruction of the residential group in Madrid.

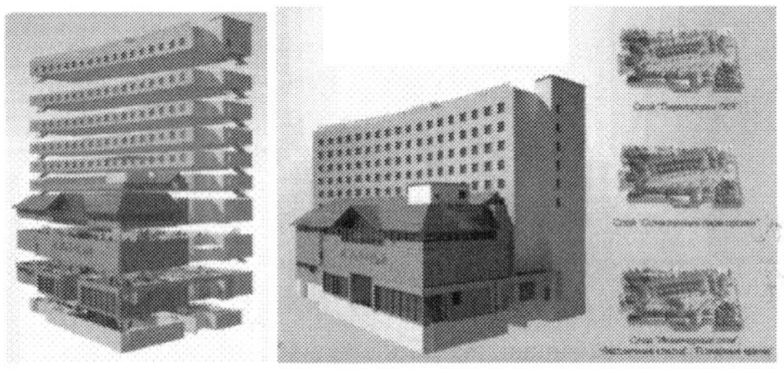

Fig 3. The layer digital model of the trade centre.

On the fig. 3 is the digital model of the trade centre in Ekaterinburg, developed for the Ministry of Emergency Situations to educate members of the organization to rescue operations in case of emergencies. The model allows to disassemble the building in layers and to show all communications.

In the thesis work "The use of energy-efficient standard Green Zoom for design of residential buildings" (fig. 4) the conceptual design of the high-rise apartment building was developed, its evaluation with the Russian system of energy efficiency and sustainability Green Zoom for civil and industrial construction projects has been made.

Fig. 4. Visualization of the "green" apartment building.

The strategic goal of the Green Zoom standard is to create a favourable living environment with affordable and comfortable housing. The evaluation of buildings according the Green Zoom system is carried out at 9 divisions: architectural and planning solutions, the project location and the organization of transport provision, environmental sustainability territory, water efficiency, energy efficiency, reduction of harmful emissions into the atmosphere, environmentally rational choice of building materials and waste control, the ecology of the internal environment of buildings, innovation in design and regional characteristics.

In the Fig. 5 the futuristic project of an individual apartment house, designed to meet the requirements of environmental friendliness and energy efficiency, designed to accommodate the automation system "Smart House" is shown.

Fig. 5. Visualization of the "green" futuristic smart house.

4 Conclusion

Thus, in the article the effective modern technologies in construction industry are presented. The digital modeling of buildings (BIM-technology) and the results of its application in real projects are described.

For more extensive use of the BIM-technology it is necessary to develop educational standards for universities. The department of Applied Informatics in the Ural State University of Architecture and Arts prepares a multi-disciplinary specialists in information technologies in the architecture, and teaches students BIM-technologies. We developed new educational program "Artificial Intelligence in Capital Management" in the Federal University. We also have developed and shortly described here the system that allows us to comprehensively teach the principles of environmental design and information modeling through the practice-oriented methods.

References

1. V. Talapov. *BIM Fundamentals: Introduction to Building Information Modeling* (Moscow: DMK, 2011)
2. BIM Project Execution Planning Guide. Available on http://bim.psu.edu/
3. G. Zakharova, A. Krivonogov. Building Information Modeling on the basis of "green" standards for the purpose of sustainable development of territories. Izvestiya Vuzov. Physics (№ 8/2, 37-40, 2016).

4. G. Zakharova, A. Krivonogov. Necessity of teaching Green BIM technologies in the higher school of the 21st century. II International Conference "Safety Problems of Civil Engineering Critical Infrastructures" (Ekaterinburg, Russia, 116-120, 2016)
5. E. Krygiel, B. Nies. *Green BIM: Successful Sustainable Design with Building Information Modeling* (Wiley Publishing, 2008, 268 p.).
6. W. Feist. *Gestaltungsgrundlagen Passivhäuser* (Verlag Das Beispiel, 2000,143 p.) Russian translation: W.Feist. *The main provisions on the design of passive houses* (Moscow: ACB, 2011, 144 p.).

New approach to decision making by multiple criteria analysis

Tomas Vanicek[1,*], and *Jana* Kucerova[1]

[1]Czech Technical University, Faculty of Civil Engineering, Prague, Czech Republic

Abstract. Many decision processes in technical and economical sciences require multiple criteria decision making. The most widely applied methods for multiple criteria evaluation of alternatives are based on the evaluation of alternatives in terms of an additive preference function. All of them require the estimation of weights of usually conflicting criteria. There are several methods how to find the weights of the criteria and how to find the evaluation of each solution in each criterion. The decision process based on simple weighted sum of values may not be the best approach in all situations. This paper contains a new approach of the evaluation of measured value set by different mathematical operators than the usually used multiple criteria evaluation methods. The approach was applied in a case study for multiple criteria evaluation. Generally, this new decision-support tool can help in various situations where different types of effects caused by a construction or reconstruction can occur. This is a very frequent situation in dealing with building defects, too.

1 Averaging operators

In different situations, various aggregating operators can be used to obtain a value we call an average.

Let R be the set of all real numbers and m the number of measurement of the excavated value. Then, the general averaging operator can be defined as a mapping Avg from R^m into R, satisfying the following conditions:

1. A is a continuous mapping of R^m into R.
2. A is monotonic in each of its m coordinates, that is: if $p_j \geq q_j$ for each $j=1,\dots,m$, then $A(p_1,p_2,\dots,p_m) \geq A(q_1,q_2,\dots,q_m)$.
3. A is idempotent, that is: $A(p, p, \dots, p) = p$ for any real number p.
4. A is internal, that is: $min(p_1,p_2,\dots,p_m) \leq A(p_1,p_2,\dots,p_m) \leq max(p_1,p_2,\dots,p_m)$

Special types of averaging operators called quasiarithmetic means are most frequently used. These operators can be expressed by an arithmetic formula.

Let $a_1,a_2,\dots,a_m$ be real numbers representing some measured values.
Such an α-quasiarithmetic mean is defined for positive values $a_1,a_2,\dots,a_m$ by the following formula:

* Corresponding author: vanicek@fsv.cvut.cz

$$Avg_\alpha(a_1,a_2,...,a_m) = \left(\frac{1}{m}\sum_{j=1}^{m} a_j^\alpha\right)^{1/\alpha} \tag{1}$$

For $\alpha=1$, we obtain the commonly known arithmetic mean, for $\alpha=2$, the harmonic mean.

In the limit $\alpha \rightarrow 0$, we obtain a geometric mean, for $\alpha \rightarrow \infty$ we obtain maximum and for $\alpha \rightarrow -\infty$ we obtain the minimum.

1.1 Level of Consensus

Of course, the measured values are not the only important piece of information for decision making. Information about the level of the consensus among the values can also be important.

Consensus level concept can be introduced for cases when the answer is *YES/NO* or for cases when the answer uses a discrete scale.

Consensus is a value from the interval $[0,1]$ that determines the level of agreement of the measured values.

Consensus is equal to 1 when all the values are same.

Consensus is equal to 0 when half of the values is *YES* and the other half have is *NO*.

Consensus can be viewed as a truth value of the proposition, "The measurements proof this situation".

This concept of consensus can be supplemented to express an agreement of the measurements results with a specified target value τ as follows:

Let X be a discrete random variable of size n with probability distribution $p(X)$ on the previously given closed interval $[X_{min},X_{max}]$ so that $p_i = P(X_i)$ for $i=1,...,n$, and let $d_X = X_{max} - X_{min}$ be the width of X. Then, the τ-agreement $Agr(X|\tau)$ of the distribution is a number defined by the following formula:

$$Agr(X|\tau) = 1 + \frac{1}{n}\sum_{i=1}^{n} p_i.\log_2(1 - \frac{|x_i - \tau|}{2d_x}) \tag{2}$$

Further generalization can be made to abandon the restriction that implies that the values express only a choice of some milestone in the discrete scale. Commonly the results of measurements can be described as a choice of any real number on the closed interval $[X_{min}, X_{max}]$. Then, the formula for τ -agreement changes to

$$Agr(X|\tau) = 1 + \frac{1}{N}\sum_{i=1}^{N} p_i.\log_2(1 - \frac{|x_i - \tau|}{2d_x}) \tag{3}$$

where N is the number of the measurements. The main difference between these formulas is that in the first formula n is the number of levels for the discrete scale (for the Likert scale, usually $n=5$, or $n=9$), and p_i is the number of measurements according to the corresponding level, while in the second formula, N is the number of measurements, and the value of each of them is evaluated separately.

Regardless of the choice of different discrete scales or different intervals for attribute value estimation, the value of τ -agreement is a number from the closed interval $[0,1]$. The value of τ -agreement equals 1 when all the measured values match value τ .

The value of τ -agreement is *0* when the values from all measurements match one of the boundary points of interval the $[X_{min}, X_{max}]$, while τ is the opposite boundary. $Agr(X| \tau)$ can be considered a fuzzy truth value of the proposition "The measurements proof that the measured value has the quantity of τ ".

1.2 Consensus optimization

Let us look to a value $Agr(X| \tau)$ as to a function of an argument τ and the set of values be an closed interval $[0,1]$. The range of the argument τ can be also considered as a closed interval $[\tau_{min}, \tau_{max}]$ where τ_{min} and τ_{max} are such values that the measured quantities are clearly between them. If we do not have any information about the foundations of measured values, we can put τ_{min} to be the lowest measured value and τ_{max} to be the highest.

The function $Cons_X(\tau) = Agr(X| \tau)$ is continuous because it is defined as a combination of continuous elementary functions. So it is a continuous mapping from a closed interval $[\tau_{min}, \tau_{max}]$ into a continuous interval $[0,1]$. Such a function must reach a maximum. This maximum is depended only on the set of measurements X. We can denote the value of τ for which this maximum is reached as $Mavg(X)$. This value $Mavg(X)$ holds four conditions for general averaging operator:

1. A is a continuous mapping of R^m into R. Clearly $Mavg(X)$ is a mapping of R^m into R. And it is continuous because it was formed as a combination of continuous elementary functions.
2. A is clearly monotonic in each of its m coordinates, that is: if $p_j \geq q_j$ for each $j=1,...,m$, then $A(p_1,p_2,...,p_m) \geq A(q_1,q_2,...q_m)$.
3. A is idempotent, that is: $A(p, p, ..., p) = p$ for all real numbers p. For the situation of an idempotent measurement the value $Agr(X| \tau)$ is equal to 0 for τ not equal to p and it is equal to *1* for τ equal to p. So the maximum of this function is reached in the point $\tau=p$ and it is equal to *1*. The value $Mavg(X)$ is equal to p.
4. A is internal, that is: $min(p_1,p_2,...,p_m) \leq A(p_1,p_2,...p_m) \leq max(p_1,p_2,...,p_m)$. Clearly the maximum of the function $Cons_X(\tau)$ is reached between the points $min(p_1,p_2,...,p_m)$ and $max(p_1,p_2,...,p_m)$. So the last condition is fulfilled.

The value $Mavg(X)$ can be used as other way how to compute an average of measured or estimated values. One big disadvantage of such averaging is the fact that it is very complicated to express such a value especially in the situation when the basic set X is extremely large. Unfortunately this is usually the situation of technical measurements. Much more useful can be this way of averaging in situations when the dimensions of set X are quite small. This is usually the situation of expert estimate of values used for example for multicriterial evaluation and multicriterial optimization. In such a situation the number of estimated values is not large, usually tens maximally hunderts of expert estimates. The *Mavg* can be a very useful way to deal with such sets of estimations.

2 Case Study

In the following text we introduce a small case study which shows the benefits of the above defined average operator *Mavg*. The input data describe the lifetime of concrete construction elements. Data obtained by classical computation of an average value from the expert estimations are given in the following table:

Table 1. Construction element.

Construction element	**Lifetime (years)**	**Average lifetime (years)**
Concrete Foundation	80-150	100
Concrete Outdoor Walls	60-80	70
Concrete Indoor Walls	100-150	120
Concrete Outdoor Steps	60-80	70
Concrete Roof	80-150	100
Outdoor Walls with Surface Finish	100-150	120
Ventilation Shafts	40-70	60
Window Sills	60-80	70
Concrete Fence	60-80	70
Concrete Sewer Shaft	60-80	70
Concrete Sidewalk	20-30	25

For closer research the distribution of input values schould be known. For the second element – the concrete outdoor walls – the expert estimations were gradually 60,60,65,70,70,70,75,75,75 and 80 years. The arithmetic mean of these values is 70 years. By computing the value $Agr(X|\tau)$ for different values of estimation τ we obtain subsequently:

$$Agr(X|60) = 0,9865,$$

$$Agr(X|65) = 0,9913,$$

$$Agr(X|70) = 0,9939,$$

$$Agr(X|75) = 0,9923,$$

$$Agr(X|80) = 0,9864.$$

Even with closer estimations of agreement value τ we obtain

$$Agr(X|69) = 0,9934,$$

$$Agr(X|71) = 0,9936.$$

So the value $Mavg(X)$ is equal to 70 years (or very close to it) and the estimation obtained by simple arithmetic mean computation agrees with the computation by the new method.

Another situation will occur in the estimation of first construction element – concrete foundation. The expert estimations were gradually 80,80,80,80,80,80,100,120,150 and 150 years. The arithmetic mean of these values is 100 years. Computing the values $Agr(X|\tau)$ we obtain:

$$Agr(X|80) = 0,9918486,$$

$$Agr(X|90) = 0,9918389,$$

$$Agr(X|100) = 0,9914760,$$

$$Agr(X|110) = 0{,}990149,$$

and even smaller values of $Agr(X|\tau)$ for higher values of τ.

More detail exploring leads to values:

$$Agr(X|84) = 0{,}9918887,$$

$$Agr(X|85) = 0{,}9918894,$$

$$Agr(X|86) = 0{,}9918886.$$

So the maximum agreement value $Mavg(X)$ is somewhere near 85 years and this should be the lifetime of concrete foundation estimated by the expert group. Such a computation describes the situation better then the simple arithmetic mean computation with the resulting value 100 years.

References

1. J. Kučerová, J. Kaiser, P. Strnad, T. Vaníček, M. Višňovský, *Data Acquisition and Data Processing in Strategic Planning* (ČVUT, Praha, 2014).
2. L. Horák, J. Kaiser, M. Sůra, L. Truhlář, T. Vaníček, L. Vaníčková, *Artificial Intelligence Methods in Building Industry* (ČVUT, Praha, 2014)
3. J. Čermák, L. Horák, J. Kaiser, M. Sůra, T. Vaníček, L. Vaníčková, *Artificial Intelligence Methods in Building Industry II*, (ČVUT Praha, 2015)
4. E. Kašša, *Životnost staveb* (VUT Brno, 2013).

The role of managing knowledge and information in BIM implementation processes in the Czech Republic

Vladimír Nývlt[1,*]

[1]The Institute of Technology and Business, in České Budějovice, Okružní 517/10, 370 01 České Budějovice, Czech Republic

Abstract. Czech construction sector is at the initial stages of implementing and assessing Building Information Management (BIM) on pilot projects. Object modeling developed over the last 20 years is seemed as a stable ground for many professionals. 3D data models are basis for further concepts associated with BIM, helping to support Co-ordinated Project Information (CPI) and Integrated Project Delivery (IPD). This paper presents the role of managing knowledge, information and critical success factors (CSF) associated with BIM implementation within the construction industry in the Czech Republic. Determining the CSF in the context of BIM maturity levels should support BIM implementation processes within the construction industry and also within the associated bodies (education, government, technology).

1 Introduction

The concept of BIM was addressed as it is perceived today in Eastman's Building Description Systems (BDS) in the 1970s [1] and the term was first used as it is used today in van Nederveen's 1992 paper [2] However, it was not until 2003 that commercial BIM tools became abundant and the industry slowly started adopting BIM in its processes [3]. During the mid- to late 2000s, many firms across the world had to work through technical and organizational issues without clearly knowing the direction in which they were moving. The industry level and organization level of this period are referred to as the early stage of BIM adoption, or the first wave of BIM adoption. During this period, some firms failed to adopt BIM, whereas others succeeded. Nevertheless, a sufficient number of BIM cases succeeded to lead other firms to adopt BIM, and many recommendations were made based on the experience and lessons learned throughout this period [1]. However, a company with limited resources cannot adopt all the recommendations simultaneously.

* Corresponding author: nyvlt@mail.vstecb.cz

2 Understanding Information

Exploring Knowledge systems, we may conclude: objects are "documents" if they meet the following four criteria [4]:
- "materiality (they are physically – including the digital form – present),
- intentionality (they carry meaning),
- development (they are created),
- perception (they are described as a document)."

This definition is exceedingly broad, and the meaning is that anything can be a document, if someone decides consider it as such.

Documents are either available in digital or non-digital form; indeed, it is possible for both documents to coexist side by side. (e.g. real wall, and its digital representation). The complete versions of documents can only be entered into information (retrieval) systems if they are digitally available.

The fact that we can distinguish between structure and data on the syntactical level of digital documents means – at least in theory – that we are always capable of automatically extracting the data contained in a document and to save or interconnect them as individual factual documents. "In principle, any quantity of data carrying information can be understood as a document", as Voss [5] writes. In order for all documents – text documents, data documents, graphical documents or whatever – to be purposefully accessible (to both men and machines), they must be clearly designated. "This is accomplished by the Uniform Resource Identifiers (URI) or the Digital Object Identifiers (DOI)." [6]. Under these conditions, it stands to reason that sets of factual knowledge which complement each other should be brought together even if they originate in different documents or in different systems. This is the basic idea of "Linked Data" [7].

In cases of automatic fact extraction in the context of Linked Data, the original document fades into background, in favor of the received data and their factual documents. It must be noted that the data must be defined in commonly agreed manner (with help of standards like RDF, the Resource Description Framework). Additionally, we absolutely require a Knowledge Organization System (KOS) that contains the needed terms and sets them in relation to one another.

In addition to extracted or received facts there are data files which cover certain topical areas, e.g. economic data, geographic data, production data, database of products, national libraries of objects, real-time data, etc.

3 Knowledge representation

Ingwersen and Järvelin [8] create model for information retrieval and knowledge representation that places the cognitive work of the actor at the foreground (see Figure 1).

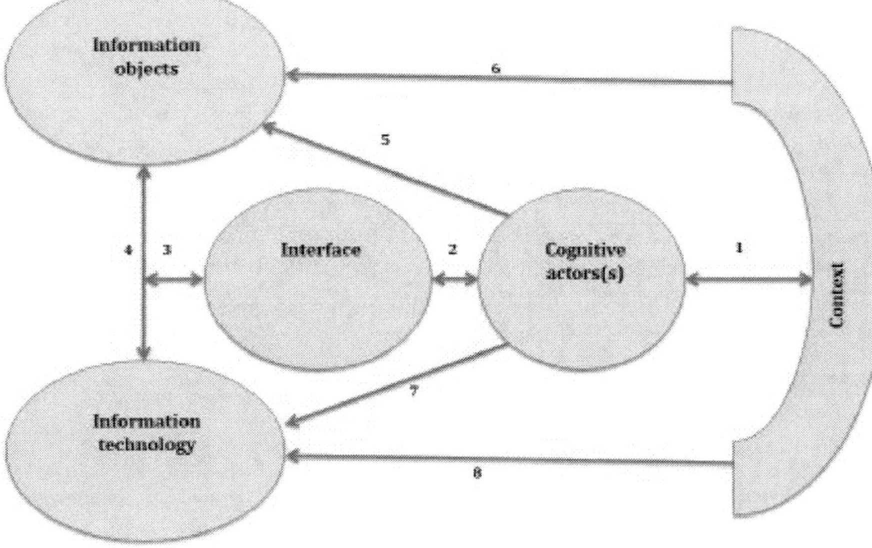

Fig. 1: The Cognitive Actor in Knowledge Representation and Information Retrival. Source: [8], p. 261. Arrows: Cognitive Transformation and Influence Double Arrows: Interactive Communication of Cognitive Structures.

The actor (or a team of actors) is anchored in cultural social and organizational horizons (context – Relation1). Actors are, for instance, authors, architects, designers, system engineers, interface designers, creators of KOSs, users of any origin. Through a man-machine interface (relation 2), the actor comes face to face with the information objects as well as with information technology (relation 3). "Information objects" are the documentary units (surrogates) as well as the documents (where directly digitally available) that are accessible through methods and tools of knowledge representation and that have been indexed by form and content. "Information technology" summarizes the programming-related building blocks of an indexing and retrieval system (database, BIM database, retrieval software, Industry Foundation Classes (IFC), etc.) as well as retrieval models. Since the information objects can only be processed through information technology, there is another close link between these two aspects (relation 4).

The actor always interacts with the system through the interface, but he requires additional knowledge about the characteristics of the information objects (relation 5) as well as the information technology being used (relation 7) additionally, information objects and information technology do not work independently of the social, cultural and organizational backgrounds (relations 6 and 8). As opposed to the interactive relations 1 through 4, relations 5 through 8 feature cognitive influences "in the background", which – as long as an information system (Building Information Model) is "running" – must always be taken into consideration, but cannot be interactively influenced.

Together with (Stock & Stock [8] try to look at "all objects in real world putting into information systems as surrogates of these objects. Then call all these surrogates as documents in information systems, what helps us to unify how to handle with all different objects in different level of granularity." So the indexer (the information system engineer) is an active and important part in the process of maintaining the institutional (BIM) memory. He decides whether a document should be stored or not, and which metadata are used to describe the document. Such decisions are of great importance for the preservation [9] and retrievability of documents. "Thus the information system engineer as indexer is embedded in social context" ([10], p. 13) of an institution that may be represented also as Knowledge Organization System (openBIM).

The object of knowledge representation is – without any restriction – explicit objective knowledge. In the case of subjective knowledge, this knowledge must be externalized in order to make it accessible for further processing. This externalization is clearly limited by implicit knowledge. Explicit objective knowledge is always contained in documents.

In knowledge representation, the documents from which the respective knowledge is to be determined are available. The object at hand is the documents` content. What is "in" a database, a website, a building model, an image, a patent, an architectonic exhibit.

4 Managing knowledge sharing

In information science, knowledge is regarded as something static, which is fixed in a document and stored on a memory. This storage is either digital (such as the World Wide Web), material (as on library shelf) or psychical (like the brain of a company employee). Information, on the other hand, always contains a dynamic element; one informs (active) or is informed (passive). "The production and the use of knowledge are deeply embedded in social and cultural processes; so information science has a strong cultural context" [4].

Burita et al [11] conclude: "When referencing to a knowledge system, we have to realize that knowledge cannot be separated from human beings. Therefore, we have to think in terms of "biological systems", where functions define structures, and to avoid thinking of "physical systems", where structures define functions."

In economics and business administration, information has long time been discussed in the context of entrepreneurial decisions. "Information always proves imperfect, and so becomes a motor for innovative competition" [8]. "With conceptions for learning organizations, and, later knowledge management the subject area of industrial economics and of business administration on the topic of information has broadened significantly from around 1980 onward" [12]. "The objective is to share and safeguard in-company knowledge" [13] and "to integrate external knowledge into an organization" [14]. Knowledge is employed, information signals can be observed more closely from three points of view [8]:
- "in their relations to each other (syntax)
- in the relations between signs and the objects described (semantics)
- in the relations between signs and their user (pragmatics)"

Disciplines of pragmatics and semantics investigate the meaning and usage of signs. Stock & Stock [8] also propose knowledge to be:
- "partly a skill of conceiving of an object as it really is on the one hand, and that of successfully dealing with objects of knowledge on the other,
- partly the epistemic state that a person occupies as a consequence of successfully performing his or her cognitive tasks,
- partly the content that cognitive person refers to when doing so."
- Looking at this description of knowledge, we can see that knowledge falls into two aspects of skill and state:

1. Knowledge as the skill
a. of correctly comprehending an object (TO KNOW THAT)
b. of correctly dealing with an object (TO KNOW HOW)
2. Knowledge as the state

a. of a person that KNOWS (something)
b. that which is known itself, the CONTENT
c. its linguistic expression

Knowledge has two fixed points: the knowing subject and what is known, in which aspects (1a,b and 2a) belong to subjective knowledge and (2b,c) are objective knowledge [17]. Of course, (2c) is of our special interest, because it can only be objective if the linguistic expression is permanently fixed on a physical carrier, a document. Document may be any "surrogate" of real object. In civil engineering industry, what is of our major focus here, it can be, as an example:
-drawing or any unique part of drawing
-description of an building element with static calculations,
-graphs,
-contracts
-etc...

In further text, word "document" has the meaning of a "digital surrogate" in information system of any kind of real object in the real world, including "paper" documents.

According to Brookes [15], p.131), "the connection between knowledge and transmitted information can be expressed through a (pseudo-mathematical) equation. Knowledge (K) is understood as a structure (S) of concepts and statements; the transmitted information (Δ I) carries a small excerpt from the world of knowledge." The equation goes:

$$K[S] + \Delta I = K[S + \Delta S] \tag{1}$$

The knowledge structure of the receiver is modified through Δ I. This effects a change to the structure itself, as signified by the Δ S. The same Δ I, received by different receivers with different K[S] than each other, can effect different structural changes Δ S. The process of information differs "from person to person and from situation to situation" [15]. So the "understanding of a person`s knowledge structure" [16] is essential for information science.

Information activities process knowledge, present it in user-friendly manner, represent it through condensation and the allocation of "information filters", and prepare it for easy and comprehensive search and retrieval. All aspects that go beyond the original knowledge are informational added values, which can be found in private as well as public information sources. In addition to "knowing how" and "knowing that" the informational value added is "knowing about". Information activities lead to knowledge "about" documents and "about" the knowledge fixed in the documents. For Buckland [4] "knowing about" is more concerned with information science than "knowing how" and "knowing that".

5 Conclusion

Managing knowledge and information seems to be the most crucial point in implementation of BIM within any country specific industry. It is obvious that we are challenged by how to transfer basic thinking about processes within building industry. It is not only about BIM in meaning of Model or Modeling, and we need to look even further that to BIM in meaning of Management. There is big potential, supported by technology development, to employ LIM (Land Information Modeling) – geographical data together with BIM-Model and to step towards "Better Information Management". Managing knowledge sharing and to change and/or improve processes within whole industry becomes challenge we are facing just now. It seems to be a way, how to deal all barriers, BIM adoption must overcome.

Of course, the lack of trained personnel remains a significant barrier to BIM adoption, forcing many companies to retrain experienced CAD operators in the new tools. Because BIM requires different ways of thinking about how designs are developed and building construction is managed, retraining requires not only learning but the unlearning of old habits, which is difficult. New graduates, whose entire undergraduate experience was influenced by their familiarity with BIM and its use for the full range of student projects, are likely to have a profound influence on the way companies of all kinds deploy BIM. Inevitably, a good deal of innovation in work practices is to be expected.

References

1. C. Eastman, P. Teicholz, R. Sacks, K. Liston, *BIM Handbook: A Guide to Building Information Modeling for Owners, Managers, Designers, Engineers and Contractors, 2nd Edition* (Wiley, New Jersey, 2011)
2. S.V. Nederveen, R. Beheshti, W. Gielingh, *Handbook of Research on Building Information Modeling and Construction Informatics: Concepts and Technologies*, 1-18 (2009)
3. J. Liaserin, Building Information Modelling - The Great Debate. Available on http://www.Liaserin.com/features/bim/index.php (2008)
4. M.K. Bucklan, Journal of the American Society for Information Science and Technology **63** (1), 1-7 (2012)
5. J. Voos, Libreas. Library Ideas **15**, 13-18 (2009)
6. F. Linde, W.G. Stock, *Information Markets. A Strategic Guideline for I-Commerce* (De Gruyter Saur, Berlin, 2011)
7. C. Bizer, T: Heath, T. Berners-Lee, International Journal of Semantic Web and Information Systems **5**(3), 1 -22 (2009)
8. W.G. Stock, M. Stock, *Handbook of Information Science* (De Gruyter Saur, Berlin, 2013)
9. P.S. Larsen, Journal of the American Society for Information Science **50**(11), 1020 – 1027 (1999)
10. S. Briet, What is Documentation? (Lanham, Scarecrow, 2006)
11. L. Burita, P. Gardavsky, T. Vejlupek, Journal of Systems Integration **3**(1), 19-31 (2012). Available on http://www.si-journal.org
12. I. Nonaka, H. Takeuchi, *The Knowledge-Creating Company. How Japanese Companies Create the Dynamics of Innovation* (Oxford University Press, Oxford, 1995)
13. G.J.B. Probst, S. Raub, K. Romhard, Managing Knowledge. Building Blocks for Success (Wiley, Chichester, 2000)
14. W.G. Stock, I. Peters, L. Weller, Advances in Librarianship **32**, 137-158 (2010)
15. B.C. Brookes, Journal of Information Science **2**(3-4), 125-133 (1980)
16. C. Cool, N.J. Belkin, *Interactive Information Retrieval. History and Background*, 1-14 (Facet, London, 2011)

Detecting, modeling and quantifying differences in terms meaning, in BIM environment

Miroslav Šůra[1,*]

[1]CTU Prague, Faculty of Civil Engineering, Thakurova 7, 166 29 Prague, Czech Republic

Abstract. Professional tasks in BIM involve sharing information among many participants, and require the use of the most precise terms to meet the following conditions. First, the user (reader) must not hesitate over the meaning. Second, various users should understand them alike, or better – in the same way. Less vague expressions provide more reliable basis for decision analysis and decision making, and for a joint, seamless, unquestionable consensus. Formal definitions of terms are used for that purpose where necessary, e.g. for logical or computer processing, but other terms remain or need to remain without such precise formal definitions. The principles for comparing difference in understanding the meaning of a term by different subjects are examined. We specify principles for measuring and calculating difference in, and vagueness of, meaning. Then, variables and scaling methods are proposed to measure the difference in and vagueness of meaning. The presented measuring scales can be applied for example in the process of refinement, unifying or splitting BIM terms, or group consistency in understanding a BIM-oriented terms can be assessed using the concept of vagueness.

1 Introduction

1.1 Vagueness

Vagueness [1] is a common feature of many tasks not only in the area of building, with or without BIM, and the tasks include communication. The vagueness is due to using natural languages. Building tasks as well as many other technical and professional tasks are often starting their existence from formulations in a natural language, too. Later on, however, the analysts strive to express the ideas in such a form that eliminates the vagueness. To this end, formal languages are used.

Some disciplines, like mathematics and physics, are completely relying on formal languages. Applied, and technical, disciplines also use formal tools to eliminate vagueness—to be effective and practical, to reach as exact and reliable results as possible. For some parts of the task, however, it is necessary to use natural language as a tool. This stems from the fact that these parts are not purely technical [2, 3] but take into account

* Corresponding author: sura@fsv.cvut.cz

human beings, social questions, emotional aspects, artistic views and so on [4]—and terms in descriptions using natural language are naturally vague.

In BIM environment, another additional feature applies besides the above—various professions come into contact via BIM, professions that otherwise might have less or no need to interact.

1.2 Semantic differential

The tasks in building area using BIM—or some of the sub-tasks—then combine 'formal descriptions' and 'natural (vague) descriptions'. The part with formal descriptions uses variables to express values, and ideally eliminates vagueness. Such a variable, that does not contain vagueness, has 'zero semantic differential'. Other variables, typically from the part with natural (and vague) descriptions, do not ensure the elimination of vagueness effectively; their semantic differential is non-zero, the participants need not to understand it in the same way.

The vagueness (of the meaning) of a variable need not always be a problem. Vagueness *is not* inaccuracy, incorrectness or unfaithfulness of the actual value of the variable, but vagueness also includes the awareness of the fact that other participant(s) may understand the terms in various different ways. Many practical tasks are satisfactorily tackled without strict descriptions and definitions of each relevant variable. In this case, removal of vagueness provides no substantial additional benefits. Typical case is if the solution exhibits only minor sensitivity to that variable change.

In other cases, however, the vagueness is a problem—e.g. when results are not interpreted, explained or assessed equally by every BIM participant and/or user. This may be due to semantic differential in some (both input and output) variable(s). Then locating and decreasing the semantic differential(s) is an adequate option as it reduces the danger of defects in any respective building life cycle phase (planning, construction, use…).

1.3 The case for semantic differential and vagueness

Some building associated terms like 'comfort of living' or 'road safety' are vague by themselves—in this case it is not a drawback but rather a merit and advantage. This derives from the fact that building isn't a static discipline but an evolving one, responding to new demands, scientific progress and technological innovations [5].

The content/meaning of such terms as 'comfort of living' changes in time [6]. These changes make the term more or less vague over time—and semantic differential varies over time, too. Some constituents of comfort are well defined, exhibit (almost) no vagueness, and (almost) no semantic differential (heat, humidity), others probably change (space requirements), others may be vague due to individual views, attitudes, location, culture, and also change over time (colors & lighting, furnishing, population density, amenities…) [7]. Not recognizing the related semantic differential change causes defects, in building plans, constructions…

2 Calculating semantic differential and vagueness

2.1 Dimensions, Aspect Ratings, Aspect Rating Density Function

Semantic differential is a type of a rating scale designed to measure the connotative meaning of objects, events, and concepts; and the semantics or meaning of words, particularly adjectives or featuring descriptors, and their referent concepts [8]. This is applicable also to building and BIM to encompass (some of the) variables defined for a building task. The connotative meaning is represented by a point in a space which is a Cartesian product of m *aspects* X_i (or *dimensions*) $G \equiv X_1 \times X_2 \times \ldots \times X_m$. If an individual aspect is a discrete scale, it is mapped on some suitable matching continuous scale. The aspects are defined using a pair of contrasting names (good–bad, slow–fast, elaborate–simplified) or a single name (desired population density, maximum walking distance, amenities availability, collision rate...). Viewed from the other side, the semantic differential approach helps in discovering and defining relevant variables if the problem is formulated in a natural language and a recognized or suspected vagueness is present.

A subject i rates the meaning separately for each aspect, which yields a point $\boldsymbol{a}^{(i)}$ within the *aspect space* G. The difference between two ratings $\boldsymbol{a}^{(i_1)}$ and $\boldsymbol{a}^{(i_2)}$ is the *semantic differential* in (the perception of) the meaning between the two subjects. It need not be simply a vector connecting the two points, but rather a calculated vector value:

$$\boldsymbol{d}^{(i_1\setminus i_2)} \equiv \langle d_1^{(i_1\setminus i_2)}, d_2^{(i_1\setminus i_2)}, \ldots, d_m^{(i_1\setminus i_2)} \rangle = \boldsymbol{u}\left(\boldsymbol{a}^{(i_1)}, \boldsymbol{a}^{(i_2)}\right) . \qquad (2.1)$$

For ease of calculations, to make it more general and versatile, the ratings defined as a point $\boldsymbol{a}^{(i)}$ in G are extended to continuous *aspect rating density* function† or just *aspect rating* function $f^{(i)}(G)$.

We suppose that aspect rating function satisfies factorization condition (the dimensions are handled as mutually independent, rather uncorrelated, for each subject i of N subjects)

$$f^{(i)}(G) = f^{(i)}(x_1, x_2, \ldots, x_m) = \prod_{k=1}^{m} f_k^{(i)}(x_k) , \quad i = 1, \ldots, N , \qquad (2.2)$$

and aspect rating-completeness or rating-unit-measure condition, and is non-negative

$$\int_{X_k} f_k^{(i)}(x_k)dx_k = 1 , \quad \forall x_k : f_k^{(i)}(x_k) \geq 0 , \quad k = 1, \ldots m . \qquad (2.3)$$

To compare two aspect densities we define *semantic differential density* as two subjects' aspect densities difference (for $k = 1,2, \ldots, m$)

$$d_k^{(i_2\setminus i_1)}(x_k) = u_k^{(i_2\setminus i_1)}\left(f_k^{(i_2)}(x_k) - f_k^{(i_1)}(x_k)\right) , \quad \boldsymbol{d}^{(i_2\setminus i_1)} = \langle d_1^{(i_2\setminus i_1)}, \ldots, d_m^{(i_2\setminus i_1)} \rangle , \qquad (2.4)$$

where $\mathrm{u}_\mathrm{k}^{(i_2\setminus i_1)}$ may—but if appropriate need not—be an identity.

2.2 Aspects vagueness and semantic differential

Vagueness $vag_k^{(i)}$ of subject i's rating of k-aspect (or vagueness of subject i on aspect k) should depend on the rating of aspect x_k, i.e. on the aspect density function $f_k^{(i)}(x_k)$ and can be measured using statistical or probabilistic functions. Among the first to come into consideration is variance $var_k^{(i)}$, or optionally standard deviation $\sigma_k^{(i)}$, $var_k^{(i)} = {\sigma_k^{(i)}}^2$. Using variance for measuring vagueness $vag_k^{(i)}$ of k-aspect rating we define

$$vag_k^{(i)} \equiv var_k^{(i)} = \int_{X_k} \left(x_k - \mu_k^{(i)}\right)^2 f_k^{(i)}(x_k)dx_k , \quad k = 1, 2, \ldots, m . \qquad (2.5)$$

The mean $\mu_k^{(i)}$—or expected value E—of the aspect k rating function by subject i is

$$\mu_k^{(i)} = E\left[f_k^{(i)}(x_k)\right] = \int_{X_k} x_k \, f_k^{(i)}(x_k)dx_k , \quad k = 1, 2, \ldots, m . \qquad (2.6)$$

In vector form, vagueness and mean as related to subject i's rating are

† The conversion from discrete to a continuous scale can be performed using Dirac delta functions.

$$\boldsymbol{vag}^{(i)} = \langle vag_1^{(i)}, vag_2^{(i)}, \dots, vag_m^{(i)} \rangle . \qquad \boldsymbol{\mu}^{(i)} = \langle \mu_1^{(i)}, \mu_2^{(i)}, \dots, \mu_m^{(i)} \rangle , \qquad (2.7)$$

Aspect density function mean being a counterpart of single-valued aspect rating on a discrete scale, the semantic differential can be viewed as a function based on (two) aspect rating densities means. If the function is the difference of the two means we have (in vector form)

$$\boldsymbol{d}_{\mu}^{(i_2\backslash i_1)} = \boldsymbol{\mu}^{(i_2)} - \boldsymbol{\mu}^{(i_1)} = \boldsymbol{E}\left[\boldsymbol{d}^{(i_2\backslash i_1)}(G)\right] . \qquad (2.8)$$

The above *semantic differential density* (2 .4) is 'directed'—it "points" from one subject's aspect rating to the other, and is or may be negative in some intervals. When trying to bring closer two or several participants in a building or BIM task, the 'directing' may turn to an inconvenience, it might increase aversions. To have an 'undirected' quantity ψ for semantic differential density, the following condition should hold

$$\psi\left(f_k^{(i_1)}(x_k), f_k^{(i_2)}(x_k)\right) = \psi\left(f_k^{(i_2)}(x_k), f_k^{(i_1)}(x_k)\right) \geq 0 . \qquad (2.9)$$

One possible form of ψ is

$$\psi\left(f_k^{(i_1)}(x_k), f_k^{(i_2)}(x_k)\right) = \left|f_k^{(i_1)}(x_k) - f_k^{(i_2)}(x_k)\right|^n . \qquad (2.10)$$

The mean (2 .7) are vectors of numbers (one number per aspect) and thus the semantic differential (2 .8) cannot reflect the various courses (shapes) of aspect rating functions $f_k^{(i)}(x_k)$. Other formulas for semantic differential can be used to reflect the 'shape' of aspect rating functions, for example

$$d_k^{(i_1\uparrow i_2)} = \sqrt{\int_{X_k} \left(x_k - E\left[exc_k^{(i_1\uparrow i_2)}(x_k)\right]\right)^2 exc_k^{(i_1\uparrow i_2)}(x_k)dx_k} , \qquad (2.11)$$

$exc_k^{(i_1\uparrow i_2)}(x_k)$ may be defined by (2 .10) with $n = 1$, and $E\left[exc_k^{(i_1\uparrow i_2)}(x_k)\right]$ is its mean

$$exc_k^{(i_1\uparrow i_2)}(x_k) = \left|f_k^{(i_1)}(x_k) - f_k^{(i_2)}(x_k)\right| . \qquad (2.12)$$

3 Comprehension within an expert group

Mutual comprehension within a group of experts [9] extends the concept of *semantic differential* to a group of N subjects, each subject i within the group being assigned a weight $w_k^{(i)}$ for aspect k, and having his own aspect ratings $\mathbf{f}^{(i)}$, associated into *group ratings* $\mathbf{F}$, $\mathbf{F}_w$, and *group rating density functions* $\mathrm{f}_k^{(\mathbf{F})}(\mathrm{x}_k)$, $\boldsymbol{f}^{(\mathbf{F})}$.

$$\boldsymbol{w} = \langle \boldsymbol{w}^{(1)}, \dots, \boldsymbol{w}^{(N)} \rangle, \quad \boldsymbol{w}^{(i)} = \langle w_1^{(i)}, \dots, w_m^{(i)} \rangle , \quad \sum_{i=1}^{N} w_k^{(i)} = N, \; k = 1, \dots, m , \qquad (3.1)$$

$$\boldsymbol{f}^{(i)}(G) = \langle f_1^{(i)}(x_1), \dots, f_m^{(i)}(x_m) \rangle , \quad \boldsymbol{F} = \langle \boldsymbol{f}^{(1)}(G), \dots, \boldsymbol{f}^{(N)}(G) \rangle , \quad \boldsymbol{F}_w = \langle \boldsymbol{F}, \boldsymbol{w} \rangle , \qquad (3.2)$$

$$\boldsymbol{f}^{(F)} = \langle f_1^{(F)}(x_1), \dots, f_m^{(F)}(x_m) \rangle , \quad f_k^{(F)}(x_k) = \frac{1}{N}\sum_{i=1}^{N} w_k^{(i)} f_k^{(i)}(x_k) , \; k = 1, \dots, m . \qquad (3.3)$$

3.1 Group rating

Using the above, *weighted group mean rating* $\mu_k^{(\mathbf{F})} \equiv E_w^{(\mathbf{F})}[f_k(x_k)]$ on aspect k is defined as

$$E_w^{(F)}\left[f_k^{(F)}(x_k)\right] = \frac{1}{N}\sum_{i=1}^{N} w_k^{(i)}\, E\left[f_k^{(i)}(x_k)\right] = \frac{1}{N}\sum_{i=1}^{N} w_k^{(i)} \int_{X_k} x_k\, f_k^{(i)}(x_k)dx_k \ , \qquad (3.4)$$

or in vector form:

$$\boldsymbol{E}(\boldsymbol{F_w}) = \langle E_w^{(F)}\left[f_1^{(F)}(x_1)\right],\ E_w^{(F)}\left[f_2^{(F)}(x_2)\right],\ \dots, E_w^{(F)}\left[f_m^{(F)}(x_m)\right]\rangle\ . \qquad (3.5)$$

Group semantic differential Gd may be defined as an extension of semantic differential of two subjects (2 .11) and may be defined in various ways, e.g. as an average of semantic differentials of all pairs of experts in the group

$$Gd_k^{(F)} = \frac{1}{N(N-1)}\sum_{i_1,i_2=1}^{i_1,i_2=N} d_k^{(i_1\text{T}i_2)},\quad i_1 \neq i_2\,, \qquad (3.6)$$

or as an average of semantic differential of group mean rating density $f_k^{(F)}(x_k)$ and each individual rating density $f_k^{(i)}(x_k)$ (see (2 .11), (2 .12))

$$exc_k^{(F\text{T}i)}(x_k) = \left|f_k^{(F)}(x_k) - f_k^{(i)}(x_k)\right|\ , \qquad (3.7)$$

$$d_k^{(F\text{T}i)} = \sqrt{\int_{X_k}\left(x_k - E\left[exc_k^{(F\text{T}i)}(x_k)\right]\right)^2 exc_k^{(F\text{T}i)}(x_k)dx_k}\ , \qquad (3.8)$$

$$\widetilde{Gd}_k^{(F)} = \frac{1}{N}\sum_{i=1}^{N} d_k^{(F\text{T}i)}\,. \qquad (3.9)$$

In addition to mean and semantic differential we can define vagueness in aspect k in various ways similar to (2 .5)

$$GVag_k^{(F)} = \int_{X_k}\left(x_k - \mu_k^{(F)}\right)^2 f_k^{(F)}(x_k)dx_k = \int_{X_k} x_k^2\, f_k^{(F)}(x_k)dx_k - \left(\mu_k^{(F)}\right)^2, \qquad (3.10)$$

$$\widetilde{GVag}_k^{(F)} = \frac{1}{N}\sum_{i=1}^{N}\int_{X_k}\left(x_k - \mu_k^{(i)}\right)^2 w_k^{(i)} f_k^{(i)}(x_k)dx_k \leq GVag_k^{(F)}\,. \qquad (3.11)$$

4 Conclusions

In collaborative building processes, mutual comprehension forms a substantial part of success. The concept of semantic differential offers a tool to identify potential or real threats of miscomprehension and to foresee and prevent defects in building processes as a consequence thereof. Data for semantic differential evaluation can be collected, or may be already available. For example, if an expert panel on some MCDA task evaluates the alternatives, the criteria play the role of aspects and each alternative is one of the objects rated on aspect scales. Calculating the vagueness and/or semantic differentials may provide hints of possible miscomprehensions.

- High semantic differential may be due to different opinions of experts, or to perceiving the underlying item in divergent ways. Or it may be due to different methods used by experts, or to novelty of the expert area and its development.
- Low semantic differential may be not due to equal comprehension, but to traditional views or to equal methods used by the participating experts. Or it may be due to lack of creative thinking.
- Finding a subgroup of experts exhibiting substantially lower semantic differential or vagueness may be a sign of some dissension, and unifying the views may be desirable. Or it may simply be a sign of several extant paradigms or schools.

Higher semantic differential or vagueness may indicate misunderstanding, although it may have other causes, too, and in the field of interest it may lead to discovering, besides miscomprehensions, also other features. It may also indicate the intention of some participants to keep the term vague ('social housing' being an example).

Funded as well as unfunded support from various European projects and foundations is gratefully acknowledged.

References

1. B.A.W. Russell, *Vagueness*, Australian Journal of Philosophy **1**, 84-92, (1923)
2. L.H. Dreyfus, *What Computers Still Can't Do: A Critique of Artificial Reason*. (MIT Press, New York, 1992)
3. R, Penrose, *The large, the small and the human mind*, (Cambridge University Press, 1997)
4. D. Kahneman, A. Tversky, *Prospect Theory*, Econometrica, **47(2)**, 263-291 (1979).
5. J.A. Laitner et al., *The Long-Term Energy Efficiency Potential: What the Evidence Suggests*. Available on http://amca.org/assets/member-document/6ACEEE-Vision-Report-2012.pdf (2012)
6. D. Madland, R. Teixeira, *New Progressive America: The Millennial Generation*. Available on https://cdn.americanprogress.org/wp-content/uploads/issues/2009/05/pdf/millennial_generation.pdf (2009)
7. J. Rifkin, *The zero marginal cost society: the internet of things, the collaborative commons, and the eclipse of capitalism* (Palgrave Macmillan, New York, 2014)
8. C.E. Osgood, G. Suci, P. Tannenbaum, The measurement of meaning (University of Illinois Press, Urbana, Illinois, 1957)
9. S. Plous, *The psychology of judgement and decision making* (McGraw-Hill, Inc., New York, 1993)

Building defects and methodology of partial risk assessment as a step towards improving quality

Jaroslav Synek[1] , and *Miloslava* Popenková[2*]

[1]Faculty of Civil Engineering, Department of Construction Technology, Czech Technical University Prague, Thákurova 7, Prague and Metrostav, a. s., Koželužská 245, Prague 8, Czechia
[2]Faculty of Civil Engineering, Department of Construction Technology, Czech Technical University Prague, Thákurova 7, Prague, Czechia

Abstract. This article deals with the assessment of partial risks in construction projects. Partial risks compromise the quality of the construction during preparation, implementation and use. They are the cause of defects associated with individual parts of construction and supply. These risks are also a source of increased repair costs. We need to identify partial risks and suppress them at all stages of the construction project to improve quality and reduce construction costs. The article defines the sub-risk and its impact on the complexity of buildings. It deals with the methodology of evaluation and use of analytical results for the prevention of building defects. The article also presents the benefits of applying the sub-risk assessment and the resulting preventive measures.

1 Introduction

According to the available data, building projects are very often burdened by a bad reputation resulting from exceeding the implementation time and rising costs of projects that produce poor economic results. These are not just large projects, but also medium and small projects in all types of economies. The risks associated with these projects are the cause of poor results. Risk realization leads to building defects during preparation, implementation and use. Participants in building projects (investors, banks, designers, implementation firms, operators and users) should therefore be interested in improving the project's results and should therefore try to limit the realization of risks, the causes of the project's failure. In the long term, prevention is the most effective risk mitigation, ie procedures that prevent its implementation. However, routine risk management for the global risks of the whole project is not effective enough to ensure the quality of the individual parts of the project where individual risks to the quality of the buildings are realized. The work focuses on the limitation of operational risks, namely technical, technological risks associated with the development and use of technologies and manufacturing risks associated with production processes, not with other risks (contractual, financial, social, local, etc.). Therefore, in order to determine effective preventive

* Corresponding author: miloslava.popenkova@fsv.cvut.cz

procedures for individual parts of a building project, the risks and circumstances of their implementation must be analysed on the basis of defects analysis of parts of the construction (individual types of constructions, crafts and embedded technologies). The results of analyses use taxonomy, a list of defects, determining causes, as well as determining the magnitude of the consequences of a defect. Depending on the frequency and cost of defects in the evaluated production system, it is necessary to determine the risk values and subsequently determine the severity and extent of risk prevention. This is an effective process leading to the necessary reduction of potential losses and damage to construction projects.

2 Targets

The aim of the paper is to describe the already used methodology for identifying and evaluating the partial operational risks associated with the construction project and to clarify the procedures that can be used for effective prevention from the point of view of the contractor of the construction project. To prevent partial risks, it is necessary to focus on their identification and analysis, the procedures for their evaluation and the identification of the most serious of them.

3 Definitions and characteristics of risk

The general definition of risk is considered according to [1]: *"The possibility that something unfavourable will happen in the future"*, also *"Failure, Threat, Damage or Loss"*.

The risk for production processes can be understood as a threat, danger and uncertainty of the result. In the case of operational risks of construction projects, we consider the so-called technical risks as defined [2]: *"The probable value of the loss incurred by the wearer, Risk recipients by realizing the hazard scenario, expressed in cash or other units."*

This definition allows for an analytical, mathematical formulation describing the possible consequences for the implementation of the risks for the project. Risks can be defined as the probability of a threat or other negative event associated with a project endangering the achievement of its goals. If the overall objective of the project is the economic success (project profit), the completion of the required construction period and the proper functioning of the project, i.e. the quality (safety), the implementation of the risks threatens to achieve them. The risk is generally characterized by the probability of occurrence of a negative event (defect), the level of risk (high, medium, small), impacts (consequences), predictability, degree of responsiveness, relation to subject, rankings (primary, secondary, residual), size, acceptability, etc. (see taxonomy of risk [5]) and tables, as originals of good quality and well contrasted, are to be in their final form, ready for reproduction, pasted in the appropriate place in the text.

4 Risk management, identification and classification and mathematics

In order to achieve the objectives of the project (profit), it is necessary to actively control and control the risks based on their knowledge. This assumes the ability to evaluate them in relation to the goals and implementation practices used in the project. Own risk management includes their identification and classification (assessment), management plan, monitoring, risk response and risk communication.

4.1 Overall global risks

Normally, the project's risks are generally assessed in relation to the project, site and conditions of implementation. These stem from an open environment of construction projects that require uninterrupted responses to the changing situation.

4.2 Partial risks

If we examine in detail the procedures needed to meet the project objectives, focusing on global general risks is not sufficient. For direct risk and quality management, it is necessary to focus on the quality / risks contained in the individual parts of the project, the components, and the technologies whose systemic partial quality / risks create or threaten the resulting quality of the project system. The project is to be understood as a system in accordance with the general theory of systems, i.e. as a system of elements and their interrelations, relations (see e.g. [3]).

Therefore, the basic task in risk management and ensuring the required quality of buildings is knowledge and way of eliminating partial quality / risks associated with the individual components (components) of buildings within the building - system. Construction is always a set, system of individual components and their elements. Only as a complex system they create the construction and its quality, respectively the quality of its use. For each partial part of the complex system - buildings are laid specific requirements, each part has specific features, requires specific requirements and conditions for assembly and processing, has specific conditions of use. The specific (partial) properties of the component are a condition for the proper functioning of the global building system as it is an indispensable part of the building system. The required functions of the system (the construction project) can be filled only if the complex system is working on the principle of conjunction, all required inputs must be filled in to meet the output requirements of the system, (quality) function of the building (e.g. maintaining the internal environment for the function)) At the same time. If they are not filled, the response of the global system (output) does not meet the specified quality requirements, the system function is defective.

In order to achieve the specified quality, it is therefore necessary to focus on fulfilling the partial requirements for the quality of individual components, parts of the construction. This quality is necessary to require at all stages of the construction, from the formulation of the project, through the design, the realization to the construction. If this requirement is not met, the quality of the building as a system is jeopardized. Therefore, it is necessary to continuously check:

- Determination of partial component quality requirements from design to implementation
- Check for compliance with partial component quality requirements
- Running the building as a system in accordance with partial component quality requirements

The quality of the use of the building is a decisive criterion for quality assessment because the highest costs during the building life cycle are spent on building use, user maintenance and repairs. In order to achieve the necessary quality of use, it is therefore necessary for all members of the project team (investor, designer and contractor) to know and be able to work with the partial risks associated with the individual parts of the construction and their elements both in design and implementation. Risks need to be identified based on the knowledge of the results of the quality evaluation of previous projects. For identified defects, it is necessary to investigate and determine their causes (e.g. by category - ground, engineering, transport, etc.), to determine defective elements of the construction and their causes, the cost of their removal.

The methodology should be based on current practices used to identify and assess risks [1] and use the specialized knowledge and information available to investors and building operators, designers and developers. This information should be shared by all participants in the construction project. Information on defects and disturbances is collected by participants in construction projects but rarely can be processed, archived and, most of all, actively used in the following projects. In some cases, such information is used systematically to improve new designs and practices. The aim of this methodology is to use the feedback possibilities to reduce design and implementation risks.

5 Current status

Risk analyses are currently used to assess the global risks of the entire construction project. Therefore, the risk is perceived as a risk of aggregation and such risk assessment does not provide sufficient information for the preventive measures of the implementing company related to the individual parts of the construction and the implementation processes. Therefore, there is no basis on how to prevent the realization of partial risks associated with individual parts of buildings and implementation processes. Methods of identification (risk assessment) of individual parts of buildings and processes (partial risks) are not well known and used.

The proposed procedures are aimed at suppressing and eliminating the individual risks directly related to parts of the construction and processes. These risks describe the differences in the threat to a certain part of the structure or process that may cause serious failures leading to permanent damage to structures and may require a reduction in load (structural requirements) or a change in design use as a result of reduced requirements. Permanent damage to the structure may result in additional reinforcement or rebuilding of the structure. Targeted prevention focused on the risk parts (processes) in preparation and implementation suppresses additional changes and repairs that are usually significant and costly. Therefore, it is necessary to analyse in detail the risks associated with individual parts and to implement these practices in the implementation practice. Ignoring this option brings companies involved in the endangered project a loss of reputation and loss of position on a highly competitive construction market.

6 Methods

The basis of the method of partial risk assessment is the use of feedback from the preparation of the construction and its implementation. The default information is the number of defects already encountered in the production system. The occurrence, frequency of failures is monitored for individual types of buildings (residential, administrative, production, transport, etc.) and their parts (supporting structure of the lower and upper structure, protection of the building against water, internal non-bearing structures, individual technological parts etc.) category. It is also necessary to identify the costs necessary to repair the defect (damaged structure or part of the building) and assign it to the individual categories (components). These costs are determined as an average at specified time or other intervals (e.g. calendar year, project), and in time series. The risk is then determined as the result of the observed frequency of occurrence and the long-term costs of removal of the defect.

Part of this method is to determine the causes of defects that use the information found to correct the defect. The causes are investigated at all stages in the design, implementation, even during the use of the building. Using partial risk assessments effectively reduces the high risks associated with design and implementation (e.g. large underground constructions). The method uses the information the contractor gains in evaluating the quality of the design of the building, the implementation processes and the information obtained during the use of the construction.

7 Results

The results of the partial risk assessment, in the context of a particular construction system of the building company (used construction techniques and technologies), are able to look for effective ways of limiting the risks associated with certain types, parts of buildings and building technologies used. Finding the critical parts of the building (construction) associated with critical risks, where it is necessary to focus on the risk reduction, increasing the resistance to defects allows the application of Pareto rule (80/20) [4]. Its application identifies those parts whose improvement will bring significant overall risk reduction and quality improvement. Preventive practices focus on individual components of the solution, identified parts, their formulation and parameters, technical solutions (design), manufacturing process techniques. Also included are the procedures used during the inter-operational and final inspection of the construction, the use of the construction, its maintenance and repairs.

Detailed analysis of the parts and their relation to the complex function of the construction system is usually associated with feedback, sufficient information from the implementation and operation needed to find the method necessary to reduce the risk of defect to an acceptable boundary. Using Pareto rule on the most endangered parts of buildings will usually significantly reduce defects in the short term.

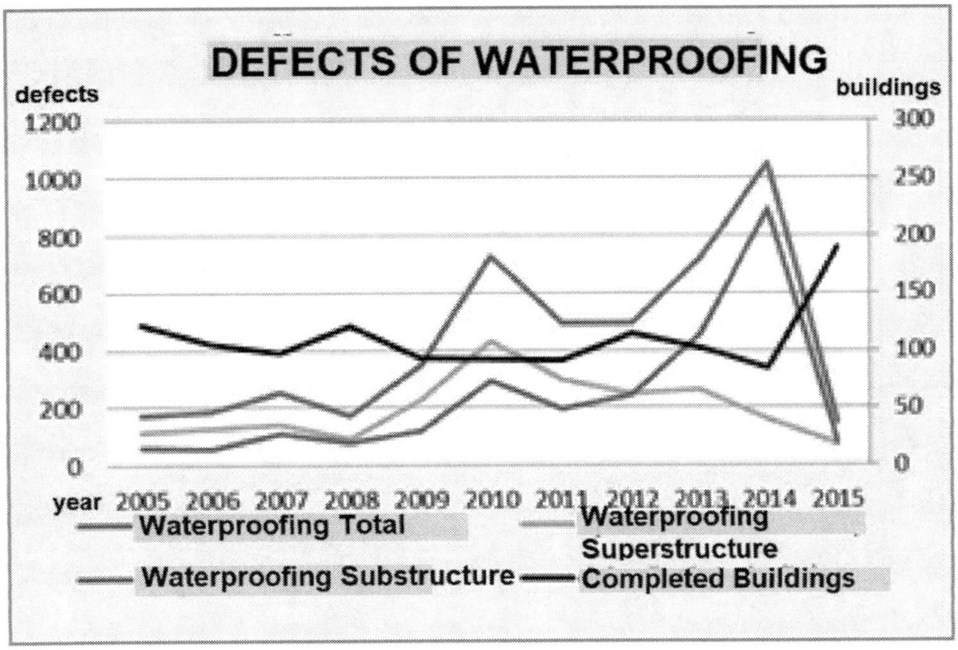

Fig. 1. Disturbance of Waterproofing of the Substructure – (number on right) red global waterproofing, green waterproofing of top structure, blue waterproofing of substructure, black number of completed building (on left).

An example is the application of this procedure to improve the quality of the substructure, water tightness. Insufficient water tightness of the substructure to groundwater impacts the construction mainly due to defects in the internal environment (leaks reduce the quality of the environment) and the formation of corrosion. The development of corrosion can mean, in the long term, a reduction in the bearing capacity and stability of the structure and thus its usefulness. It is therefore essential that the underground parts of the buildings are adequately protected from the negative effects of groundwater. Coating waterproofing usually applied to the outer surface of the structure is very sensitive to the correct design (overall water tightness), including penetrations (piping, etc.) and expansion of the structure. Coatings are susceptible to imperfect design and mechanical damage by other building activities (sharp fittings are installed just after a thin waterproofing from asphalt or plastic foils and can easily break through waterproofing). Water leaks often cause extensive defects in the indoor environment. These defects are difficult to locate and, above all, difficult to repair. Even after costly and extensive repairs to the substructure, they often carry permanent traces of damage, water leaks (even if they are limited) and, above all, the corrosion of the reinforcement.

For these reasons, procedures have been sought to significantly reduce these defect risks. All components of the solution have been examined and tested in relation to the manufacturing system. New specifications and requirements for substructure (parameters, proposed technical solutions, materials and products), technological procedures and control procedures for continuous and final quality control were formulated. As a result of the changes made to design and implementation, significant reductions in construction deficiencies, reduction of risks associated with permanent damage to the structure of corrosion and reduction of repairs were made.

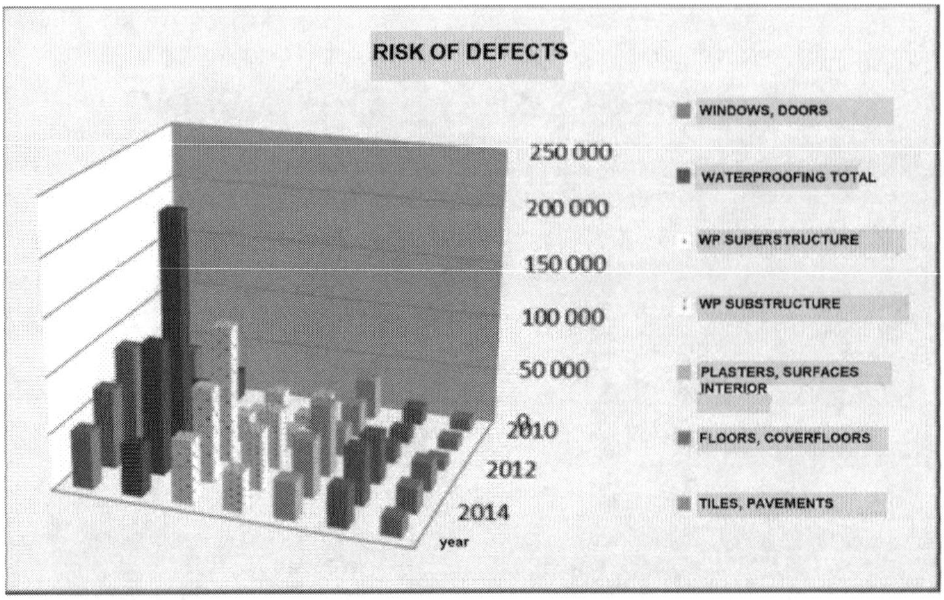

Fig. 2. Partial risks of structural defects y. 2010 – 14 (green windows and doors, blue waterproofing global, white and one blue point waterproofing of top structure, white and two blue point waterproofing of substructure, yellow surfaces, interior plaster, purple floors and floor coverings, grey tiling and paving). Volume of risks on right.

8 Conclusion

Applying the described method of selecting parts of buildings with a high risk of damage allowed the focus on detailed analysis of defects of the part of the construction in all the components of the design, preparation and realization stages. The results of analyses identified the partial risks associated with the concept of the problem and the finding of effective preventive procedures to eliminate defects, i.e. to increase the resistance of the structure before the risks are realized. The application of the methodology brought a reassessment of the approach to evaluation of project proposals and methods of implementation and reduction of the number and extent of defects.

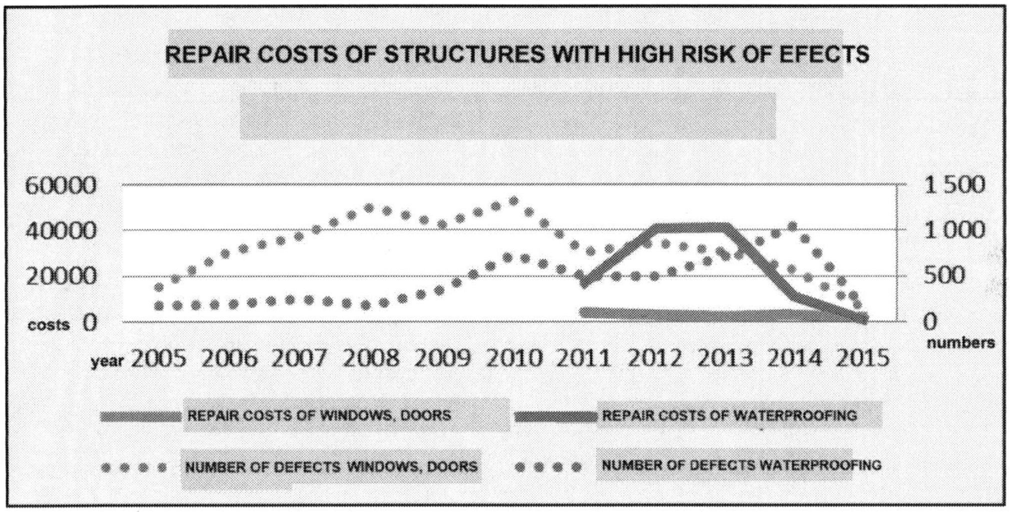

Fig. 3. Decline of development of defects and cost of repairs of high-risk structures y. 2005 – 2015 (red dots numbers of waterproofing defects, blue dots number of windows and doors defects /on right/, red costs of waterproofing reapirs, blue costs of windows and doors reapairs /on left/).

References

1. Oxford Learner's Dictionary. Available on http://www.oxfordlearnersdictionaries.com/definition/english/risk_1?q=risk
2. M. Tichý, *Ovládání rizika: analýza a management* (C.H. Beck, Prague, 2006)
3. A.D. Hall, *A Methodology for Systems Engineering* (Van Nostrand Reinhold, 1962)
4. H.F. Cervone, OCLC Systems & Services: International digital library perspectives **31** (3), 108-111 (2015)
5. D. Baloi, A.D.F. Price, International Journal of Project Management **21,** 261-269 (2003)

MATEC Web of Conferences **146**, 01006 (2018) https://doi.org/10.1051/matecconf/201814601006
Building Defects 2017

Building renovations in BIM systems

Marian Smutný[1] *

[1]CTU Prague, Faculty of Civil Engineering, Thákurova 7, 166 29 Praha 6, Czech Republic

Abstract. Renovation of buildings is a demanding challenge for computer systems. Nowadays, complex 3D models of historic buildings can be scanned and modelled, including sculptural filigrees and stucco decorations with possibility to generate different projections, cross-sections and elevations according to descriptive geometry rules. However, all of this falls within the field of "3D Modelling". BIM systems work with elements that have to be classified according to their purpose, or according to the way they are realized. The system is then able to sort, filter or group them. Also, it is able to compute their bill of quantity and quality based on their attributes. Last but not least, it has to be able to schematically display civil engineering drawings on the basis of classification of the elements which differ considerably from the display of descriptive geometry. In regular constructions, repeatability and predictability of the links between the elements suit computer systems. Thus, the issue of building renovations in BIM systems is to find a balance between modelling irregular constructions and irregular space arrangements, while maintaining the ability to interact with other BIM elements.

1 Introduction

BIM programs are based on 3 different engines:
1. object-oriented (1, 2) engine - each object serves a separate function. It is defined by its properties' - what it is and what it can do. Examples of such BIM programs are *ArchiCAD, VectorWorks*.
2. database (3, 4) engine - all information is stored and maintained in a relation database from which visual representation of drawings and models are generated. Examples of such BIM programs are *Speedikon, Revit.*
3. geometry modeller (5) engine - all information is stored as a 3D model with attached attributes. It may be based on a parametric modeller as *Parasolid* or *ACIS*, or one can use a specialised modeller. Examples of such BIM programs are *AutoCAD Architectural, TriForma, and Allplan.*

Each of these approaches gives certain advantages and certain disadvantages. In this article the *Allplan* BIM system is described - how it is able to create a virtual building with

* Corresponding author: marian.smutny@fsv.cvut.cz

irregular constructions (walls, rooms, openings and vaults) of historic buildings and their modifications while keeping their interactivity with reasonable accuracy and high returns.

2 The Modeller

Allplan BIM modeller is not based on a robust parametric modeller but on a **specific simplified modeller** which is particularly **suited for building renovations** mainly because of two features:

Modeller is based on simple 3D solids consisting of a polygonal **plan shape extruded** with top and bottom faces of a **spatial envelope** (in *Allplan* terminology the term "*Planes*", or "*Roof Planes*" is used) which constitutes a **storey**. Top and bottom shapes of 3D solids can be referenced to top or bottom "*Planes*" in various distances, or can be fixed independently from "Planes". The advantage of this approach is that the architect can focus on the floor plan, and the modeller ensures that constructions will fit to the storey height especially when it is formed by vaults.

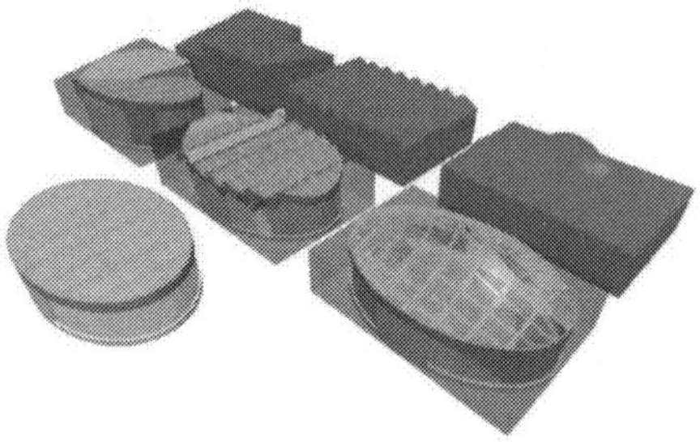

Fig. 1. Violet colour represents a virtual spatial envelope "*Roof Planes*" where BIM elements continuously adapt to each of them.

The modeller uses both **positive and negative 3D solids**. Positive 3D solids are the basis for walls, beams, slabs, etc. Negative 3D solids are the basis for openings such as doorways, windows, niches, etc. The big advantage with the renovation of buildings is it can be placed **independently** or can be overlapped. Thus, the window can be **traversed through a folded wall** or **wall connection**. Equally, it is possible to freely add and remove a plan vertex to floor plan contours of the BIM element in order to form **irregular walls** or openings with **rabbeted jambs**.

The undisputed advantage of working with the BIM system is to modify the schematic representation of BIM elements which is reflected in its 3D geometry. Nevertheless, the ability to interact remains as well as ability to list elements. After these low-end modifications, 3D solids still remain "intelligent" where in the floor plan it is displayed schematically, keeping interactions and listing the bill of quantities.

3 Modelling Process

To illustrate this procedure, the Town Hall renaissance building in Mělník is selected. Survey drawings were delivered in MicroStation DGN files. Survey drawings consisted of the shapes of rooms, elevation dimensions, schema of vault geometry, ceiling heights, vault impost, directions of ceiling beams, and test pits.

DGN files were imported into individual drawings as the base drawings for individual floors. Properties of elements (layers, linetypes, colours, fonts ...) are unified and elements is sorted and cleaned for easy orientation in the base drawings.

3.1 Walls and Openings

It starts with creation of walls. The content of the survey drawing is also copied as the base drawing for creating BIM elements. All layers except walls are switched off. In the base drawing, lines are trimmed to borders, therefore individual areas are closed. Lines of door jambs and window reveals are deleted and unified into intact borders of walls.

The "*Polygonal Wall*" function is selected to create irregular walls. At the command prompt, "*Automatic Geometry Examination*" feature is turned on. The BIM elements of the walls are created by clicking inside the closed areas. Because the walls are set same parameters (material, graphics, height), the joined edges are not displayed in connection to the other walls.

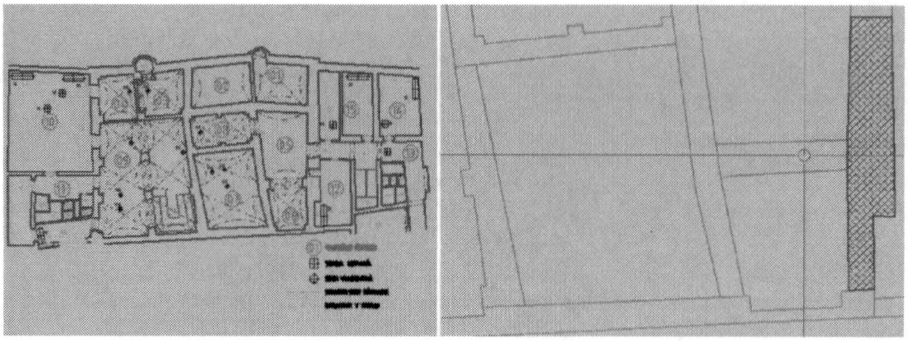

Fig. 2. Lines of survey drawings is used for creating individual BIM "*Walls*".

After having created the walls, then all layers of base drawing are displayed, so it is possible to snap on the edges when inserting openings. The "*Door*", "*Window*", "*Niche*" functions are used depending on the type of the opening. The location and width of the opening is determined by snapping on the background lines. All walls and openings are now completed.

Door architraves are modelled using 3D solids and stored as parametric objects are referred to "*Makro*". Also, it is possible to define how the object will be altered when changing the dimension. These objects can be assigned to openings so that they can move, rotate and adapt to the dimensions of the openings, and they are also the billing of quantity depending on various dimensions.

3.2 Vaults and Ceilings

After the walls and openings have been created, the vaults are created. For these purposes, the function "*Roof Plane*" is used which makes a virtual element. The "*Roof Plane*" function is more general in use. It can be better imagined as a **spatial envelope** to which BIM elements (walls, openings, ceilings, beams, rooms, roofing...) are adapted. BIM *Allplan* elements are usually height-related to either the top and/or bottom pairs of "*Planes*" which is usually represented as the height of the storey. For example, **walls** are referenced by the **lower edge** to the **lower plane** and the **upper edge** to the **upper plane**. The **ceiling slab** is usually referenced to the **top plane** with the **thickness**.

This function can be used to define **irregular segmental vaults with lunettes** over individual rooms. First, an **auxiliary room** is created **outside the room** boundary with the same rotation as the room and the average room width. This is because the *Allplan* "*Arc Roof*" function can create a segment arc only between strict parallel edges (called "*Eaves*"). Finally, the parameters of the arc are taken.

Subsequently, the "*Roof Plane*" will be created **inside the room**. The contour will be taken by clicking inside the room with the "*Automatic Geometry Examination*" feature turned on. The **arc** from the auxiliary vault is **transferred and taken** to the setting of the "*Mansard Roof*" function which is not limited to parallel eaves, and the segment arc will apply to even divergent edges. To create a **groin vault or a lunette**, the "*Dormer*" function is used. Same span, arc height and "*Eaves*" setting for groin vault are used as before ("*Eaves*" is a substitute name for "*Impost*" in this case).

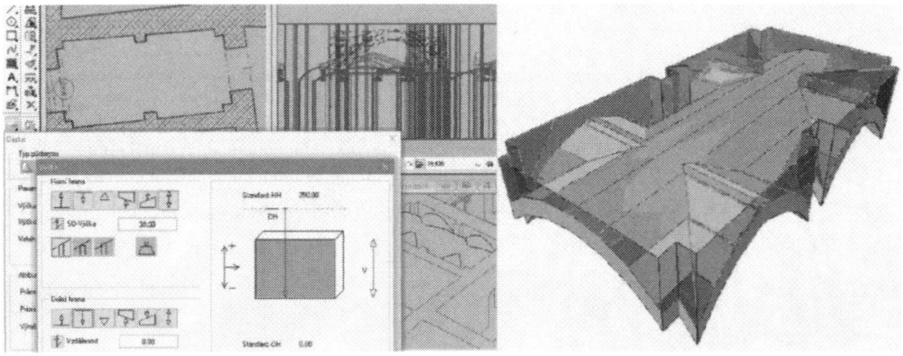

Fig. 3. Creating irregular vaults and settings of BIM elements are referenced to "*Planes*".

The **ceiling plate** is created with a **bottom edge referenced to the upper "*Plane*"** and the **upper edge with a thickness** of 30cm. As the ceiling plate is adapted to the "*Roof Plane*", a 30cm thick vault is created.

The procedures described above are made for the remaining vaults and rooms and all storeys including the attic. The advantage of the virtual spatial envelope is that it adapts not only **vaults slabs** with **infilling**, but also to **plasters** (room surfaces) or to new **dividing walls**.

The procedure for creating a truss in an attic is not documented here, as it uses procedures and terminology that are identical to the documentation provided with *Allplan*.

3.3 Stairs

Stairs are built with some irregularities. The base drawing is set to display only layers with stair and wall outlines. The "*Free Stairs*" function is used with "*Automatic Geometry Examination*" turned on which detects the stair outline. If a mistaken detection occurs, it is advisable to edit lines in the base drawing and trim the walking line from connection to the outline.

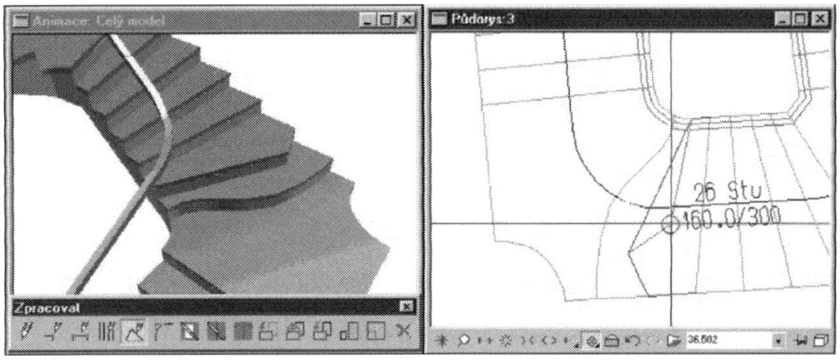

Fig. 4. Modification the schematic representation of BIM elements is reflected in its 3D geometry.

Subsequently stair parameters are **fine-tuned** such as the number of **steps**, the **height** of the stairs, the insertion of the **half-landings**, and the degree of **rotation**. When the steps are rotated, it is enough to rotate only the steps bordering the landings, the other intermediate **steps are interpolated**.

Each step can be modified with **low-end modification** functions directly in the floor plan, such as "*Add vertex*", "*Remove vertex*", "*Modify point*" **without any disruption** of the interaction. All modifications change the 3D representation and it is listed in the bill of quantity.

4 Conclusion

The ability of BIM programs used on historical irregular construction depends on their engine architecture. Therefore, we can conclude that the geometry modeller engine is a more advantageous approach than database or object-oriented engines.

References

1. Object-oriented design. Available on https://en.wikipedia.org/w/index.php?title=Object-oriented_design&oldid=79350684
2. H. Nguyen, A. Oloufa, *Proceedings of the ICCCBE Conference*, 615-620 (1997)
3. Database design. Available on https://en.wikipedia.org/w/index.php?title=Database_design&oldid=800547877
4. G. Lee, J. Jeong, J. Won, C. Cho, S. You, S. Ham, H. Kang, Journal of Computing in Civil Engineering **28** (2), 210-222 (2012)
5. Geometric modeling kernel. Available on https://en.wikipedia.org/w/index.php?title=Geometric_modeling_kernel&oldid=805124377

MATEC Web of Conferences **146**, 01007 (2018) https://doi.org/10.1051/matecconf/201814601007
Building Defects 2017

Defects search during the inspection of civil and industrial buildings and structures on the basis of laser scanning technology and information modeling approach (BIM)

Anna B. Galieva [1,*], *Vladimir N.* Alekhin [1], *Alexey A.* Antipin[1], and *Sergey N.* Gorodilov[1]

[1]Ural Federal University, Institute of Civil Engineering, Ekaterinburg, Russia

Abstract. This paper presents the issues of the inspection, creation of drawings and 3D model of the building in the absence thereof a technical documentation for the object of reconstruction. An approach based on the use of laser scanning technologies and information modeling of objects is proposed. The laser scanning of facades of residential buildings, objects of cultural heritage, industrial enterprises is considered. The proposed approach makes it possible to improve safety during the work on the inspection of building structures, to execute a construction project drawings and specifications with the maximum degree of detail. On the basis of the created three-dimensional model of the object it is possible to subsequently make a static calculation of the building's structures.

Reconstruction of industrial and civil facilities in recent years has become more widespread. When reconstructing architectural monuments design organizations face the challenge of preserving the historical appearance of the object while replacing or strengthening structural elements [1 – 3]. For industrial objects, it is important not only to determine the basic geometric characteristics of load-bearing structures, but also to obtain information on the presence of base, deflections, defects and damages, the features of the connections of elements, and also to reveal the actual location of pipelines, communications for inventory, modernization and technical re-equipment. When reconstructing the housing stock, the task is to perform accurate measurement work for the subsequent development of design solutions to restore operational reliability, improve energy efficiency, determine the cost of restoration work.

When preparing working documentation for the reconstruction or re-equipment of the facilities, designers need accurate taking off drawings, executive documentation, which are often either simply missing or do not coincide with reality. The technology of laser scanning makes it possible to obtain a three-dimensional model of the object with a sufficient amount of details for a subsequent processing with minimal time.

* Corresponding author: a.b.galieva@gmail.com

The newest laser scanners can produce up to 1,200,000 measurements per second. The result of scanning is a cloud of points with 3D coordinates X, Y, Z. Modern software systems make it possible not only to process three-dimensional point clouds, but also to compare them with a three-dimensional model of the object to detect deformations [4 – 12].

The stage of real-time scanning of the object is followed by a stage of cameral work, where the data processing is performed with the help of specialized software and an information model of the building or structure (BIM) is created. BIM and GIS are key sources of data for modeling and analyzing the energy efficiency of buildings when developing solutions for optimizing energy consumption. BIM models are not just graphic objects, it's information that allows you to automatically generate drawings, perform project analysis, and even manage the engineering systems of an object [13 – 15].

Three-dimensional laser scanning to determine the exact geometric parameters of the elements of facades of buildings and structures is widely used both in Russia and abroad. However, little is known about the implemented projects for the reconstruction of housing stock based on laser scanning.

An interesting example of an implemented solution for the renovation of a residential building is the Innova project in Peltosaari, Riihimaki district, Finland [16], initiated by PAROC company.

Thanks to the use of laser scanning technology, the task of reconstructing an object with the maximum preservation of the appearance of the building is solved.

Realized solutions for the reconstruction of the building to bring it to the level of a "passive" house can reduce the actual emissions during operation by 75 tons per year.

Works on scanning of residential buildings are also carried out by specialists of the CAD Department of the Institute of Civil Engineering of Ural Federal University. An example of scanning an apartment house in the town of Rezh, Sverdlovsk region, is shown in Figure 1. The data obtained can be used to prepare a report on measurement works, estimate the cost for repair of facade structures, reconstruction and renovation of facilities, etc. [17].

Fig. 1. Result of scanning an apartment building in Rezh, Sverdlovsk Region.

In the summer of 2016 specialists of the Department made a survey of the building structures of the visual part of the cultural heritage object Ekaterinburg State Academic Opera and Ballet Theater for subsequent restoration works. During the work, laser scanning technology was used to determine the altitude marks and geometry of the dome of the theatre's audience (Fig. 2).

Fig. 2. The result of scanning the Opera and Ballet Theater, Ekaterinburg. General view.

Scan data was transferred to Revit software for further processing and determination of elevation marks (Fig. 3).

When reconstructing industrial facilities, specialists have to solve a number of specific problems: searching for optimal architectural and structural solutions, increasing the energy efficiency of the enterprise and the building, including the replacement of technological and engineering equipment, dismantling and erection of new structures. However, the most difficult task is the task of linking all, both new and existing, engineering networks. Crossing networks of different purposes during the construction phase entails not only the rework of the documentation, but additional financial costs for materials and products. The situation is often complicated by the fact that there is no documentation (both design and executive) for old industrial facilities, and the information compiled on the basis of the survey and measurement results is not exhaustive. In this case, the use of laser scanning helps [17 - 19]. The result is a three-dimensional cloud of dots, illustrating the exact location of all communications. Based on the results of the scanning, a three-dimensional information model is created, which allows obtaining a complete picture of the object and developing the most effective solutions for its reconstruction.

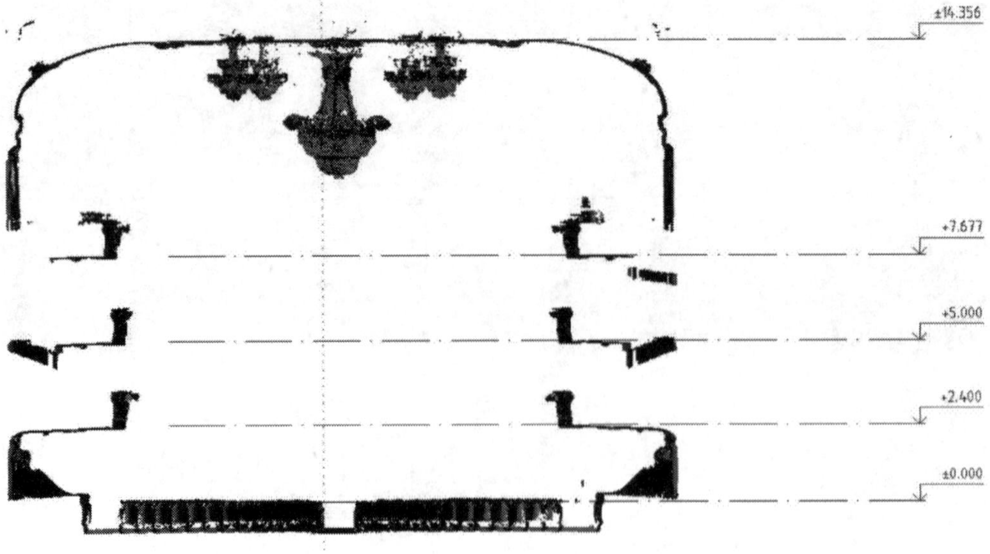

Fig. 3. Processing point cloud in Revit software.

Figure 4 presents a panoramic image of the reconstructed workshop in Ekaterinburg. The building was built in the 40s of the XX century. There is no documentation for structures of the building.

Fig. 4. Panoramic photograph of part of the workshop.

Laser scanning of the object (the fragment of the scan is shown in Fig. 5) was produced by students S. Pugach and V. Leshukov of the CAD Department of the Institute of Civil Engineering, as part of their final qualification work.

The use of laser scanning technology greatly reduced the time of measurement, helped to identify deformations of structures, inaccuracies in erection of bearing structures, peculiarities of the location of engineering communications, and allowed the creation of a detailed three-dimensional model of the object for subsequent work on reconstruction of the facility. On the basis of the existing model, the installation of energy-efficient equipment and the reconciliation of new communications with existing ones was designed (Fig. 6).

Fig. 5. The result of scanning the workshop.

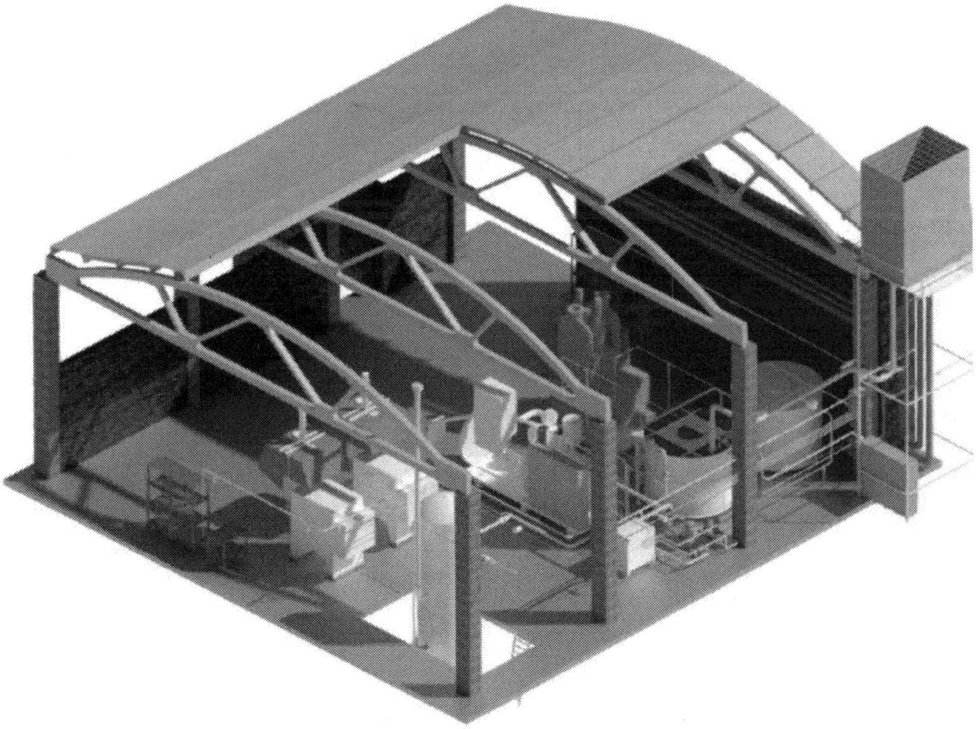

Fig. 6. Fragment of the information model of the object for the development of the reconstruction project.

As a result of using the "laser scanning - BIM-technology" bundle, possible numerous collisions are eliminated even at the design stage, the process of creating the estimate documentation for the dismantling of structures, replacement of equipment, construction or erection works, etc. is automated [17, 20]. The presence of an information model allows you to determine the number of equipment, replacement and maintenance schedules for each particular device, the number of materials and products for overhaul, and much more. Thus, the use of the information model is not limited only to the design stage, but can also be successfully applied at the stage of operation of the facility.

References

1. V.D. Olenkov, D.S. Popov, A.A. Pronina, E.A. Samarina, Design. Art. Industry **1,** 93-98 (2012)
2. S.G. Sheina, D.K. Upennikov., Internet-journal "NAUKOVODENIE", **7**, 3 (2015)
3. T. Kozlova, V.V. Talapov, AMIT: network journa*l* **3**, 12, (2010).
4. A.V. Seredovich, *Materials VII International. Sci. Congress «GEO-SIBERIA-2011»* (, Novosibirsk, 19-29 April, 2011).
5. D.V. Komissarov, A.V. Ivanov, *Materials of scientific. Congress «GEO-SIBERIA-2005»* (Novosibirsk, 25-29 April, 2005)
6. V.A. Seredovich, A.V. Ivanov, *IX Intern. Sci. Congress «Interexpo GEO-Sibir-2013»* (Novosibirsk, 15 - 26 April, 2013).
7. A.P. Maneev, Electric stations **9** (986) 1, 33-37 (2013)
8. A.V. Radziukevich [et al.], *VIII International Scientific Conference "Geodesy, Geoinformatics, Cartography, Mine Surveyor"* (Novosibirsk, 10-20 Apr 2012)
9. K.B. Rysbekov, J.I. Amirov, *Achievements of the higher school-2011*, (Czech Republic, Prague, 2011)
10. A.A. Kuznetsova, D.A. Gura, T.E. Alkachev, Scientific works of the Kuban State Technological University **4**, 77-83 (2014)
11. N.N. Alekseenko, Vestnik MGSU, **2**, 62-73 (2016)
12. I.K. Kiyamov, R.Kh. Mingazov, A.F. Muzafarov, R.A. Ibragimov, A.A. Sibgatullin, Exposition Oil Gas **7** (32), 41-43 (2013)
13. N.Sh. Mustafin, A.A. Baryshnikov, A.M. Skrizhkov, Regional development: an electronic scientific and practical journal **8** (12) (2015)
14. I.O. Orelyana Ursuo. CAD and graphics **7**, 78-86 (2008)
15. Pokalnis V.A., Zemlyakov G.V. *Innovations in concrete science, construction production and training of engineering personnel: a collection of articles on the materials of the International Scientific and Technical Conference dedicated to the 100th anniversary of the birth of I.N. Akhverdova and S.S. Ataeva* (Minsk, 2016)
16. http://www.paroc.ru/campaigns/проект-«иннова»-(innova)- date of the application 13.03.2017
17. V. Kupriyanovsky, S. Sinyagov, D. Namiot, J. Kupriyanovsky, International Journal of Open Information Technologies **4**, 5, 14-24 (2016)
18. D. Furaev, CAD and graphics **2**, 24-27 (2016)
19. A. Kalimulina, CAD and graphics **6**, 24-28 (2015)
20. V.D. Astrakhantsev, I.I. Zolotarev, *XII Intern. Sci. Congress «Interexpo GEO-Siberia-2016»* (Novosibirsk, 18-22 April, 2016)

Energy consumption during the building life cycle – influence of investment activities and operations

Dalibor Vytlačil[1,*]

[1]Czech Technical University in Prague, Faculty of Civil Engineering, Department of Engineering Informatics, Thákurova 7, 16629 Prague, Czech Republic

Abstract. The paper describes the dynamic model of maintenance and investments of a building structure and HVAC systems. The aim of the research is finding the time dependent curve for energy consumption and also the cash flow that depends on the investments to energy saving arrangements and operations. The solution is based on the system dynamics method. The method makes possible to interconnect technical and economic parts of the problem. The main parameter in the model is the energy consumption in the building per floor square meter and year. This parameter is influenced by a deterioration of the building structure and the components of the active elements. The investments realized with the aim to decrease the energy consumption is another influence. The example of the computer simulation of the building parameters during the life cycle is presented in the paper.

1 Introduction

The prediction of maintenance and investments in the buildings requires more attention. The reason is the amount of financial resources spent for this purpose. The costs of maintenance works are higher than the costs during construction because it is small scale work, it needs a preparation of the site, it influences the operations in the building with possible financial losses [1-2]. The facilities management (FM) deals with described problems. FM is now the important part of the maintenance management domain with the holistic view of problems concerning the buildings.

It is necessary to introduce an objective function for the evaluation of the process quality. In this paper it is annual energy consumption per quadrat meter of the floor space. About 50% of the cost is spent for energy systems in the buildings. These costs are divided in the ratio 2:1 for manpower and materials.

Another important factor in the problem is the investment activity that can substantially change the output parameters, in this case to reduce energy consumption. The aim of the maintenance is to keep the buildings in the same conditions but investments want to improve the functionality of the building structure. This activity is focused on the passive as

* Corresponding author: vytlacil@fsv.cvut.cz

well as the active elements of the buildings. The model links both factors to one system of interconnected activities.

2 Description of dynamic model

The solution is based on system dynamics method. This approach allows us to describe the whole system with all relevant elements, [3-4]. The main part of the model is depicted in Figure 1. The stock parameter is annual energy consumption per square meter that is influenced by increasing flow derived from the deterioration of the building structure and by decreasing flow derived from the investments to the structure.

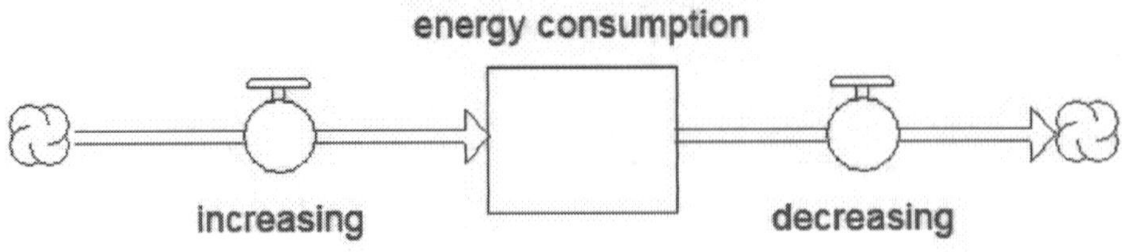

Fig. 1. The main subsystem of the model for the described problem.

2.1 The structure of the model

The deterioration depends on costs spent for planned maintenance and ad hoc maintenance (repairs). The decreasing rate depends on the investment costs.

The model includes the financial subsystem that is represented by the cash flow stock. The cash flow element does not include all incomes and outcomes but only relevant items. The income is calculated as saved money from the energy saving arrangements in the building, [5]. The outcome is derived from the maintenance and investment costs, see Figure 2.

2.2 Input values

The example was calculated for the family house with floor space 150 m^2. The simulation was performed for two levels of planned maintenance 20 $CZK.m^{-2}.year^{-1}$ and 40 $CZK.m^{-2}.year^{-1}$. The time period was 40 years. The heat source is a gas boiler. The initial energy price is 2,12 $CZK.kWh^{-1}$ and can be changed every year by energy price rate. During the life span two investments are considered, in the fifteenth and twenty-fifth years. The investment cost is 2000 $CZK.m^{-2}$. All input parameters are in Table 1.

Table 1. Input values of the variables.

Initial price of energy	[$CZK.kWh^{-1}$]	2,12
Initial consumption of energy	[$kWh.m^{-2}.year^{-1}$]	120
Annual change of energy price	[%]	3
Planned maintenance costs	[$CZK.m^{-2}.year^{-1}$]	20/40
Repair costs	[$CZK.m^{-2}.year^{-1}$]	20

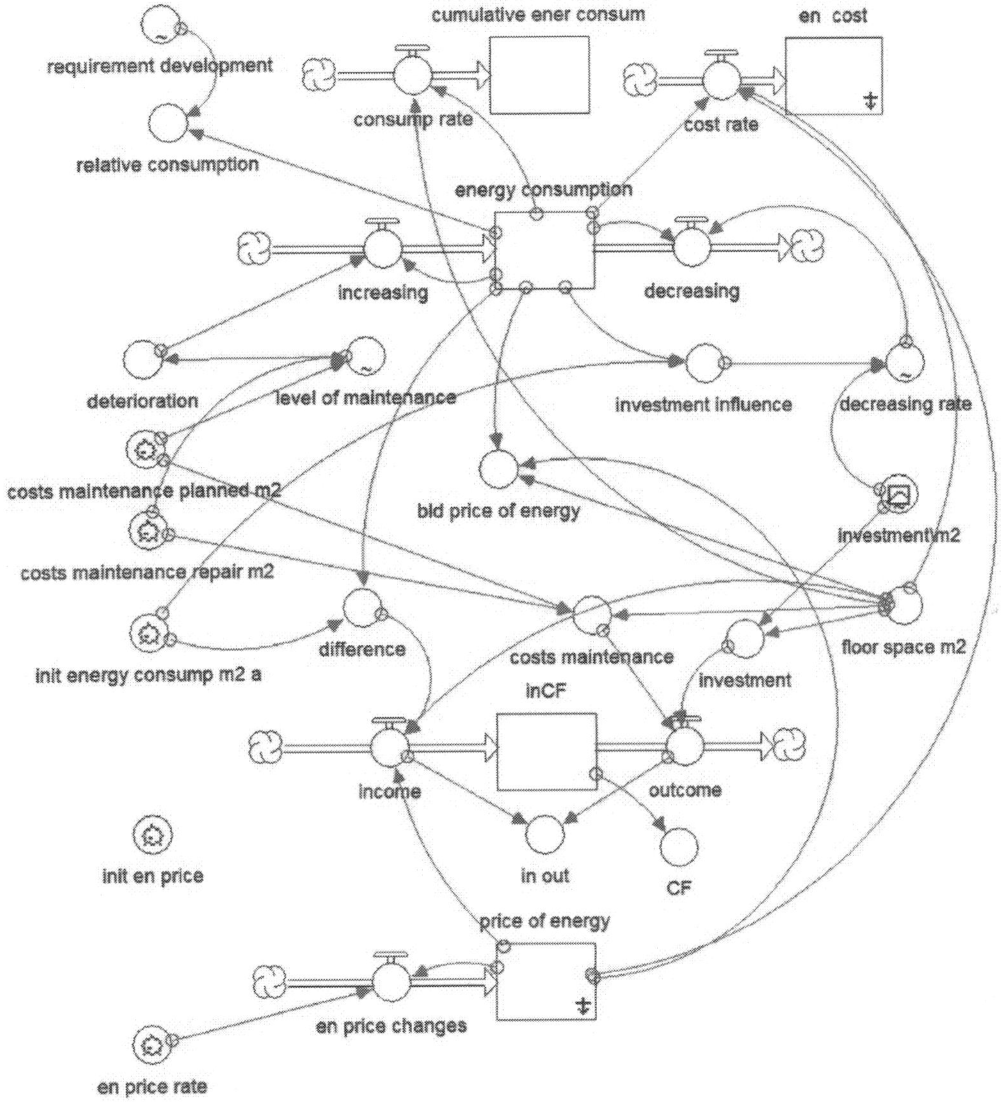

Fig. 2. The dynamic model of the investigated system.

3 Results

Output parameters are depicted in the graphs in Figure 3 and Figure 4. The selected parameters are written in Table 2.

To find the influence of certain parameters the sensitivity analysis was performed. The resultant values are drawn in Figure 5. The graph shows the influence of the planned maintenance to the cash flow. It should be compared to the energy consumption. The higher maintenance values improve the energy consumption but the cash flow parameter is worse. It means it is necessary to find the optimal value of the maintenance level.

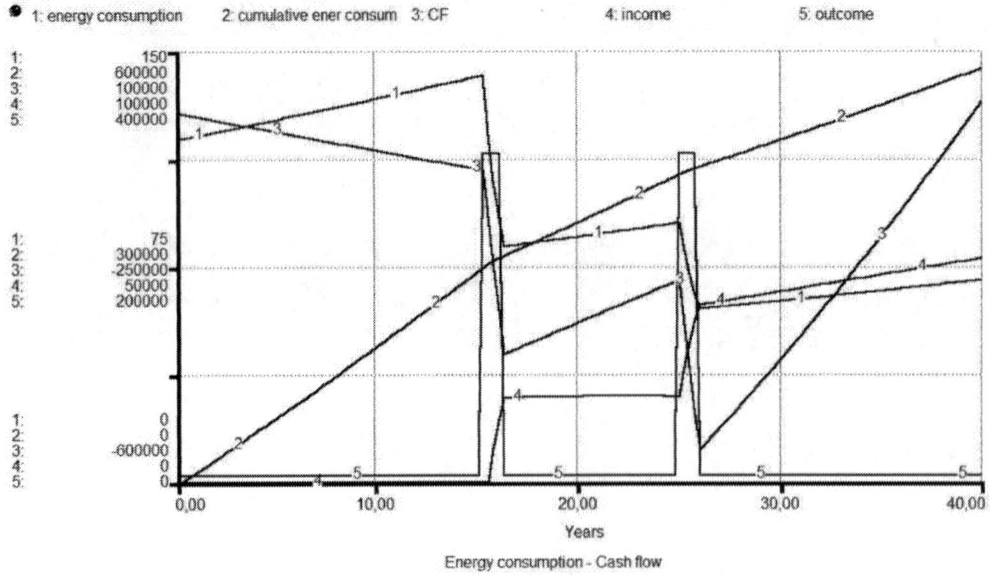

Fig.3. System parameters for planned maintenance 20 CZK.m^{-2}.year^{-1}: 1-energy consumption [kWh.m^{-2}.year^{-1}], 2-cumulative energy consumption [KWh], 3-CF [CZK], 2-income [CZK], 3-outcome [CZK].

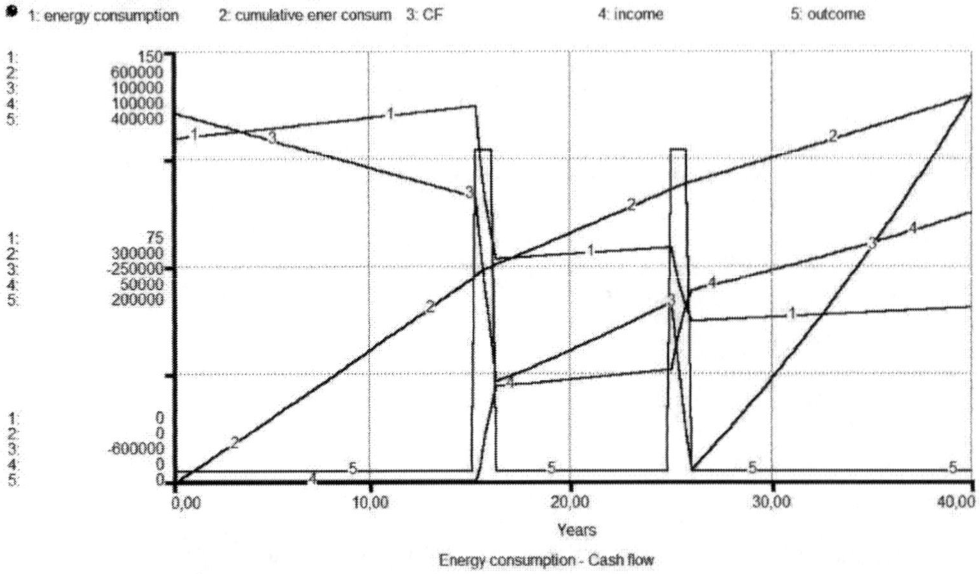

Fig. 4. System parameters for planned maintenance 40 CZK.m^{-2}.year^{-1}: 1-energy consumption [kWh.m^{-2}.year^{-1}], 2-cumulative energy consumption [KWh], 3-CF [CZK], 2-income [CZK], 3-outcome [CZK].

Table 2. Output values from the simulation for two cases.

Parameter / Case	Consumption of energy	Final consumption of energy	Energy costs
	[KWh]	[kWh.m^{-2}.year^{-1}]	[CZK]
C 1	577356	70,5	2144640
C 2	539607	60,5	1978895

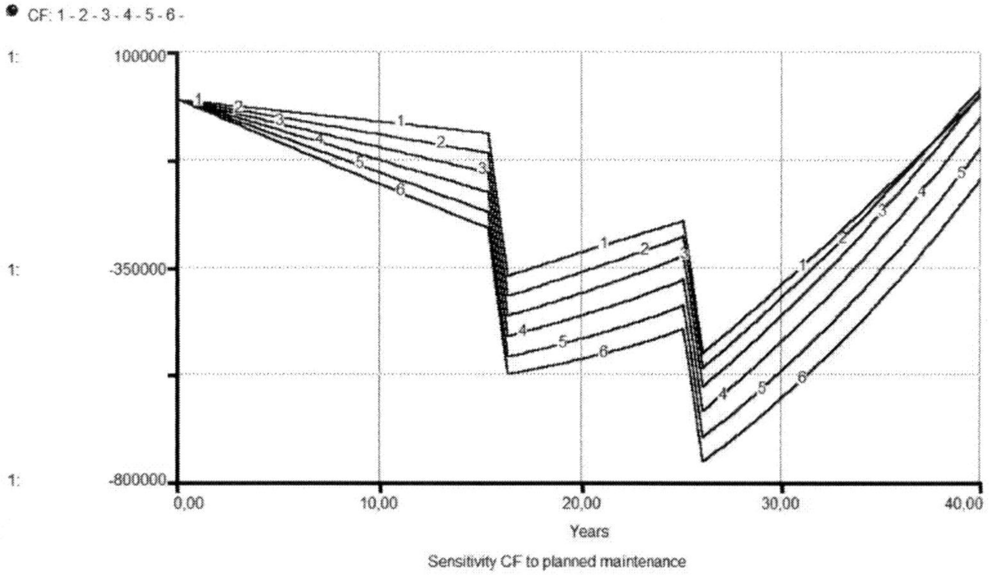

Fig. 5. Sensitivity of cash flow [CZK] to planned maintenance - from 10 to 100 CZK.m^{-2}.year^{-1}: 1-10 CZK.m^{-2}.year^{-1}, 2-28 CZK.m^{-2}.year^{-1}, 3-46 CZK.m^{-2}.year^{-1}, 4-64 CZK.m^{-2}.year^{-1}, 5-82 CZK.m^{-2}.year^{-1}, 6-100 CZK.m^{-2}.year^{-1}.

4 Conclusions

The model deals with two kinds of parameters. The hard parameters describing the building structure and HVAC systems and soft parameters that is difficult to predict, e.g. changes in the energy price. The variability of the parameters is influenced also by the time period of the simulation – the life span of the building is long compare to other technical systems.

The aim of the research was to describe complex dynamic system with technical and economical parameters influencing the dynamic behavior. The designed model is possible to use for the calculation of the different parameters during the life cycle of the buildings. The model is not the replacement of standard FM programs but it is the tool for strategic decision making.

Another investigation has to be done in long time parameter changes. In this topic, there are the changes in the thermal properties of used materials (due to material moisture, mechanical stress), boiler efficiency, sealing compounds etc. The rate of the changes is influenced by physical properties of the materials but also by the quality of the construction and maintenance works.

Uncountable benefits are the important attribute of the problem. The material and component replacements bring often better functionality and therefore the simple comparison of the building states in certain time is rather difficult. The replacement of the windows is the typical example.

The solution of the problem needs to combine the knowledge from many fields – operations management, project management, structure engineering etc. The system dynamics is the good approach that can connect all these knowledge.

This research has been supported by SGS grant SGS17/122/OHK1/2T/11 Dynamic models of changes in building industry investments.

References

1. B. Chanter, P. Swallow, *Building Maintenance Management* (Blackwell Publishing, Oxford, UK, 2007)
2. H. Plat, *The Economics of Property Management – The building as a Means of Production* (Butterworth Heinemann, Oxford, UK 2001)
3. J. Morecroft, *Strategic Modelling and Business Dynamics* (Wiley, Chichester, UK, 2008)
4. J.D. Sterman, *Business Dynamics: Systems Thinking and Modeling for a Complex World* (Irwin/McGraw-Hill, Boston, MA 2000)
5. D. Vytlacil D, *Proceedings of the 11th IBPSA Conference* 2092-2097, (International Building Performance Simulation Association, Glasgow, 2009)

Reducing failures rate within the project documentation using Building Information Modelling, especially Level of Development

Kristýna Prušková[1,*]

[1]Faculty of Civil Engineering, Czech Technical University, 16629, Prague, Czech Republic

Abstract. Paper´s focus is on differences between traditional modelling in 2D software and modelling within the BIM technology. Research uncovers failures connected to the traditional way of designing and construction of project documentation. There are revealed and shown mismatches within the project documentation. Solution within the Building information modelling Technology is outlined. As a reference, there is used experience with design of specific building in both ways of construction of project documentation: in the way of traditional modelling and in the way when using BIM technology, especially using Level of Development. Output of this paper is pointing to benefits of using advanced technology in building design, thus Building Information Modelling, especially Level of Development, which leads to reducing failures rate within the project documentation.

1 Introduction

At current times, there are several ways of project documentation´s construction used. It is a mixture of traditional ways and modern technologies. We can find architects who still design with a pen and a paper and, on the opposite site, architects using the most advanced technologies for design.

1.1 Different approaches to project documentation´s construction

The aim of this paper is to reveal the research of differences between using traditional way of building design in 2D software (or simply drawings) and using an advanced technology called BIM – Building Information Modelling from the building designer´s perspective and consequences connected to using these different technologies. As a reference for the comparison of these two different approaches to the construction of project documentation is used experience from construction of project documentation of specific building.

* Corresponding author: pruskova@mail.vstecb.cz

1.2 Processes needed within the construction of project documentation

In the traditional way, there are drawings in 2D. Every drawing should correspond to other, for example, elements shown in the floor plan must be also shown and exactly shown in the same place in other drawings as section plan, etc. This develops high demands on the designer´s imagination and precision of his work. This way of modelling building design, unfortunately, gives a space for the human failures that can be revealed in other phases of building life cycle.

The other part of the building design is a coordination of processes that must be done between the participants of special parts of documentation. Because of project documentation is usually constructed by more participants, it is necessary to provide effective communication between them. These participants are entering the design process in different phases of project documentation´s development and with different request for information (RFI). Every participant who enters to the design process makes changes in the design according to his profession. At first, there is an architect, who gives the main idea of a building design. Then there is a building designer who is usually responsible for the whole project documentation and who provides all the coordinating communication and information transfer necessary for the good comprehension and conjunction of all the professional parts of documentations. The responsible building designer needs to develop a plan for his coordinating activities. The main part of these activities is project work division between other participants and determining their accountabilities.

1.3 Project documentation required legislation in Czech Republic

Project documentation of building contains a lot of information about the design of building or a construction in a widely way. There are generally several participants, who contribute to design of the building. Each participant has responsibility for different part of this design and for his part of project documentation´s construction. Current Czech Republic´s Legislation demands the project documentation in the form of file with paper documents. The file must be in prescribed structure. There are no demands for the digital form of documentation within the legislation. The structure divides documentation to many parts. These parts are usually elaborated by different project participants according to their specialization and professional orientation. The documentation parts need to correspond to each other. That´s why there is huge time-consuming and demanding thinking process, which must coordinate all the participants.

During the design process, there are a lot of phases, when each participant do the proposal, then the assessment of the proposal and all proposed data are analysed. When the proposal´s analysis is sufficient to all aspects of the design project, after that the participant is able to finish his work. But when there are conditions, not satisfying project design, it´s necessary to do the changes in order to satisfy all conditions. These project changes should be announced to all other participants. Participants have to accept these changes and there is a probability that several changes would be unacceptable for some participants and that´s the reason for other changes. This process seems to be very complicated.

1.4 Traditional way of building design – modelling in 2D software

Using 2D software means construction of drawings by the lines and texts. Construction of documentation using traditional technology demands great coordination between participants and each participant is highly loaded by the processes connected with the design.

The main problem in a traditional way of modelling in coordination of all participants´ work is in data sharing. In current times, there are project participants with different knowledge and experience and there are a lot of software possibilities. That´s why there is used lot of different platforms and data formats used within a single project. Every participant works with different software, which is the most suitable for their own knowledge and experience and for their profession. It means that every participant has different approach to create a documentation, which makes him feel comfortable to use for his part of work. Shared data are transformed into different software and with different information contained. That means that every participant works on his own data model. Consequences are different request for information (RFI), because every participant uses different information relevant for their part of the work on a project design. Next to these requests for information there is a high probability in disinformation are misunderstandings, which may cause the undesirable changes in the project. These aspects are the main causes of failures within the project documentation. Because every single change needs to be inscribed into each data model used with all participants in order to create a collection of suitable documentation parts, which correspond to each other.

2 Using BIM Technology

A simple misunderstanding or disinformation in the project documentation may cause the huge and especially expensive reparation in the following phases in building life cycle. Using BIM Technology not only in building design leads to reduction of failures within the project documentation. According to the Podmanický and Nývlt (2015), the BIM technology is becoming a useful tool not only for its attractive visualization and presentation outcomes [1].

2.1 Comparison of effort and effect according to the Graphic Curve by Patrick MacLeamy from HOK

The Patrick MacLeamy´s graphic curve on the Effort (and Effect) expended on the phases of building life cycle says two important things:
- 1) The better is project developed, the less costs are invested into design changes.

That means that well-prepared pre-design, when the ability to influence the costs and performance is on the highest level, decrease the possibility of needed costly design changes in the following life cycle phases. On the opposite site, pre-design that has to be developed with changes or modification in further life cycle phases is more demanding for efforts and, of course, cost.
- 2) The efforts made on the project documentation created by using traditional technology and by using BIM technology are approximately the same (according to the area under the Mac Leamy´s curve. But it differs in the amount of efforts made in different phases of building lifecycle.

Traditional technology demands the most efforts in the phase of construction project documentation. The BIM technology, compared to that, demands the more efforts earlier in the building life cycle. It´s in the phases of schematic design and mainly in the phase of design development, when the ability of influence costs and performance of the building is on a higher level than in the phase of construction project documentation.

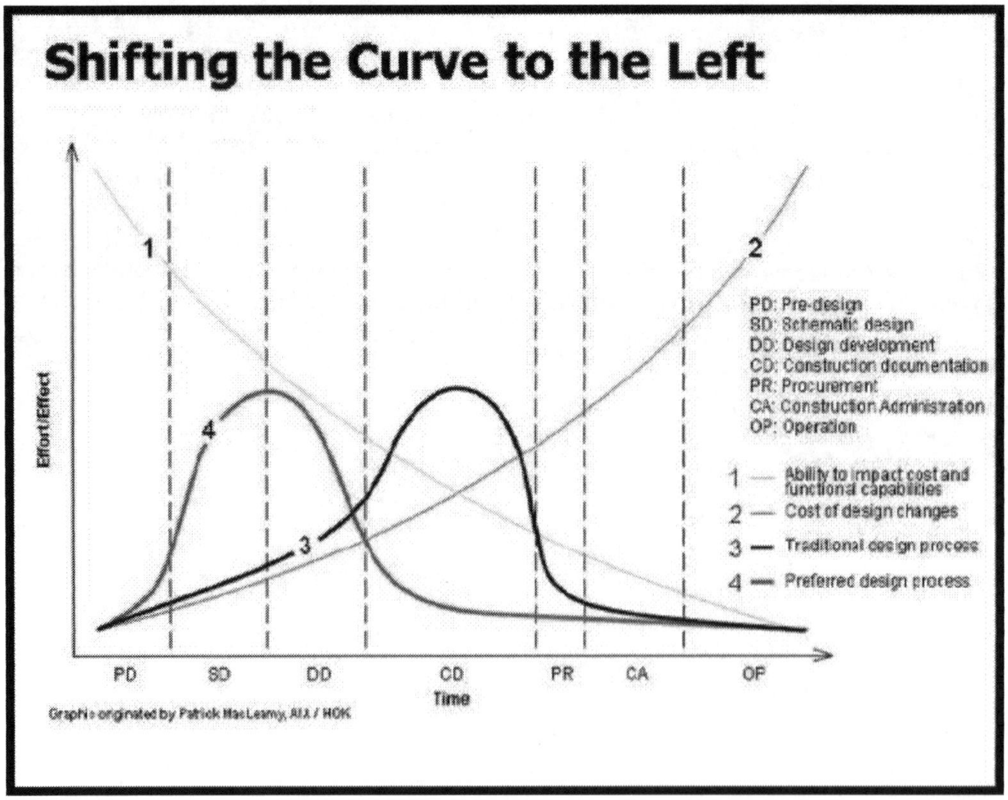

Fig. 1. Graphic originated by Patrick MacLeamy, AIA / HOK [2].

That's why the BIM technology is considered to be a better way for building design in many aspects. It's reflected in decreasing rate of failures that revealed during the building life cycle thus also in cost of project modifications and, of course, in time that is necessary to provide demanded changes or modifications of the project. The knowledge management is the integral part of the processes needed to perform using BIM Technology. "The need of knowledge management primarily comes from BIM's lack of tolerance to erroneous or missing information" [3].

3 Using of Level of Development as a tool for definition

There are many terms used in connection to the levels and project phases. We can talk about Level of Detail, which informs how much detailed information is included in the element model, or about the Level of Information. The Level of Development (LOD) helps to define the amount of information given to element or group of elements in different part of the building life cycle [4].

Every model contains a lot of elements in different phases of progress within the building life cycle. LOD is a tool, which helps determine these phases a thus the reliability of information contained about the element [5]. The preciousness of elements' definition in the LOD language is secured by the Level of Development Specification, which is at least annually actualized according to the new experiences with this technology and the progress of using this technology. The LOD is divided into levels LOD100, LOD200, LOD300, LOD350, LOD400 and even LOD500 according to the project's progress and the amount of information requested in the different phases [6]. LOD contributes to better orientation and visualization of information contained in project model.

4 Management with using BIM Technology

BIM Technology itself doesn´t contain the management of knowledge, but it can be a very useful tool in a connection of well-prepared plan of processing of knowledge. Extracted knowledge from building information model can be useful not only for design and build processes but also for all other processes that may use information about building [7]. The management of processes within the using BIM Technology can be ensured by BIM Project Execution Plan. Definition of BIM project scope can be done using definition of priority of requirements [8]. "Principles of creation of BIM Project Execution Plan is based on definition of BIM Goals that are achieved through the application of BIM Uses" [9]. This plan of processes ensured the most suitable plan for using BIM Technology that finally helps to all participants within the design process to be well oriented in the model and its project phases and to be able to carry out the most efficient work done. Such process models are usually based on Business Process Model and Notation standard or its adaptations [10] and offers mainly behavioural view of the process. This view can be extended by adding more information about structure of participants [11] and states of process that allows better process effectivity measurement [12, 13].

5 Conclusion

From the building designer´s perspective, BIM Technology is a very useful tool for better coordinated work of participants within the whole building life cycle. BIM model itself has high-level demands on information. It ensures the higher quality of not even the project documentation, but of the project design itself. Using this technology lighten the difficulty of processes that has to be carried out by people´s thought in a traditional way of modelling. Thu human factor in traditional way of modelling my cause many failures of different scale due to the high demands on human´s imagination and experiences. The BIM Technology doesn´t allow most of these failures, because it can ensure the precious sharing a developing of information used within the whole project life cycle. Widely using of enhanced BIM technology will leads to the higher quality of project design and, without dispute, to the sustainable development of buildings and structures.

This work was supported by the Grant Agency of the Czech Technical University in Prague, grant No. SGS16/200/OHK1/3T/11

References

1. P. Podmanický, V. Nývlt, *International Multidisciplinary Scientific GeoConference Surveying Geology and Mining Ecology Management, SGEM 2015* **2**(6), 377-384, (2015)
2. AIA. Available on https://www.aia.org
3. V. Nývlt, K. Prušková, IOP Conf. Ser.: Mater. Sci. Eng. **245**, 042070 (2017)
4. K. Prušková, *International Multidisciplinary Scientific GeoConference Surveying Geology and Mining Ecology Management, SGEM 2017* **17**(62), 821-826, (2017)
5. C. Eastman, *BIM Handbook*. (2011)
6. Level of Development Specification. Available on http://www.bimforum.org/lod
7. R. Pinka, J. Kaiser, *International Business Information Management Association (IBIMA)*, 2602-2616 (2017)
8. J. Myslín, J. Kaiser, *International Business Information Management Association (IBIMA)*, 4002-4013 (2017)

9. Building Information Modeling Project Execution Planning Guide and Templates: Version 2.0. Available on http://bim.psu.edu/
10. J. Kaiser, *CESB 2016 - Central Europe Towards Sustainable Building 2016: Innovations for Sustainable Future,* 781-788 (2016)
11. J. Myslín, J. Kaiser, J, *Multi-role Activities in BIM Execution* Plan, Advanced Science Letters (will be published)
12. J. Myslín, J. Kaiser, *International Business Information Management Association (IBIMA)*, 3598-3606 (2017)
13. J. Kaiser, J. Myslín, J. Rádl, *Process Management in Building Industry(ČVUT*, Praha, 2016)

Dynamic model of implementation efficiency of Building Information Modelling (BIM) in relation to the complexity of buildings and the level of their safety

Martin Hotový [1,*]

[1]CTU, Czech Technical university in Prague, Faculty of Civil Engineering, Czech republic

Abstract. The aim of the work is to present the efficiency of the establishment BIM into the building industry on the created dynamical model related to the extent to the complexity of buildings, in which Building Information Modeling (BIM) has been used as „a tool of the industry revolution 4.0 in the CZE". The work shows the rank of the influence in establishment of new technologies 4.0 and Internet issues on increasing demands and complexity of buildings and their safety. A result of simulation in the dynamical model emerges the conclusion why we should be interested in the rise of the complexity extent of processes of the buildings in relation to corresponding cost of BIM. It shows that the establishment of BIM achieves the necessary extent of the efficiency, i.e. in what assumptions and parameters the significant reduction of the direct or indirect costs has been achieved in the construction, management of the buildings and the facilities in their life cycle.

1 Introduction to efficiency issues

Implementation of BIM is currently a worldwide trend and with its significance, it moves the current building industry towards the modern and system-intelligent "*Building 4.0*". BIM processes are implied by the increasingly expanding possibilities of information technologies (ICT). "BIM is a digital presentation of physical and functional building characteristics. BIM is a source of shared information about the building, creating a reliable foundation for decision-making in the course of its life cycle from the primary design to its demolition." [1,7]

There is a number of attributes that currently prevent wider BIM implementation. Apart from such as the traditional human reluctance to implement new things when they

* Corresponding author: martin.hotovy@fsv.cvut.cz

are not required, such as, for example, the missing need to create projects in the BIM environment for public contracts or unfamiliarity with the benefits of the BIM use or with the BIM in general (according to a survey, up to 70% of respondents in the UK aren't even aware of what BIM is exactly [1,7]). Therefore, only the "incomplete" fact whether the investment into BIM will pay off for them economically is essential for the companies.

After the introduction of BIM, the survey [2] states a growth of labour productivity by 30% and decrease in queries and information or other conflict misunderstandings in the course of the building and construction works up to 50%. In the survey with BIM users, the interviewees identically cite an increase in productivity. Up to 17% of interviewees talk about a 100% increase [2]. However, efficiency is dealt with only from a limited viewpoint of the primary design and projection. For example, it does not include the costs of further periods such as licences, library and database creation costs, in-service training, training of new workers and fluctuation of employees etc. In this concept, ROI (1. , Table 1.) in the BIM solution also does not comprehensively include parts of the whole building's process: subcontractors, contractors and other parts of the construction process and sustainability, that necessarily have to be transferred to the BIM system. Since the decisive element of the BIM efficiency are costs of the building operation in its entire life cycle, that makes up 65-80% of the total costs [3, 9] from which almost 1/3 consists of management and maintenance. Frequent reason for BIM application: "*The buildings owners want to decrease the direct running costs in the buildings because of the rise in the prices (for example) of energies during past years*" [12].

$$\frac{(B - (\frac{B}{1+E})) \times (12 - C)}{A + (A \times C \times D)} = \text{ROI, first year} \qquad (1)$$

The formula variables are:

Table 1. Taken from - REVIT Building Information Modeling - (ROI).

A = cost of hardware and software
B = monthly labor cost
C = training time
D = productvivity lost during training
E = productivity gain after training

1.1 Return rate of investment into BIM in the entire complex

Therefore, within the entire complex of the construction process and the life cycle of the buildings, the economic return rate and the profitability of the BIM implementation can be different. On the contrary, according to [4], it isn't possible to fully establish ROI in the BIM implementation along with its economic efficiency / return rate of the solution from the viewpoint of the existing unavailability of a sufficient number of demonstrable data (Table 2.). However, the implementation of BIM has unquestionably many advantages.

Contractors and subcontractors: the extent of transfer to BIM as an important element for the efficiency of savings in all the phases of LCC (Life cycle cost) [3, 9].

Table 2. The extent of transfer to BIM – CZE [9].

Rather YES	16%
YES	21%
Rather NO	42%
Strictly NO	21%

The National Institution of Building Sciences (NIBS) states, that the building industry can increase its performance by 4% in new buildings and by 1.5% in building reconstruction when using BIM [4]. Some other foreign studies and authors talk about a savings ratio of up to 20%. However, we are still talking about the economic profitability only, i.e. purely the financial aspect. But we also have to take other benefits into account that the BIM implementation brings and in which it is possible to estimate a positive benefit such as, for example, the limitation of risks and the safety of buildings. [5].

- reduction of the risk of a wrong analysis of the total price of the project (limiting extra costs)
- reduction of administrative burdens – print costs, user and operational documents etc.
- increase in cooperation of the involved parties in later stages
- significant increase in the quality of checks with a positive impact on the safety of the building during its use
- improvement of logistics planning and works in the BIM environment with an effect on error reduction in the construction part of the project – especially in complex projects
- reduction of the time necessary for project execution and reduction of risks that stem from technical-economical conflicts
- increase in safety (system warnings of current dates of regular repairs, revisions or checks in every building, including the placement of the technological system)

2 Dynamic models

The ability to understand very complex dynamic systems under uncertain conditions – that is, precisely in the situation when there isn't enough relevant information. The possibility of simulating both operational processes and complex aggregate strategic issues [10].

Essential advantages: explicitness, reliability of logical consequences, complexity
It is necessary to find the "weak" spots in which a small change brings about great consequences to the result of the behaviour of the entire system and to analyse them.

"The syndrome of 90%": 90% of the project are created in accordance with the project plan. The subsequent 10% will cause a delay or fundamental problems. For example, a delay in the schedule or financial plan in extensive investment projects by 100-200 and even more percent is not an exception. This issue exists also, for example, in strategic projects in the building industry. It is often caused by a wrong estimate of the expected size of the project. Another reason is a big delay in the detection project and process errors – before we're able to detect the weak spots, it's already too "late".

2.1 The places where errors in projects arise with which the system dynamics in relation to the BIM can help:

- wrong mental model (limited perception, errors in consequence)
- information and their sharing (we need: fast, comprehensible, adequate..)
- project processing (building on “unsuitable” paradigms..)
- analysis of feedback in the course of the project (test)
- underestimation of small changes with a subsequent extensive influence
- long duration of the project (effects of development cycles of the sector)
- necessity of cooperation and communication between the sectors

2.2 Determining the aim of the model

- Theoretical vs. practical model (ratio on the representation scale)
- Interpreting the situation (understanding it) vs. predicting its behaviour?

Model assembly:

Mental model → Conceptual model → Diagram, preliminary model → Model → Evaluation [11].

2.3 Basic Dynamic BIM implementation model (Without implementation of the subset employess)

The model Basic Dynamic BIM implementation (Figure 1.) is based on the system dynamics method [10]. The structure of the model is created by means of stock elements and flow elements [11].

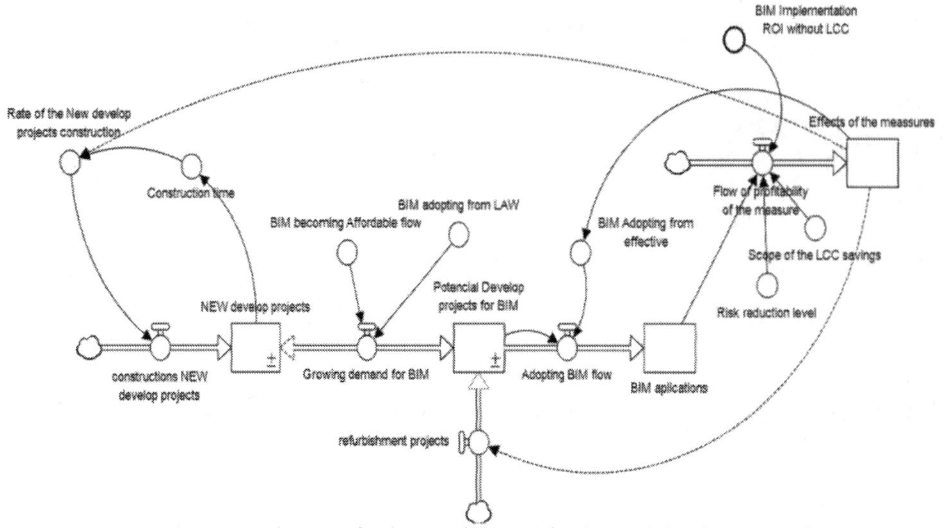

Fig. 1. Demonstration Dynamic BIM implementation model (Created in the SW Stela© Architect).

2.4 Mental description of considerations coming from the dynamic model

Employment model, training (Figure 2.) shows the dynamic model as one of the key partial sub-models. The transfer to BIM concerns all the parts of this process in its coherence and complexity, and the process must respond to this. The extent of the effect from a BIM implementation is significantly influenced by two more attributes.

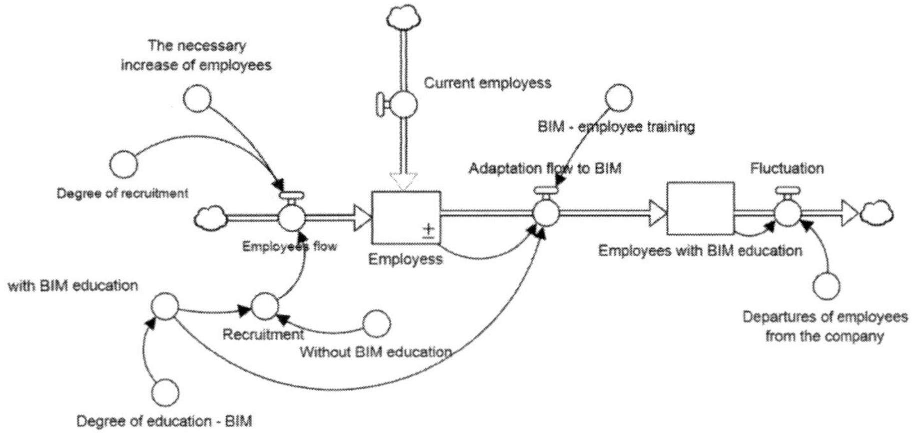

Fig. 2. Employment model, training (Created in the SW Stela© Architect).

-The extent of BIM implementation
-Costs / return rate for the period of the whole LCC, i.e also in the operational phase
-Human resources

The actual rate of return of the SW and HD purchase and employee training included in the ROI calculation for the efficiency calculation of BIM implementation isn't decisive and it is one of the elements of efficiency measurement. The model exhibits a strong feedback in case the conditions for positive BIM effects are met, i.e. the crossing of the BIM implementation extent in projects (that is, in case of a support from the state for these measures – the condition of a BIM use in public contracts that make up more than 50% of the volume in the building industry/year in investments [3,9]).

However, when a significant positive BIM effect on the building industry is achieved, the model shows a strong pressure on the sub-model ***Employment model, training*** (Figure 2.) again, since the need of an in-service training or recruitment of new employees who are capable of working with BIM significantly increases. Maintaining balance and positive effect is possible to achieve with a level of work education with BIM in the educational process at secondary schools and technical universities.

BIM processes: Planning → Execution → Management and operation → Demolition

This way, there are many benefits in the use of the BIM model from advanced visualisation methods including the simplification of the construction management to the reduction of costs for the operation of the building structure [6]. Therefore, BIM also demonstrates a visible effort to be the BI "M" (Building Information „*Management*") in its concept and application during the whole LC (Life cycle).

It is important for the people who work in the building industry across-sectors to also familiarise themselves with the BIM environment as users, along with people in the whole contractor and subcontractor chain. This is due to the complexity and long-term nature of the projects. Information that is provided by BIM must be understood by all the persons involved, so that they can efficiently use the information in connection to other elements and data of the processes. This requires a high degree of cooperation.

So all the more, the last word in the BIM abbreviation – modelling (M) – is being replaced by *"management – information management"* and that is the so-called knowledge management that has a unique place in such an advanced system integration such as BIM. Integration of processes and their source information coherence, which creates backward and mutually influencing dynamic bonds, shows that it is a developing dynamic process. Therefore, it is possible to apply ***System dynamics*** to describe it and to understand it. Especially in a situation of uncertainty, when data are not available in the necessary extent (since BIM is still a very young platform).

This work was supported by the Grant Agency of the Czech Technical University in Prague, grant No. SGS17/122/OHK1/2T/11.

References

1. National Building Information Modle Standard Project Committee. *BIM FAQ*. Available on http://buildingsmartalliance.org/index.php/nbims/faq/
2. REVIT® Building Information Modeling. *BIM's Return on Investment Autodesk*. Available on http://www.autodesk.com/revit
3. M. Slanec, Význam metody BIM pro stavební praxi v ČR. Available on http://www.bimfo.cz/Aktuality/Vyznam-metody-BIM-pro-stavebni-praxi-v-CR.aspx (2016)
4. S. Mordue, P. Swaddle, D. Philp, *Building Information Modeling For Dummies* (John Wiley&Sons Ltd., 2016)
5. Consulting Alinea. BIM. *Cost vs benefits*. Available on http://www.building.co.uk/bim-cost-vs-benefits/5081845.article (2016)
6. S. Kalinischuk, Journal of Systems Integration **6**(3) (2015)
7. National Institute of Building Sciences. N*ational Building Information. Modeling Standard*. (National Institute of Building Sciences, USA, 2012)
8. Underwood J., Isikdag Umit. *Building Information Modeling and Construction Informatics* (IGI Global, 2010)
9. A. Anisimova. *Základy Implementace BIM na českém stavebním trhu* (Fineco, 2012)
10. J.D. Sterman, *Business Dynamics Systems Thinking and Modeling for a Complex Word* (McGraw-Hill Companies, 2000)
11. D. Vytlačil, *Systémová analýza a Syntéza* (Nakladatelství ČVUT, Prague, 2007)
12. D. Vytlačil, Prediction of a building energy conservation market. CESB 2007 Prague International Conference - Central Europe Towards Sustainable Building (2017)

System description of the defect creation and elimination in the buildings

Dalibor Vytlačil[1,*]

[1]Czech Technical University in Prague, Faculty of Civil Engineering, Department of Engineering Informatics, Thákurova 7, 16629 Prague, Czech Republic

Abstract. The paper deals with the dynamics of the system of defects creation in the buildings and also with the possibilities of the defects elimination. The problem is described by the dynamic model that includes key elements influencing the system behaviour. The model is based on the system dynamics method that uses the stock elements and the flow elements which change the stock element values. In the described problem the main stock is the number of the equipment defects. This parameter can be increased by the flow element that depends on the time of using and the load. For the defect elimination are two possibilities: the planned maintenance and the repair of broken parts. Other elements are number of workers available for the elimination of defects and the running cost depending on the maintenance effort. The case of the main system parameters calculation is presented. The model can be used for understanding the importance of the maintenance in the field of facility management.

1 Introduction

The houses as well as all structures became in past years high complex and it has big influence on the desired quality of the design and construction works in sites. Higher complexity it is possible to see in HVAC systems and other active elements but a similar development is also in the building structure. These factors feature an operation phase. In this phase of building use we perform operations that ensure the quality of the internal environment and all processes related to the building user. Facility management is nowadays common management method which is used for planning and conducting these activities. It makes use of planning tools that are focused on the detail, e.g. scheduling systems and databases, see [1]. Important part of the processes is related to the reduction of the defects.

The correct description of the real behaviour of the system requires the holistic approach. The system approach for describing the system of interdependent elements is necessary. The main elements are the maintenance and the repairs of broken parts. It is dynamic problem that needs dynamic model for the description of the time changes of the variables.

* Corresponding author: vytlacil@fsv.cvut.cz

2 Method

The method for understanding the influence of all relevant variables that affect the system is system dynamics. This method uses the stock elements for the description of the variable changes during the time period and the flow elements for the description of rates that change the stock values, [2-3].

The stock value changes are described by Equation 1.

$$Stock(t) = \int_{t_0}^{t} [Iflow(s) - Oflow(s)]ds + Stock(t_0) \quad (1)$$

The design of the cause diagram that presents how entities in the problem are influenced is the first step for creating the system dynamics model. The arrows include also polarity explaining the direction of the change for the next element, see Figure 1.

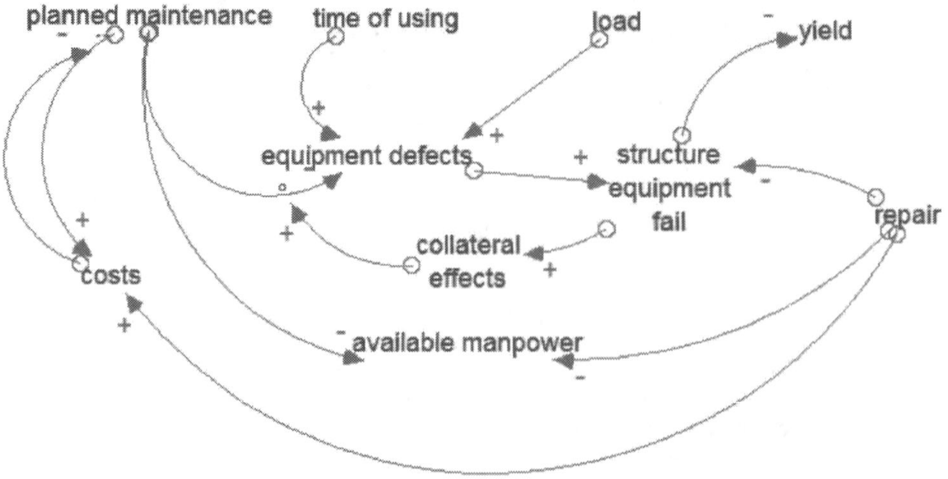

Fig. 1. Cause diagram for described problem.

2.1 Dynamic model

The key elements of the system dynamics model concerning the investigated problem are depicted in Figure 2. The main stock is *Defect* which is influenced by defect creation and two possibilities of the defect elimination: planned maintenance and the repair in case of the broken component, [2]. Presented basic model is the core in the broader model. The model for the real situation has to include subsystems dealing with financing, resources planning and supply line of spare parts for the repairs and the maintenance activities.

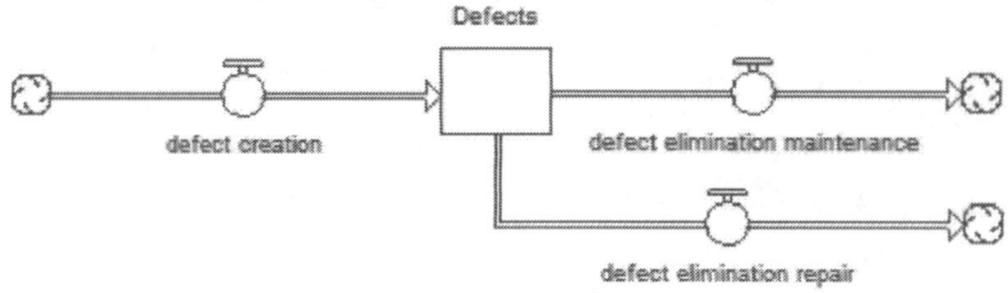

Fig. 2. Basic model of defects creation and elimination.

The model including described subsystems is presented in Figure 3. The considerable elements are used for manpower dynamics. *Active workers maintenance* and *Active worker repair* are numbers of workers used in certain time for planned maintenance or solving ad hoc problems (repairs). *Available workers* are at this time free for the next work in both kinds of defect elimination. The stock can be negative to demonstrate the situation when the system suffers from manpower shortage.

The elimination is based on the number of workers needed for one defect elimination by means of maintenance and repair (WMD and WRD) and on the time needed for the finishing the task. The defect creation depends on the time of using the component (in some cases the whole building) and on the load (intensity of the building use).

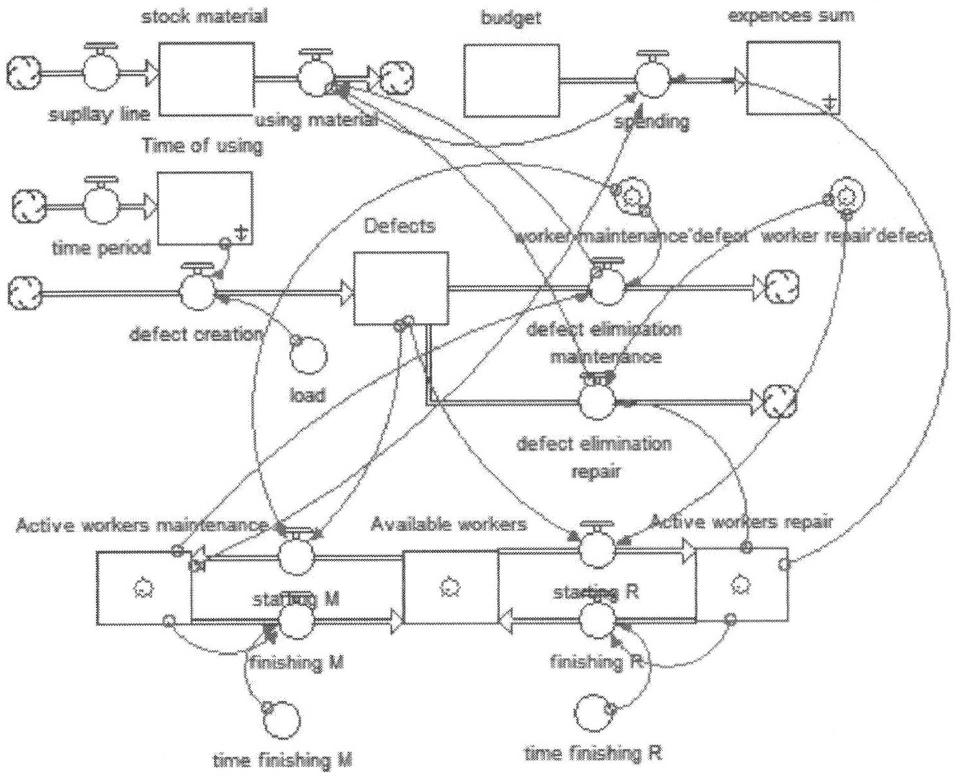

Fig. 3. System dynamics model.

2.2 Input data

The simulation is performed for the case of the office building with the floor space 11000 m^2. The initial number of workers for ensuring maintenance and repair activities is 4. The time period is 12 months. The goal of the simulation is to find out the relationship between defects and number of workers in the dynamic environment.

The parameter value *Defect* is the initial value for the simulation. The initial values for all parameters influence also the dynamics of the system particularly in the case of short time period. The parameter values are in Table 1.

Table 1. Input values.

Parameter	Value
Active worker maintenance	2
Active worker repair	2
Available workers	0
Defects	3
WMD	0.2
WRD	0.6

3 Outputs from the simulation

The output parameter are in Figure 4 and Figure 5. The first group of parameters is focused on the number of defects and flow parameters influencing the stock *Defects*. In the beginning the high efforts in the defect elimination is evident. In the end of the time period has the influence the defect creation because of the time factor.

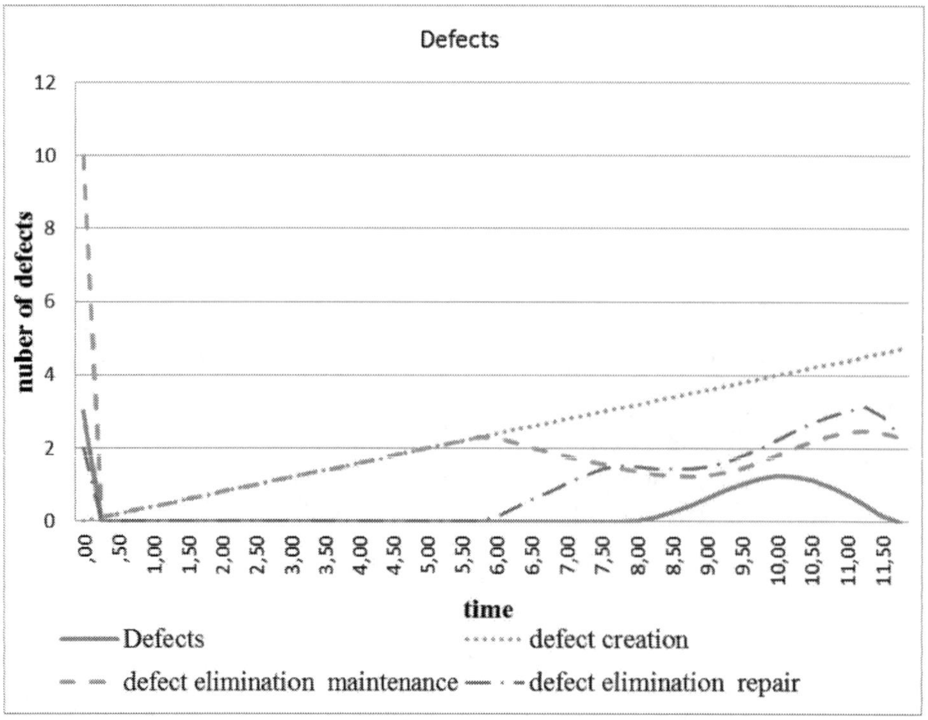

Fig. 4. Defect dynamics.

The manpower dynamics is presented in Figure 5. In the beginning the result shows the worker shortage and the need for more than 4 workers. It means in this time the company providing facility management has to add 1 worker or to use the capacity from the external environment.

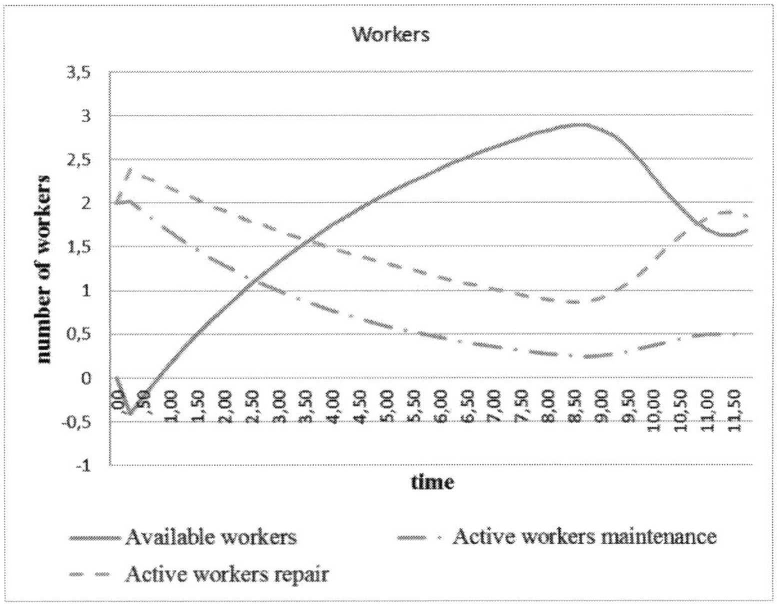

Fig. 5. Manpower dynamics.

4 Conclusions

The paper presents the system approach for the calculation of the defects and the description of the dynamics of considered parameters. The model is focused on the dynamics of manpower and does not use the restriction from component supply system or budget.

From the model it is possible to find out:

- Dynamics of the system with interconnected elements;
- Number of workers for maintenance and repair works;
- Dynamics of manpower in the case of the time dependent defect creation;
- Demand for outsourcing.

The next development will be devoted to the design of other subsystems as the stock management for spare parts and the influence of the operating cost on the financial results of the company. It is necessary also consider the changes in revenues in case of repairs that will influence the business outputs of customers (tenants in the building). The broad goal of model using is to find adequate policy for resources planning and company performance improving.

This research has been supported by SGS grant SGS17/122/OHK1/2T/11 Dynamic models of changes in building industry investments.

References

1. B. Chanter, P. Swallow, *Building Maintenance Management* (Blackwell Publishing, Oxford, UK, 2007)
2. J.D. Sterman, *Business Dynamics: Systems Thinking and Modeling for a Complex World* (Irwin/McGraw-Hill, Boston, MA 2000)
3. J. Morecroft, *Strategic Modelling and Business Dynamics* (Wiley, Chichester, UK, 2008)

Using BIM model for Fire Emergency Evacuation Plan

Adam Stančík[1,*], *Roman* Macháček[2] , and *Jiří* Horák[1]

[1]Brno University of Technology, Institute of Computer Aided Engineering and Computer Science, 602 00 Brno, Veveri 331/95, Czech Republic
[2]Czech Technical University in Prague, Department of Steel and Timber Structures, 16636 Prague 6, Czech Republic

Abstract. Is it possible to use building information modelling (BIM) to create a fire emergency evacuation plan in the national condition of the Czech Republic? This article focuses on the advantages and possibilities of using BIM for fire protection.

1 Introduction

When there is a fire, every second counts. Fire and smoke spread very fast. To avoid unnecessary loss of human life, it is important to have a well-thought out fire evacuation plan. But that is not the only requirement. For better orientation within a burning building, a floor plan with marked escape routes that everyone is familiar with is created. This plan is the graphic part of a fire evacuation plan. For a fire evacuation plan to be effective, it must be kept up to date throughout the service life of the building. This is what building information modelling (BIM) can do for fire emergency evacuation plans.

2 Preparation for BIM model

Currently there are no specifications for BIM in the Czech Republic. This has resulted in there being a big difference in the quality of various models. These models are frequently created for investors for future use, for example for facility management; however the investors often do not know themselves what to expect and see nothing more in the BI model than a 3D view of the final state of the object. Those with more experience know that they should focus more on the information aspect of the model. The basic elements (in Autodesk Revit© BIM software these elements are called 'families') of these models can sometimes be found in several world languages, from common languages such as English, German and French to Chinese, which is harder to translate. Unfortunately, this linguistic diversity appears not only in non-reported parameters, but also in reported parameters. The model designers themselves are usually familiar with the model, however there are situations where

* Corresponding author: stancik.a@fce.vutbr.cz

the creators revisit the model after a period of time and are unsure what specific items were intended for. Linguistic consistency should, therefore, be one of the basic features of a model. The language for the whole project should be clearly defined at the beginning of a project. In the case of BIM, this should at least be the case for the reported parameters.

3 Cooperation

Many professions – and people who are forced to communicate with each other – enter the lifecycle of a building. To facilitate communication, BIM sharing information using a single building information model, as shown in Figure 1 on the left. In the Figure 1 on the right, one person has been added. This person is often not spoken about within the BI model, but this person is an inherent part of every building, whether it is a family home or the tallest building in the world. That person is the designer of the building's fire safety solution – the fire engineer. It is the fire engineer who ensures the safety of the people using the proposed building. The designer of a building's fire safety solution is involved in several stages of the building's lifecycle:
1. Planning – the creation of a fire safety solution (PBR), including the design of escape routes and the calculation of evacuation time;
2. Implementation – the creation of necessary reports and the location of fire safety equipment (PBZ);
3. Building operation (control) – PBZ revision and control, and possible future use of data for rescue units in case of a fire in the building.

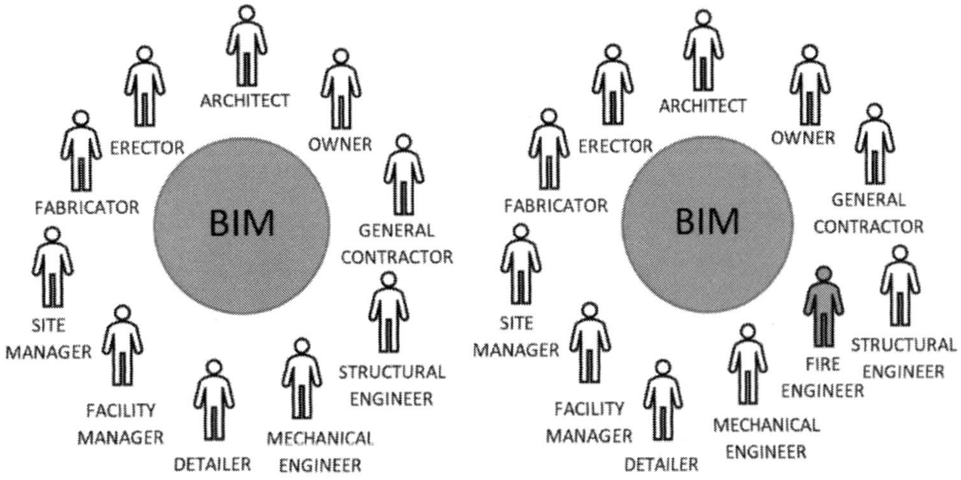

Fig. 1. The professions involved in the lifecycle of a building.

4 Parameterization

The information BIM provides is much more valuable than the actual 3D interpretation of objects in the model. The 3D view can enable us to know that there is an extinguisher in the building, however this 3D view can also be enhanced by the addition of an information folder that provides the localization, exact type, manufacturer, price, fire class, type of dispenser, temperature range, fire capabilities and amount of extinguishing media, as well as a picture of the device and, last but not least, all the information that is generally stated on the label of a fire extinguisher, such as the year of production, and the inspection and revision dates. Each bit of information is understood as a parameter for the element or 'family'. The number of element parameters is unlimited. All information is reportable and can be used not only in project documentation or construction, but also in the management of the building. How detailed the information is depends on the requirements of the investor and the authors of the model. Parameters can be further sorted into groups for better clarity. Parameters can be numeric, integral or of the YES/NO type, among other options. Most of the parameters we create and use are text-based. Other than using them for reporting, it is also possible to use these parameters to create a graphic display of a fire plan.

5 Families

As mentioned earlier, the model consists of basic elements called 'families'. The model created in the BIM Revit© tool normally consists of all three family types: system-based, model-based and embedded. Creating families is one of the advanced controls of the above-mentioned software. In the beginning, the goal was to process the signs used to create the fire evacuation plans of a building and then use them to create these plans. Example of some of these signs are shown in Figure 2. The following basic rules are used in the formation of families:
- It is a combination of 2D symbol and 3D object.
- The insertion point is located at the one edge of the symbol.
- The family is 'cleaned up' and all unnecessary items, such as unused materials, lines and hatches, are removed as they affect the resulting file size.

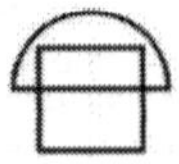

2D symbol for Fire Alarm Call Point

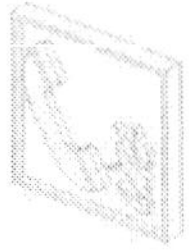

3D object for Fire Alarm Call Point

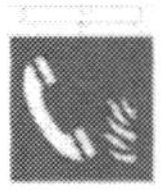

Floorplan view in coarse detail of family Fire Alarm Call Point

Fig. 2. Example of one of signs used for evacuation fire plans.

6 Evacuation fire plans

Article 33 of Decree No. 246/2001 Coll. on fire prevention states what must be included in a given fire evacuation plan. A fire evacuation plan consists of both text and graphics, to which we are obligated by the following point of the above-mentioned decree: "f) graphical representation of the direction of escape routes on individual floors" [1]. The graphical portion of an evacuation plan is prepared in the form of a floor plan on which the individual floors of the building are plotted out, as well as escape routes and directions, fire extinguishers, fire hydrants, fire alarms, the fire assembly point and the safety instructions in case of a fire or accident. The graphical map of the fire evacuation plan must also include the position in which the observer is currently located. When structural changes are made to a building, fire evacuation plans can become outdated, and this can have fatal consequences. This problem could be solved by having a maintained BI model as the basis for processing the graphical part of a building's fire evacuation plan itself. On the graphical portion of a fire evacuation plan, apart from the floor plan, which can be updated, there are parts of the plan that do not need to be changed and are, thus, valid for all future plans. These still-valid elements include the instructions in the event of a fire, the instructions in the event of an accident and the explanation of what the signs in the fire evacuation plan mean. This is done by means of legends, which are copied on to the drawings of the evacuation plans for each individual floor.

7 Creating the graphical portion of the fire evacuation plan

The actual creation of the graphical portion of the fire evacuation plan is partly dependent on the data used by the architect or civil engineer. With a few adjustments, we can accomplish an extremely clear layout of an evacuation plan in a short space of time, while maintaining the maximum information necessary for the safe evacuation of building occupants. These adjustments include: setting a single solid hatch line for walls, leaving only the numbers and names of rooms, turning off the detailed display of the fillings of openings such as windows and doors, and leaving only empty openings and a number of other modifications. To complete the graphical portion of the fire evacuation plan, it is necessary to fill in the ground plan with the above-mentioned legend. Additionally, a sign with the exact location of the fire evacuation plan must be added to the plan.

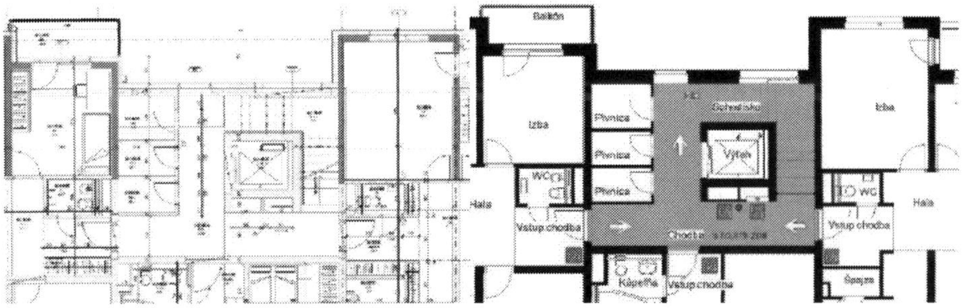

Fig. 3. Part of the evacuation plan before (left side) and after (right side) editing the ground plan of the building.

8 Add-on application and plug-in for BIM software

Almost all tasks designers of fire safety solutions undertake when designing project documentation can be programmed in order to reduce the time the work takes, especially when it comes to repeat processes. In this case, there are a few simple commands that, in their entirety, make up a complex code. BIM software Revit© provides a rich and powerful .NET API that can be used to automate repetitive tasks, and extend the core functionality of Revit in simulation, conceptual design, construction and building management, and much more. It is possible to program Revit's .NET API with any .NET compliant language, including VB.NET, C# and C++/CLI. The Software Development Toolkit (SDK) provides extensive .NET code samples and documentation to assist the beginning stages of development with Revit's API [2]. In addition, an application that would save fire safety solution designers a lot of work, and in the best cases managed to generate 'drawings' for the fire evacuation plan by itself, is in the development stages.

9 Information model and visual programming

For those who are interested in creating their own tools, but are limited by the programming language, the standard BIM parameters of the software can be extended by an add-on for visual programming that uses an algorithmic graphics editor environment. The add-on has been developed as an open source project, allowing developers and users to actively participate in its development and continual improvement. The main difference between code writing and visual programming is that visual programming uses the associated nodes to perform the given operation.

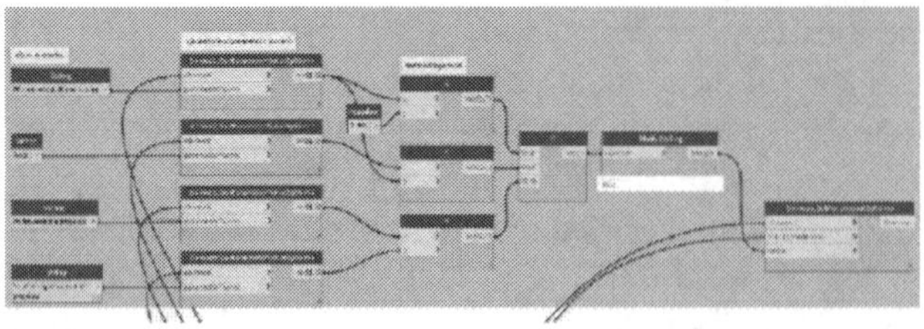

Fig. 4. A script assigning the correct values using the logic function.

There is an unlimited number of possible uses for visual programming. Designers of fire safety solutions can be assisted with seemingly easy steps, as well as complicated ones, such as calculating the length of an escape route or the number of evacuated people.

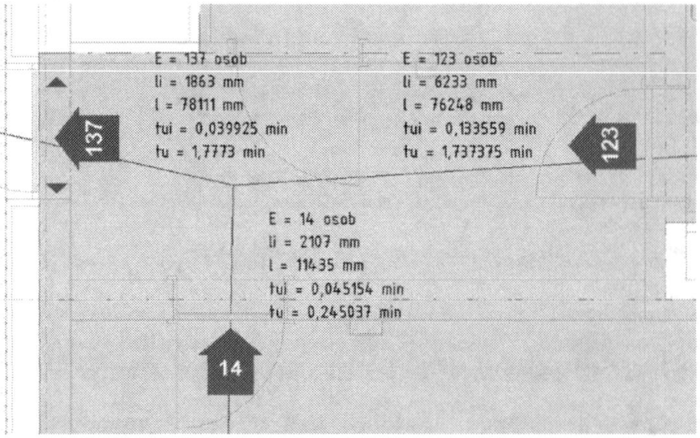

Fig. 5. Counting values from multiple consequential segments.

In this case, the designer would follow several steps. The first step for a designer using the tool is to create a new project that is joined to the building model that is linked to the original file. This ensures that the changes made in the original file will take effect in the new project as well. The next step is to select the function of individual rooms in the building. This automatically assigns values in order to help calculate the number of building occupants. Subsequently, each room is assigned to a fire compartment, where other values can be added for calculation and the persons in the individual rooms belonging to it can be added up. Finally, a network of escape routes, which represents the streams of exiting people towards the open area, is formed in the model. After defining the escape routes, the final step is to run a script created using the visual programming plug-in. This script will run a number of algorithms that primarily assign the number of people from the fire sectors, following which the number of evacuated persons is assigned to the first segment of the escape route. In case of the collision of multiple segments in one place, the script will decide on a greater value of evacuated persons. The algorithm is repeated several times before adding up the people in all of the segments. The script proceeds analogically when calculating the length of the escape route and the evacuation time. Calculated values; such as the length of the escape route, the number of evacuees and the evacuation time; can then be reported in a simple way and provide a comprehensive overview in order to assess the requirements. Using this data, if the report is unsuitable; for example, an escape door is too narrow, or an escape route is too long; the object can easily be edited and the calculation recommenced.

10 Conclusion

When there is a fire, every second counts. It is vital that fire evacuation plans are kept up to date. Thanks to BIM models and their subsequent use, we are able to pass on information quickly and with minimal loss. There is a higher requirement for fire safety solution designers to also have knowledge of information technology, but those who are fluent in programming languages or are able to use visual programming will save a lot of time on tasks that take up tens of minutes of an ordinary designer's time. In cases with a well working code, the error rate that could arise from the need to repeat the process several times during the creation of the plan can also be minimized.

Acknowledgement: This paper has been worked out under the project No. FAST-J-17-4590 and under the project No. FAST-S-17-4275

References

1. *Act No. 246/2001 Coll., On determination of conditions of fire safety and execution of state fire surveillance (Fire Prevention Regulation)* (Collection of Laws, Prague, 2001). Available on https://www.zakonyprolidi.cz/cs/2001-246
2. Autodesk Developer Network. Available on http://usa.autodesk.com/adsk/servlet/index?siteID=123112&id=2484975

Development the tool for demand control ventilation systems elements design and their impact to building life-cycle assessment: a review

Richard Pinka[1,*]

[1]CTU in Prague -Faculty of Civil Engineering, Department of Applied Informatics, Thákurova 7 166 29, Prague, Czech Republic

Abstract. The paper describes possible methodology for DCV – demand control ventilation design process which will lead to solving the most important problems of more accurate ventilation system design, the efficiency: time and costs of design while improving its effectivity (level of design confidence, degree of uncertainty, financial parameters). For this purpose it has been started developing the tool implementing the methodology. Furthermore, it describes steps of testing and validation design process by implementation state-of-the-art algorithms for measuring the occupancy or CO2 detection for DCV system control testing sites and their comparison with computation design results. It describes also the reasons and necessity of decomposition of whole ventilation system to specific elements for the design and assessment procedure.

1 Introduction

In modern commercial and office buildings over Europe is actual trend to design more interconnected HVAC systems together. As stated in [1], systems of ventilation and climatization has a frontier position within the buildings energy consumption based on the either consumption of the cooling/heating energy (as well as energy for humidification or dehumidification for climatization systems) for maintaining the air temperature and consumption for fans operation. HVAC systems conceptually divided far apart do not use the advantages of each other for the energy efficient operation, which often leads to different concept misconduct issues. The need of energy consumption reduction tends to integrate more HVAC systems together, fancoil units for the cooling purposes are smoothly changing for induction units i.e. cooling beams, connecting the cooling system with the ventilation system. The intake air for cooling beams increase their cooling capacity based on the income cooling water and if this concept is used for building as a whole, it gives better energy performance of the cooling source, because of lowering the cooling water temperature difference used to cool the air above dew point temperature (e.g. EWT/LWT

* Corresponding author: richard.pinka@fsv.cvut.cz

18/21 °C instead of 6/12 °C) as of the case of cooling fan-coils design. Nevertheless, going to the side of the air and ventilation system, more sophisticated design brings more questions such:

- Shall be prioritized air temperature and airflow or water flow for cooling capacity regulation of the induction units for interior temperature regulation?
- Shall prevail regulation dampers based on the CAV or VAV concept?
- Similarly, shall be preferred constant water flow or variable water flow system?
- How many dampers and which position shall be defined for the best price/performance ratio? Which type of actuators shall dampers have?
- Will the system be enough flexible for future interior changes?
- Last, but not least - is the DCV design expendable or favourable for the specific project circumstances?

Each question is related to each other and together they are defining the dimensions/options of system design and assessment. First dimension is duct system. Low operating energy consumption may be reached by choosing a low duct system friction Δp by appropriate duct system design. Pressure drop is often increased by inappropriate concept of duct system due to limited space and shafts for ducts, causing high width/height duct ratio, small flow area and higher amount of duct elbows. Another dimension of the design leading to energy costs reduction is to minimize the operating demand of an air flow volume V over the operating time and appropriate design of the regulation elements and their control for air distribution: these aspects of design are the main topic of this paper.

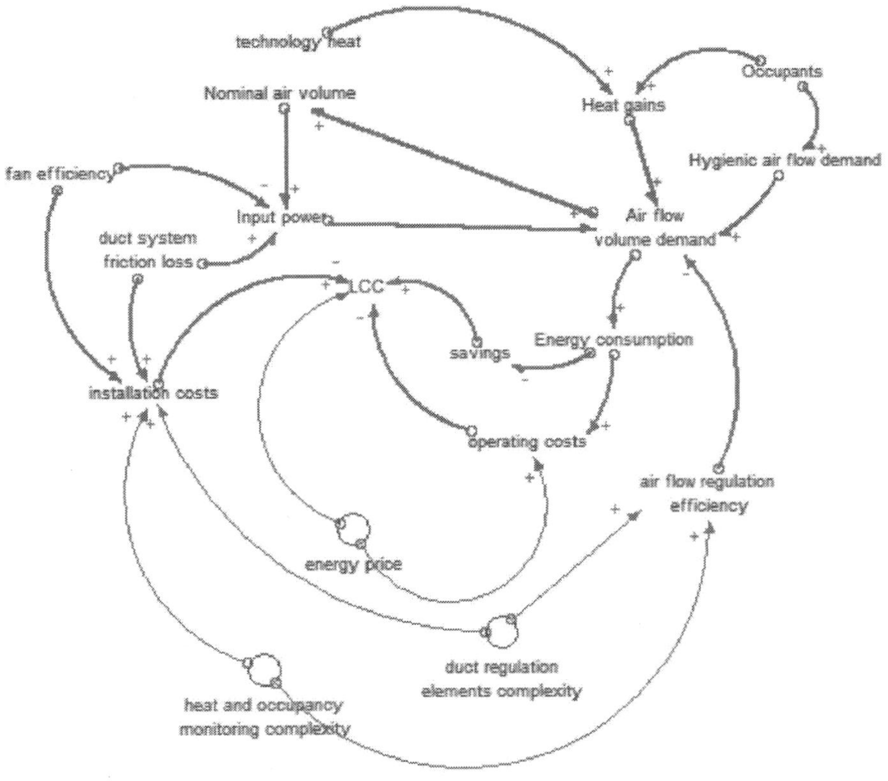

Fig. 1. Influences and relationships in the ventilation system design.

As is seen on Fig.1, the relations of the design dimensions define whether there is positive or negative impact within their relations. Installation costs, savings, operation costs may be the most important evaluation criteria for investor but may be considered by each investor differently, based on his experience and priorities. Relationships between these specific physical and intangible phenomena influencing ventilation system design is difficult to determine with high accuracy without proper tools. Nevertheless, due to the informational technologies development, such tools development is in today focus. Important aspect for the evaluation are quality of hardware installed: duct system defining the friction loss, and quality of technology: fan efficiencies. Very important role for operation air flow volume demand has the system's ability to distribute the air right to the zones where it is needed and no more anywhere else. This ability named in Fig.1 as a duct regulation elements complexity may offer high flexibility to the system in the matter of air distribution minimization. Distribution logic is described furthermore in part about the distribution and control strategies and in Fig.4. Different but essential design dimension is about data inputs for the BAS control system, to find strategy how to send right inputs about the IAQ to the system. All of these aspects are considered design stage sensitively to higher installation costs price - which are missing important data about life-cycle system costs very often due the lack of designing tools able to absorb and process data. These design options thus fall to the disgrace very often. With investor's aim to save investment money, not the operation money, may the reduction costs on regulation elements bring more issues in future than wished. Such most occurring issues not influencing only energy consumption may be:
- wrong basic concept - interfering by long-term facts not taking into account;
- overestimating the volume of cooling source´s cooling capacity;
- overloaded ventilation flow over the operating time based on the lack of conceptual regulation elements in ducting system - relatively higher energy consumption;
- insufficient ventilation capacity of AHU based on lack of regulation elements;
- wrong control design and logic - missing the proper monitoring tools or proper metric algorithms for the system control;
- unable to set up or tune up the system after some period of operation or after the interior conceptual changes.

2 Methodology definition on a case study building

The matter of the research is to firstly create the tool for base computation models used for design air distribution within building zones and its energy consumption estimation. This tool will be able to import parametric data from selected open BIM filetypes- i.e. IFC filetype. Models will further implement different control strategies based on selected system regulation element types (actuator types or whole types, e.g. in Fig.7). The matter of research moreover will include definition of the control input data strategies. The variants of strategies based on the computation model will be able to compare between each other and in the later stage will be an effort to compare computation results with measurements on appropriate model-like sites for the tool validation.

The first computational model design is developed on the data based on newly built office building with 7 storeys situated in the Prague. The computation model is based on ventilation system for the ventilation of 2nd to 7th floor. The system consists of two AHU units situated on roof. Due to similar space layouts the model was simplified by used of one 2nd storey (40 zones), two 3rd storeys (2x51 zones) and three 4th (3x52 zones) storeys for reaching the number of 7 storeys. Complex computational model includes 298 zones in total.

Primarily, an "air flow demand" will be defined by a concrete design elementary input data-sets such as occupancy numbers/rates, solar/lighting/body/notebook heat gains in each zone specified each dayhour. Based on air distribution and control strategies, the "air supply", which energy consumption will be estimated on, will be defined afterwards. Air demand and supply will take into account the occupancy ratio by occupancy curves and occupancy probabilities as another design dimensions and input variables to the model. The model will then modify regulation characteristics per each room separately, according distribution elements (cooling beams) and its regulation/control possibilities in each zone.
Axioms, which are considered for this topic are:
- volumes of air flows defined within ventilation system designs are mostly CAV and are hard to set;
- ventilation system is considered with equilateral pressures both on Supply and Outlet sides. Supply and Outlet air flows will be thus equal as well.

2.1 Practical aspects of computational model design

Various cooling/ventilation demands depends on the interior zone type (e.g. meeting rooms, open-offices, offices, kitchen, quiet rooms, corridors etc.) and their people/technology maximum loads, which has been further multiplied by the coefficients according the occupancy schedule and presence probabilities in smaller rooms (set to 0,625), because to have always 100% of maximum designed values is not proper operation profile.

Cooling load is variable input for design based on a) a zone position at the floor layout (e.g. near façade, on corner, without façade/window connection), b) daytime and month, c) actual zone occupancy- heat gains by human bodies and PC usage ([nominal value] x [multiplier according occupancy curves]), d) lighting heat gains if appropriate time. Air flows demands of all zones sums the total operation air volume demand. From this demand, the pressure drop is specified. Friction drop of system consists of duct system friction curve, (duct system friction is function of air flow). For the estimation of electric power necessary to transport the required air flow, additional pressure drops of AHU unit (531 Pa of internal pressure) and 20 Pa as of cooling beams pressure drop were included to the total operating pressure drop. Possible pressure drops of evaluated duct dampers were neglected.

Two logic setpoints for air flow regulation are defined – CO2 regulation and temperature regulation. As stated in formula (1), total air flow volume for cooling is defined:

$$[Solar\ h.gains]\ (W) + [Lighting\ h.gains]\ (W) + [Persons + laptops]\ (W) = Q_{cool,demand,time,zone} \quad (1)$$

$$[Lighting\ h.gains]\ (W) = Q_{light,max,zone}(W)\ .\ coef_{light,time}\ (\%) \quad (2)$$

$$[Persons + laptops]\ (W) = Persons_{,max,zone}\ (-)\ .\ coef_{,occupants,P}\ (\%)\ .\ [Q_{person,z} + Q_{pc,z}]\ (W) \quad (3)$$

By the logic of long-term energy estimation, sum of heat gains consists of solar heat gains in specific dayhour/month for solar gains. Heat gains data means array of at least 8 x 12 data cell structure matrix (maximum 24 x 12) per each computational part (each zone). The sum of total operating air volume demand defines its pressure loss by interpolating data within prepared pressure drop curve.

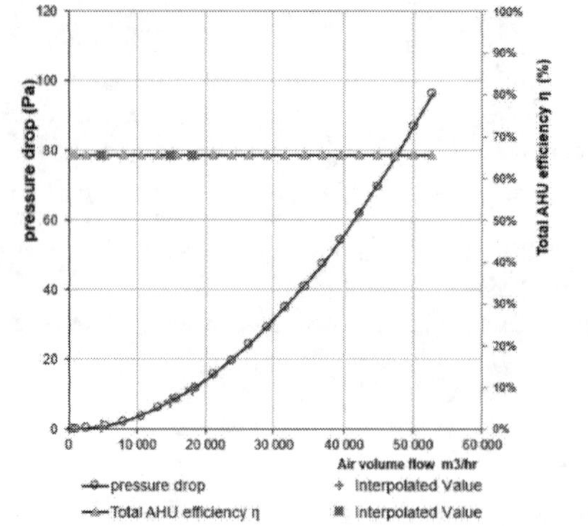

Fig. 2. Duct system friction loss curve according computed values. Interpolated values shown in figure shows example of values interpolated for the computational model operation point. Total AHU efficiency is considered for whole operating profile as a constant.

Required air volume for supply demand for cooling according defined occupancy based on operation profiles is then computed as:

$$[Persons_{,max}\ (\text{-}).\ Air\ volume\ per\ person(m^3/hour.person)]\ .\ coef_{,occup,P}(\%) = V_{hyg,demand,zone}\ (m^3/hour) \qquad (4)$$

Specific situation with small privacy and phone rooms where random factor of occupancy is needed, hygienic volume and cooling heat gains has been determined as:

$$[Persons_{,max}\ (\text{-})\ .\ Probability(0{,}625)\ ]\ .\ 100(\%) = V_{hyg,demand,zone}\ (m^3/hour) \qquad (5)$$

$$[Persons_{,max}\ (\text{-})\ .\ Probability(0{,}625)\ ]\ .\ 100(\%)\ .\ [Persons + laptops]\ (W) = Q_{cool,demand,time,zone}\ (W) \qquad (6)$$

When heat loads for cooling were determined, appropriate volume of demanded air flow volume is known by the fact, cooling capacity of each induction unit is conditioned by specific air flow. In design were defined three types of cooling induction units, with the names CB1, CB2, CB3, with nominal designed cooling capacities 1255, 928, 503 W and with specific nominal design air flows 155, 120, 80 m3/hour. These induction units may be regulated differently according selected actuators for regulation dampers (if installed), according Fig.3. Each of 298 zones has either its own set of cooling beams or hasn't none and thus is not considered in model. Air flow demand for actual operation is thus a ratio of actual heat gain and maximum cooling capacity in zone multiplied by nominal air flow for cooling in zone as seen in formula:

$$Q_{cool,demand,time,zone}\ (W)\ /\ Q_{cool,design,nom,zone}\ (W)\ .\ V_{cool,max,zone}\ (m^3/hour) = V_{cool,demand,zone}\ (m^3/hour) \qquad (7)$$

A required operating ventilation needs for further computation are higher values from $V_{hyg,demand,zone}$ or $V_{cool,demand,zone}$ named as $V_{demand,zone,time}$. When demand side of evaluation process is finished, supply side shall define how to supply the air more efficiently. As was already mentioned, there is more options to let the air distribute. That is the reason, why the final air flow demand will not be the same as air flow supply, finally implemented in operation (computational model simulation). No matter the fact, supply side will result in many different results based on the system controller types and their locations, the electricity consumption needed for air transport based on specified concepts, will be defined by firstly estimation the required mechanical power, P_{mech}, necessary to pass the required volume flow through duct system by defined pressure drop function as in :

$$P{,}mech = V.\Delta p \qquad where \qquad \Delta p = f(V) \tag{8}$$

To estimate electric input power for the mechanical movement of fans, there shall be included total AHU efficiency:

$$P{,}el = V.\Delta p \,/\, \eta_{tot,AHU} \tag{9}$$

Where V is nominal air flow volume (m3/h; m3/s; cfm), Δp is specific nominal pressure drop (friction drop or head loss plus internal pressure drop of AHU and induction units) of the system (Pa) and $\eta_{tot,AHU}$ is efficiency of AHU unit. V and Δp are outputs based on the system design. $\eta_{tot,AHU}$ is the matter of the technology efficiency and manufacturer selection. $\eta_{tot,AHU}$ of AHU unit, acc. eq. (9). Efficiency $\eta_{tot,AHU}$ is combination of more distinguished efficiencies:

$$\eta_{tot,AHU} = \eta_{drive} \cdot \eta_{VFD} \cdot \eta_{FAN} \cdot \eta_{BELT} \tag{10}$$

Where η_{drive} is electric motor efficiency (88,7% - EFF2 type), η_{VFD} efficiency of variable speed drive (98%), η_{FAN} is fan efficiency (75,6%) η_{BELT} is belt efficiency (in this case is 0 as for direct drive fan).

As already mentioned, control elements options gives the designer another dimension for the design and evaluation. In prepared computational model, damper locations in duct system hasn`t been changed and stayed the same as they were designed in real building design. Only two parameters will be changed in the computation models:

- Damper regulator types between a), b) or c) according Fig.3.
- Whether to use or not the "hydraulically joined containers" principle for DCV minimizing air flow volume control, described in and around Fig.4.

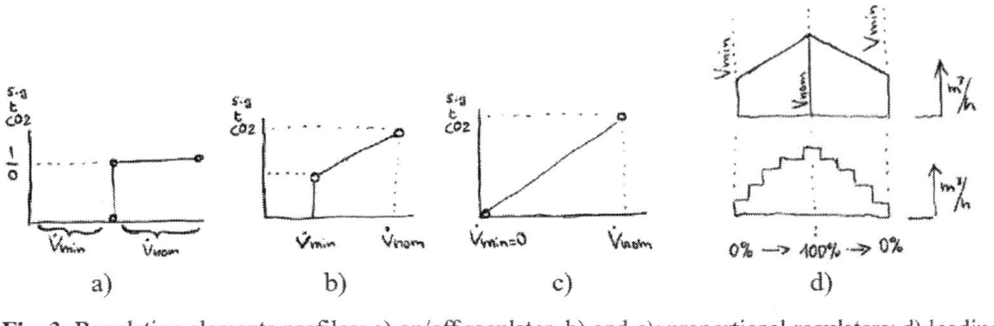

Fig. 3. Regulation elements profiles: a) on/off regulator, b) and c): proportional regulators; d) loading profiles of zones.

Regulation elements profiles are described by linear dependence (less often by non-linear description) as of a), b), c) in Fig. 3, a) on/off regulator, most simpler, knows only 2 positions; b) and c) are type of same proportional regulator, difference between them is that b) minimum value does not start in 0 but in some minimum value - operation in "off" position allows to pass minimum air volume through regulator. d) shows possible loading/unloading profiles of zone from minimum to maximum and backwards while unloading: upper is with b) and c) regulators, lower with a) regulator type, which is less accurate.

3 The problematics of air distribution and control strategies

After definition of the operating air flow demand is complete, specification of the regulation elements with their drives comes to place and without sophisticated data management it may be difficult to evaluate wide options in the design itself. Systems based on lowest installation costs demand may have lack of necessary regulation dampers for efficient operation. Case is described on Fig.4. Ventilation system consisting of main shaft duct route directs from fan in AHU F to specific floors with zones with air flow demands. Main open space zones defined as M ("mother zone") are biggest zones with demand for air. To M are connected other smaller zones such small offices D zones ("daughters"), utility rooms and meeting rooms R1. Let's say that R1 is not straight connected to the M and can have different occupants that are in other spaces (guests). In state Fig.4 a) is first operating schedule with demand of 138 air flow units, based on specific daughter zone needs and R1.

The state b) from Fig.4 shows different air demands, let's say, occupants from M and other zones moved accordingly the new air demands and in R1 arrived few more guests. Is system able to bring +30 more air flow units to the floor and redistribute the air due to different local needs? The answer depends on set of regulation elements, in this example main dumper B and possible regulation places for dampers (V1,V2,V3,V4,V2.1). In the case, there would be only manual damper or CAV regulator on B position, there would not be possibility to flow more air into floor without an impact to whole building. If B would be VAV controller (in case other floors would have the same type), fan could supply 30 more air flow units to the floor but without the required effect - give the right amount of air to the right spaces - especially for new arrivals in R1 it wouldn't be possible. In the case, where total amount per floor would not be changed, but floor demand does as c) in Fig.5, the troubles would not bring possible manual damper in B but absence of regulation elements on V positions.

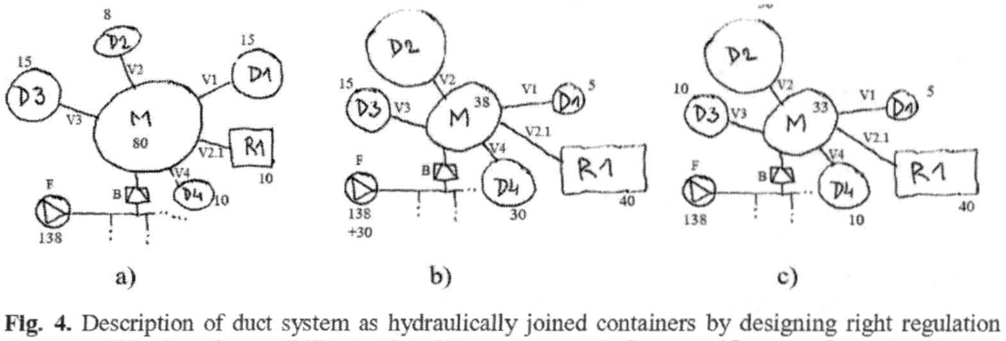

Fig. 4. Description of duct system as hydraulically joined containers by designing right regulation elements. This gives the possibility to "detach" unnecessary air from specific zones for saving it.

Either the b) or c) case, is not possible to supply the right amount to the right places and carry off unnecessary air from zones where is wasteful. Most of systems today is not able to trespass from operation a) to b) without decreasing the required flow in M room from 80 to 38 and is able only to pump even more air (42 unit in this case) to the floor to fit the needs of daughter rooms. DCV approach leads us to necessity of maximum utilization of the air, where occupants will pass from mother room to the daughter rooms and same ratio of air will be added to daughter rooms and furthermore reduced from the mother room as well. This may has similar idea as hydraulically joined containers.

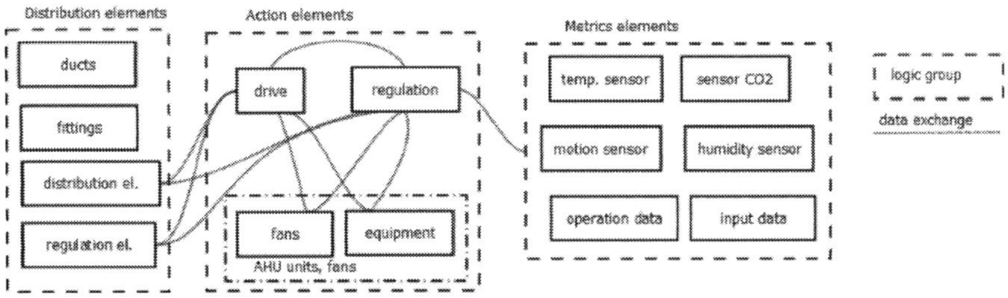

Fig. 5. Main elements of ventilation system grouped by their functionalities.

Not only because of possible issues described above, for creation of the computational model is necessary to decompose a whole system into the subsystems and their elements. By the elements division according the Fig.5 together with the definition of operating air flow volumes, control algorithm will be created. This has to consider the possible control options, which means, if fan is required to supply air to specific places, it needs already been determined, how many regulating points the air passes through till its final destination and also it needs to be determined how much each regulating element shall be opened. As result the algorithm made in design will send to fan exact data for control if implemented in operation. Based on this logic the exact regulation element profile as well as each room loading/unloading profile has to be determined.

Developing tool for design & assessment will allow to import data from BIM parametric models and allows designer made its estimations on low costs on the basis design-analyse-redesign process in the acceptable time. Explicit definition of system elements and their impact on the total air flow distribution flexibility will be then in the results very clear.

3.1 Next steps in research and validation of the models and discussion

Much has already been written about the benefits and possibilities of DCV systems. These systems works properly if they obtain proper data for their control. Data are acquired by measurement the conditions in the room for is used for calculating the amount of energy actually required and for control the distribution routes for efficient air distribution.

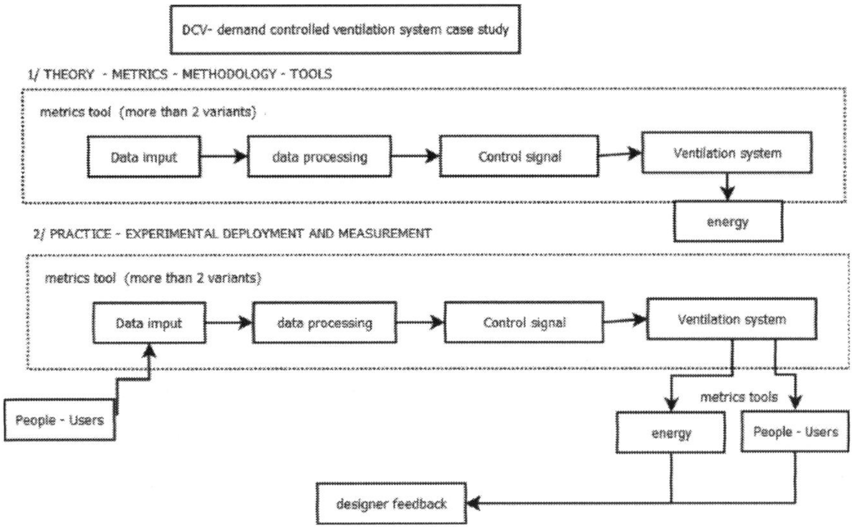

Fig. 6. Conceptual description of research topic steps.

As stated in [2, 3], there are besides CO2 other measuring technologies which are experimentally tested, according [4, 5] were tested even their combinations for their complementing, however, each technology has its own restrictions. Today´s most used CO2 sensors occupancy estimation is relatively exact but may interfere with long responsive time and air circulation patterns must be known.
The more developed metrics technology for obtaining exact and actual input data for system control, the higher chance to design and minimize air flow thus energy consumption as possible. Therefore the exact metric strategies for exact demand definition are nowadays developing and testing [5 - 9] and new are developing.

Further steps in the metrics technology research is to prepare possibilities to develop and test sophisticated metrics strategy based on an image-based processing, where continual occupants position monitoring and their movements can be achieved as stated in [7, 10] obtain exact values of occupancy rates within zones hand in hand with control logic defined in design model. As seen on Fig.6, computational model designed in developing tool is assumed later to be evaluated on site by implementation new metrics strategy to real control system and compare energy consumption differences with previous state and with computational model.

References

1. K. Ahmed, J. Kurnitski, P. Sormunen, Energy Build. **109**, 115–126 (2015)
2. E. Naghiyev, M. Gillott, R. Wilson, Energy Build. **69**, 507–514 (2014)
3. T. Labeodan, W. Zeiler, G. Boxem, Y. Zhao, Energy Build. **93**, 303–314 (2015)
4. B. Dong, Energy Build. **42**, 1038–1046 (2010)
5. H.B. Gunay, A.F. Fuller, W. O'Brien, W. Beausoleil-Morrison, I. Detecting occupants' presence in office spaces: A case study. *eSim 2016* (2016)
6. J. Yang, M. Santamouris, S.E. Lee, C. Deb, Energy Build. **123**, 192–204 (2016)
7. Y. Benezeth, H. Laurent, B. Emile, C. Rosenberger, Energy Build. **43**, 305–314 (2011)
8. Resources, D., Features, D. & Applications, F. People Counting for Demand Controlled Ventilation Using 3D Time-of-Flight (ToF) Reference Design. 1–58 (2016)
9. T. Kwevugbe, N. Brown, V. Pakka, D. Fan, Int. Conf. Enhanc. Build. Oper. (2013)
10. H.C. Shih, Energy Build. **77**, 270–280 (2014)

MATEC Web of Conferences **146**, 02001 (2018) https://doi.org/10.1051/matecconf/201814602001
Building Defects 2017

Elastic-plastic load-carrying capacity of steel members

Pavol Juhas[1,*]

[1]Institute of Technology and Business in České Budějovice, Okružní 517/10, 37001 České Budějovice, Czech Republic

Abstract. The load-carrying capacity of steel structures and members subjected mostly to bending depends in large measure on local stability of their compressed flanges and bending webs in decided cross-sections and areas. Depending on local stability, the elastic or plastic, eventually the elastic-plastic analysis and design can be applied. The actual standards for design of steel structures contain relatively detailed rules for elastic analysis by the elastic theory and for plastic analysis by the plastic theory. The elastic-plastic analysis and design of steel members are meanwhile still problematic from theory, standard and application point of view. The real elastic-plastic behavior of the steel members is complicated strength and stability problem. Therefore, the representative experimental knowledge and results about the real elastic-plastic behavior and mechanisms of failure of the steel members have been very important from the scientific and applied aspects. The author of paper realized in previous time the wide experimental research within the range of this intention. The paper contains selected knowledge and results of the previous experimental-theoretical investigation of the elastic-plastic local stability of the steel members subjected mostly to bending. The adequate slenderness – strain relation and methodology for practical calculation of the elastic-plastic bending load-carrying capacity of the steel cross-sections and members are presented in the paper.

1 Introduction

The load-carrying capacity of steel structures and members subjected mostly to bending depends in large measure on the local stability of their compressed flanges and bending webs in decided cross-sections and areas. Depending on local stability, the elastic or plastic, eventually elastic-plastic analysis and design can be applied. The actual standards for design of steel structures contain relatively detailed rules for elastic analysis by the elastic theory and for plastic analysis by the plastic theory. The elastic-plastic analysis and design of steel structures and members is meanwhile still problematic from theory, standard and application point of view, [1-3]. The paper contains selected knowledge and results of the previous experimental-theoretical investigation of the elastic-plastic local stability of the steel members subjected mostly to bending. The adequate slenderness – strain relation and

* Corresponding author: juhas@mail.vstecb.cz

methodology for practical calculation of the elastic-plastic bending load-carrying capacity of steel cross-sections and members are presented and compared with procedures that are applied in some selected standards.

2 Local stability and classification of steel cross-sections

The local stability, compactness and slenderness of the steel members and their cross-sections are relative. They depend on the loading level or material utilization, and on the buckling resistance of cross-sections and their compressed flanges and bending webs. The buckling resistance and the real behavior of compressed flanges and bending webs depend on more parameters (material properties, geometrical dimensions, material and geometrical imperfection, boundary conditions). Therefore, the steel cross-section compactness or slenderness judgments are complicated stability problem that has been theoretically analyzed through respective differential equation systems [1, 2]. For usual cases of the steel cross-sections slenderness β (width to thickness ratio) of their webs and flanges is the characteristic parameter of buckling resistance ($\beta = d / t_w$, b / t_f, c / t_f), Fig.1.

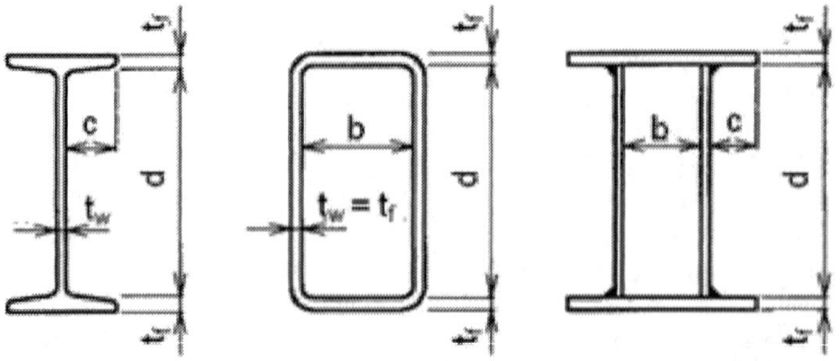

Fig. 1. Usual steel cross-sections and their dimensions.

The new European standard EN 1993-1-1:2005 and transformed national standards for the design of steel structures (CSN EN 1993-1-1, STN EN 1993-1-1 and others) contain the specific classification of the cross-sections due to dimensions and slenderness of their compression and bending parts. In these standards four classes of the cross-sections are defined.

Class 1, 2 and 3 can be considered as the compact or semi-compact cross-sections for the defined level of the material utilization. In this case of the cross-sections the full sectional dimensions can be considered in calculation. Class 4 presents slender cross-sections. In this case the effective part of cross-sections must be considered in calculation.

The standards define maximal values of the slenderness β for individual webs and flanges of the cross-sections subjected to compression or bending, otherwise combined compression and bending: β_{01} for class 1, β_{02} for class 2 and β_1 for class 3.

The web slenderness β_1 is very important for the standards and practical design of steel cross-sections. If slenderness of the web $\beta = \beta_1$, then with full elastic bending moment M_{el} of the cross-section can be calculated by simple bending theory,

$$M_{el} = f_y W_{el} \text{ or } M_{el} = f_y W_{el} \gamma_{M0}, \tag{1}$$

where f_y is the yield stress, W_{el} is the elastic cross-section modulus and γ_{M0} is the material partial factor.

The exact estimation of the limit web slenderness β_1 is problematic, because it depends on real boundary conditions and on production imperfections. Only the classic critical conception allows clearly determination of the web slenderness β_1. According to this conception the critical normal stress for the elastic limit stage

$$\sigma_{kr} = k_\sigma\, \sigma_E = 190\ 000\ k_\sigma / \beta^2 \tag{2}$$

and relative web's slenderness

$$\beta_1 = 435.89\sqrt{k_\sigma / f_y} = 28.43\sqrt{k_\sigma}\sqrt{235/f_y} \tag{3}$$

For the web with pin supporting along both longitudinal borders subjected to bending $k_{\sigma,min} = 23.88$ and if the steel S 235 with $f_y = 235$ MPa is assumed, then $\beta_1 = 138{,}95$. In accordance with the critical conception was defined the web slenderness β_1 in the previous design standard CSN/STN 73 1401:1984, [3].

The new standards take into consideration the effect of initial production imperfections (material and geometrical), therefore the limit web slenderness β_1 is smallest. According to the European standard EN 1993-1-1 and transformed national standards CSN EN 1993-1-1 and STN EN 1993-1-1 the limit web slenderness

$$\beta_1 = 124\sqrt{235/f_y} \tag{4}$$

If the web slenderness $\beta < \beta_1$, than the elastic-plastic bending moment M_{ep} or full plastic bending moment M_{pl} of the cross-section can be calculated,

$$M_{ep} = f_y\, W_{ep} \text{ or } M_{ep} = f_y\, W_{ep}\, \gamma_{M0},\quad \beta_{02} < \beta < \beta_1 \tag{5}$$

$$M_{pl} = f_y\, W_{pl} \text{ or } M_{pl} = f_y\, W_{pl}\, \gamma_{M0},\quad \beta \le \beta_{01} \tag{6}$$

W_{ep} and W_{pl} are the elastic-plastic and full plastic modulus of the cross-section.

According to the European standard EN 1993-1-1 and transformed national standards CSN EN 1993-1-1 and STN EN 1993-1-1 the limit web slenderness

$$\beta_{01} = 72\sqrt{235/f_y} \text{ and } \beta_{02} = 83\sqrt{235/f_y} \tag{7), (8}$$

3 Elastic-plastic bending load carrying capacity

The elastic-plastic bending moment M_{ep} and the full plastic bending moment M_{pl} of the cross-sections depend mostly on the actual material properties of the applied structural steel, development of plastic strains ε and buckling resistance of the bending web and compression flange.

Buckling resistance or local stability of the steel member webs subjected to bending in the elastic-plastic stadium ($\beta < \beta_1$) can be sufficient accurately defined by limit strains ε_u in the most stressed border of its compressed part ($\varepsilon_u > \varepsilon_y$).

The real elastic-plastic behavior of the steel members and their bending webs and compressed flanges is complicated. Therefore, the representative experimental knowledge and results about the real elastic-plastic behavior and mechanisms of failure of the steel members have been very important from the scientific and practical aspects. Therefore, the author of paper realized in previous time the wide experimental research with such intention. The static schemes of the tested steel beams are illustrated in Fig. 2, [4].

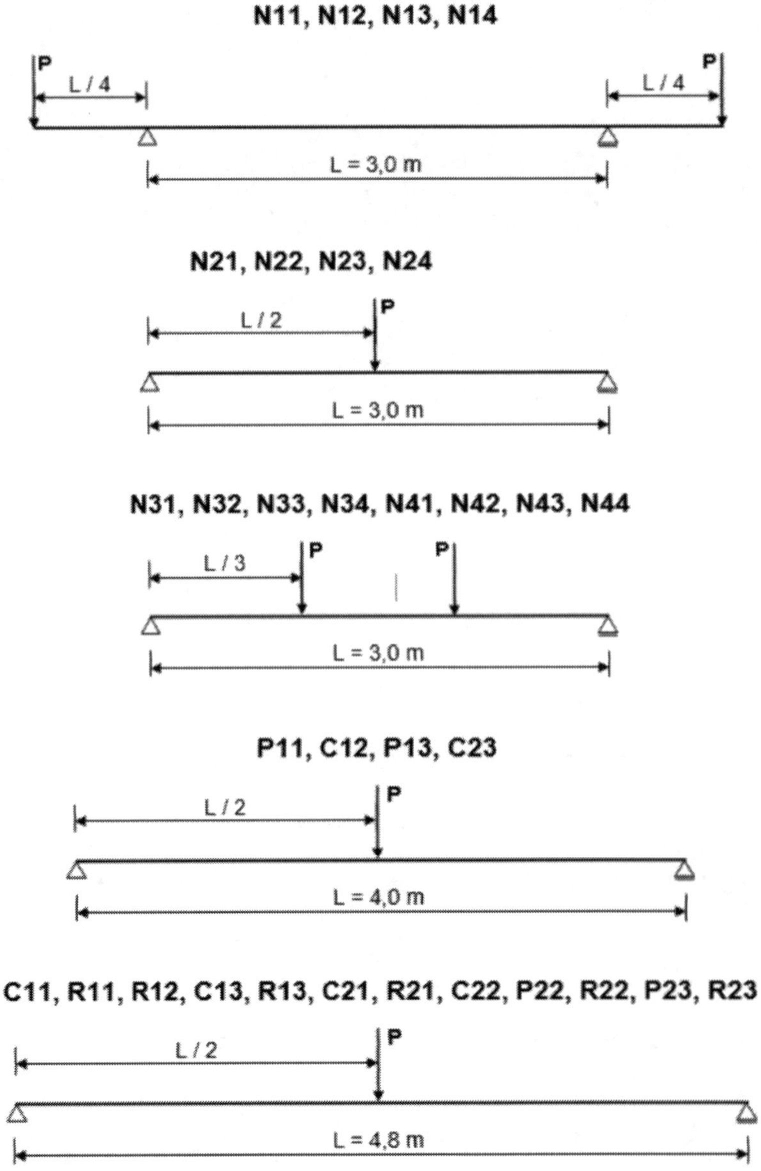

Fig. 2. The static schemes of tested steel beams.

The tested beams had different cross-section dimensions and slenderness of their webs The tests continued till elastic-plastic failure. The local tranverse buckling w and strains ε were measured and evaluated during the tests.

Following experimental and related theoretical results the author of paper worked out the original methodology for calculation of the elastic-plastic bending load-carrying capacity of the steel members M_{ep} with taking account local stability aspects [3, 4, 6-10].

This methodology is based on the development of plastic strains ε in the most stressed cross-section. This development is defined by limit strains ε_u in the edge fibers of web compressed part subjected to bending. The authors'relation $\varepsilon_u - \beta$ is presented in Fig. 3.

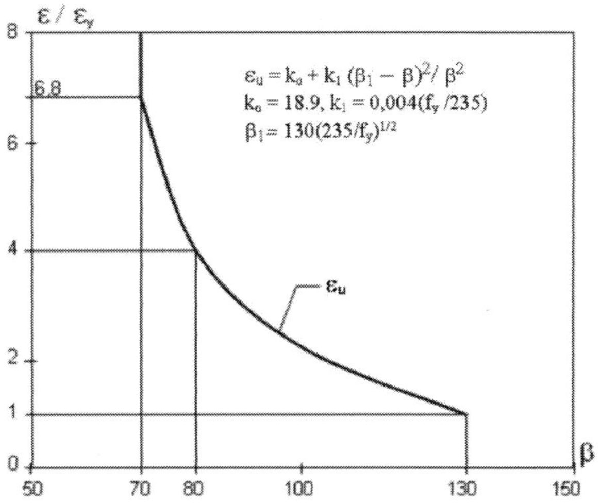

Fig. 3. The authors' relation of the limit strains ε_u to web slenderness β, steel S 235.

The presented methodology and relation $\varepsilon_u - \beta$ allow to calculate the elastic-plastic bending moment M_{ep} of the cross-section if the web slenderness $\beta_{02} < \beta < \beta_1$. For the symmetrical I cross-section

$$M_{ep} = M_{pl} - M_{el,w} (\varepsilon_y / \varepsilon_u)^2 \quad (9)$$

$M_{el,w}$ is the elastic-plastic bending moment of the web.

The previous authors' linear relation $M_{ep} - \beta$ has had form [4]:

$$M_{ep} = M_{pl} - (M_{pl} - M_{el})[(\beta - \beta_{02}) / (\beta_1 - \beta_{02})]. \quad (10)$$

The similar linear relation is also applied in Specification AISC LRFD:1999. The limit slendernesses β_0 and β_1 are, however, different.

The standard EN 1993-1-1:2005 (CSN/STN 1993-1-1) contains some procedure for the calculation of the elastic-plastic bending moment M_{ep} by the effective cross-section of class 2, which is presented in the Fig. 4.

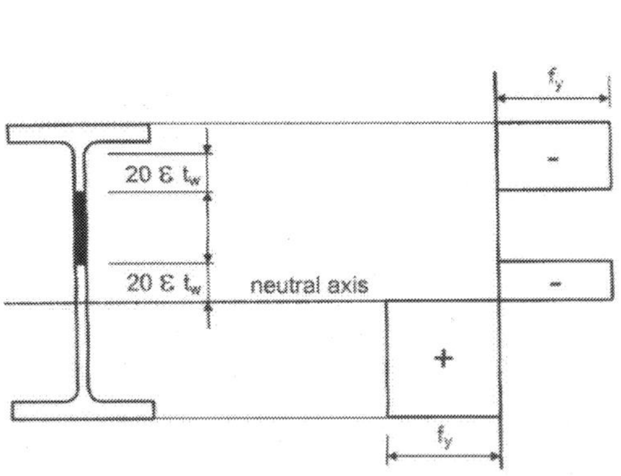

Fig. 4. The effective cross-section class 2 by standard EN 1993-1-1:2005.

It is clear from Fig. 4 that the assumed effective cross-section by EN 1993-1-1:2005 and relative national standards is in conflict with assumptions of the simple bending theory and the experimental knowledge.

4 Conclusions

The obtained and partially presented research knowledge and results have allowed following conclusions:
- The elastic-plastic load-carrying capacity of the steel members and their cross-sections at large measure depends on local stability of their webs and flanges subjected to bending and compression.
- The local stability of the steel member webs and flanges subjected to bending and compression depends on the real development of the elastic-plastic strains ε in the most stressed cross-sections and areas.
- According to the experimental and theoretical results the relation for limit plastic strains ε_u and web slenderness β has been established.
- The proposed relation ε_u - β allows common classification of the steel member cross-sections from the local stability aspects in accordance with expected or selected development of plastic strains ε.
- The proposed methodology enables to calculate the elastic-plastic bending load-carrying capacity of steel members depending on the development of plastic strains ε in the most stressed cross-sections and areas.

The presented research knowledge and results and their normative and practical application have supported the economical and reliability design of the steel members and structures.

References

1. A. Mrazik, M. Skaloud, M. Tochacek, *Plastic Design of Steel Structures* (Ellis Horwood Limited, Chichester and Prague, 1986)
2. J. Djubek, R. Kodnar, M. Skaloud, *Limit State of the Plate Elements of Steel Structures* (Birkhäuser Verlag, Basel, 1983)
3. P. Juhas and col., *Design of Steel Structures – Commentary to STN 73 1401:1998,* (Slovak Institute for Standardization, Bratislava, 2001)
4. P. Juhas, Building Research Journal **32**(3), 833-863 (1984)
5. M. A. Bratford, De Luca, Journal of Constructional Steel Research, **7**, 317-334 (1987)
6. P. Juhas, Stavební obzor **3(3)**, 88-93 (1994)
7. P. Juhas, Ocelove konstrukce, **2/2**, 13-15,43-45 (2000)
8. P. Juhas, Selected Scientific Paper, **1**, 7-21 (2005)
9. P. Juhas, *Mechanics and Material,* ASCE/ASME/SES, Baton Rouge (2005)
10. P. Juhas, Challenges, *Opportunities and Solutions in Structural Engineering Construction* (Taylor and Francis Group, London, 2010)

Cracked reinforced concrete walls of chimneys, silos and cooling towers as result of using formworks

Marek Maj[1] , and *Andrzej* Ubysz[1,*]

[1]Chair of Civil Engineering, Faculty of Civil Engineering, 50370 Wrocław,Wyspiańskiego 27, Poland

Abstract. There are presented in this paper some problems connected with reinforced concrete shell objects operation in the aggressive environment and built in method of formworks. Reinforced concrete chimneys, cooling towers, silos and other shells were built for decades. Durability of cracked shells are one of the most important parameters during process of designing, construction and exploitation of shells. Some reasons of appearance of horizontal and vertical cracks as temperature, pressure of stored material, live loads e.g. dynamic character of wind, moisture, influence of construction joints, thermal insulation, chemistry active environmental etc. reduce the carrying capacity of the walls. Formworks, as is occurred recently, are the reason for technological joints with leaking connection, imperfections of flexible formworks slabs and as result can initiate cracks. Cracked surface of this constructions causes decreasing capacity and lower the state of reliability. Horizontal, vertical cracks can caused corrosion of concrete and steel bars, decreasing stiffness of contraction, increasing of deflection and carbonation of concrete cover. Local and global imperfactions of concrete shells are increasing according to greater number of cracks...

1 Introduction

In recent years many engineers objects are renovated. In particular industry objects are not only renovated but also strengthened. Most of them lost their designed durability life. The basic reason of it is carbonation of concrete, reinforcement corrosion, loss of concrete cover, too many cracks with wide-width, too large deformations and deflections, etc. Recently, more and more often investigates the impact of object on the environment. Mechanical vibrations caused by long-term or short-term operation of equipment in their immediate vicinity may also play a role. Interesting examples of monitoring such dynamic effects on chimneys are shown in [1, 2]. But also due to the increasing standard requirements do not meet the requirements for reliable use. Another reason for renovated is change of kind of use of buildings i.e. change of live loads, change of constructions static schemas. During designing of this objects there is not expected regeneration or renovation

* Corresponding author: marek.maj@pwr.edu.pl

of parts or of the whole object. Reduced resistance due to many years of buildings exploitation is included in partial safety coefficient because of fact that phenomenon of durability is stochastic process. But due to over-exploitation and unfavorable terms such as increased chemical activity on building, structures lose their reliability before the termination of operation of the facility

2 Durability

The basic equation of reliability of each of construction element and the whole construction can by described in following equation [3]

$$A \leq R \tag{1}$$

where R – construction resistance, A – loadings acting on construction. This equation is used by engineers in stresses quantities, or resultant forces. Forces are compressive or tensile or bending moments. Generally in Limit State Design (LSD) R is regarded as a strength and A as stresses. In Serviceability Limit State (SLS) R is allowable deflection or a allowable scratch and A is static deflection or calculated scratch. R is not simply limits stresses and A is not simply loads of snow, winter, pressure etc. Equation (1) has not simple, i.e, the linear form. It involves many parameters appropriate for the loads and strength. Equation (1) can be written in the following formula

$$R - A \geq 0 \;\text{ or }\; f(R - A) \geq 0 \tag{2}$$

where function f(R – A) we can define function of safety or reliability by: R and A are treated as independent random value, but very often they are depended, particularly when loads are dynamic. Reduction of durability of construction material can began process of acceleration object destruction. According to [4], the durability of the material we mean its ability to maintain declared by the manufacturer the mechanical, functional and aesthetic material properties during design lifetime of the object. Durability of construction materials depends on the operation time of the facility, the stress number of cycles N of structures subjected to cyclic interactions and the number of exceedances of the limit states. Lowering the durability of materials reduces the computational resistance parameters and narrows the area of reliability of the whole structure. Durability of construction not only depends on properties of construction materials i.e. resistance of construction but also of properties of live loads. Such properties of load parameters as frequency, coefficient of variation, the ratio of load maximum to minimum, which characterized dynamic loads affect the behavior of the constructional material. So in this point of view acting loads - and construction resistance are not independent random values. Technological joints caused by formworks reduce the durability of the construction when they create cracks, leaks, and reduce the stiffness of the shell. Remedial work of repaired objects should restore designed reliability level of construction. But in fact during repairing there is a new construction built. New materials used to reconstruction of silo and chimney concrete cover form new multilayer silo wall. Injected or stapled cracks form walls composed of many reinforced concrete elements interconnected by network of injection. Replacement of reinforced steel bars or adding of bars change this static equilibrium which was taken into consideration in the original design. In another words repaired construction is not the same construction but only similar. As such should be re-calculated in terms of strength using new parameters of strength materials, schema static and endurance of the whole construction, i.e. ULS and SLS.

3 Constructions which lost resistance before the time of the end of design life

Some examples of such constructions presents in this paper are industrials reinforced concrete chimneys, silos and cooling towers [5]. These structures are very similar to each other. They are usually built as a shell structures, with a relatively large height and thin walls. Concrete as a material in both structures subjects to similar environmental conditions, loads, technology of object rising, destruction of the structure. In these two cases determines the safety of buildings, control of the same way. Both types of structures are similarly reinforced. They built by using formworks. The difference between them is in the prevalence of certain loads, in access to the individual elements of structure, in differentiated impact of aggressive chemicals.

4 The causes of reduced durability of chimneys and cooling towers

In designing chimney or cooling tower wall there are no place for the appearance of the larger cracks. Horizontal and vertical cracks can caused corrosion of concrete and steel bars. Cracks decreases stiffness of reinforced concrete chimney shell construction. Local and global imperfections of chimney walls are increasing according to greater number of cracks. Taking into account these facts, cracks reduce the service life of chimneys. Horizontal and vertical cracks allow condensate water infiltrates into wall. It causes corrosion and moisture of concrete wall and during winter time cracks become bigger because if glaciations water in cracks. Problems with a large number of cracks on chimneys or cooling towers walls occurred in last years in Poland. It was connected with e.g. changing chimney codes and in particularly introduction requirement of double reinforcement of concrete walls. Until the end of XX centaury reinforcement bars were located in one row, near the external surface. After receiving in Poland European Standards two rows of reinforcement are demanded. It caused several difficulties in building chimneys or cooling towers because of little space of reinforcement bars in rather thin wall. Additional problem of increasing number and widths of cracks is chimney insulation. Badly placed insulation results in an increase bending moments and the cracks are wider. Another problem is condensate which leaks through the joints and cracks. Increased temperature in the wall of the chimney insolation, chemical activity of condensate, damp thermal insulation, bad made joints causes increased corrosion of steel and concrete. The result of these are many defects of concrete cover. There are many mistakes which were made during building chimneys and cooling towers: too small distance between vertical and horizontal steel bars in concrete wall, insufficient concrete care during building, too rapid drying wall caused shrinkage in walls, using adjustable formwork in building chimneys, using joints in shell during building, non proper joint design, too long pause between subsequent concrete filling, using vibrators inappropriate for steel bars diameter and for distances between them, touching by vibrator steel bars, different distances between horizontal and vertical reinforced bars, differentials in strength of bars and in strength of concrete, different in concrete cover, diameter and area of bars of bars, imperfections in steel bars localization, height of compressed concrete area, quantity of reinforced horizontal and vertical bars, adhesion of bars to concrete, too small or destroyed thermal insulation, low quality of lap splices, type and quality of wall strengthening, imperfections, cracks, steel corrosion, technical condition of concrete, i.e. concrete carbonization, etc., destroyed moisture protecting coating, too much condensate inside chimney caused by low temperature of gas, too few steel bars calculated shrinkage for chimneys and cooling towers wall, adding loads on chimney of satellite dishes.

5 The causes of reduced durability of silos

There some reasons for reduced durability of silo walls. They arise from non-compliance of designed strength parameters. Mostly are errors of construction work, changes in operating external conditions as chemical aggression, changes of design assumptions. Specified above parameters are random variables, and during exploration of silo changed first of all coefficients of variation, i.e they increased. This resulted in a decrease of the characteristic values and computational strength parameters and increase the value of the characteristic loads. Additional parameters to those mentioned above are: This parameters are: temperature field on the silo wall caused by storage grain, insolation and daily temperature variation, temperature gradient across the silo wall, horizontal pressure caused by wind and grain pressure, concentrated loads (patch loadings), parameters connected with silo wall strength as changes in silo and thickness of the silo wall, changes in wall stiffness.

6 Some examples of repairing of chimney and cooling tower

On Fig. 1 are presented new chimney after one year of exploitation (height 180 m, diameter 7,20 m, wall thickness is 0,70 m in lower part). There are many cracks on the wall, cracked joints, destroyed concrete covers and condensate flowing from internal chimney space through cracks. Condensate formed inside the chimney flow through the cracks causes rust of steel and devastation of technological joints, which is shown on Fig. 2. Too long pause between subsequent concrete filling, using vibrators inappropriate for steel bars diameter and for distances between them and touching by vibrator steel bars caused destroyed area of joint. Cracks, separated steel bars from concrete wall changes monolithic wall to few interconnected pieces.

Fig. 1. New chimney after one year of exploitation with many cracks and condensate flowing from internal chimney space. Visible traces of replacement of concrete cover, injection of cracks, repair horizontal joints.

Thermal insulation protects chimney wall against high temperature and high temperature gradient resulting in an increase of the tension force and of the bending moment.

Additional non uniform distribution of temperature on chimney wall causes increasing of deflection of chimney and increasing vertical inner forces. This is the reason that vertical and horizontal cracks appear additionally. Thermal insulation can be destroyed or may fall down because of excessive number of condensate penetrate to surface of chimney wall. Very similar factors as in the case of chimneys cause the destruction of concrete cover and reinforcement in cooling towers (Fig.3). Using similar methods as in the case of chimney cooling towers shown on Fig. 3 was repaired (Fig. 4).

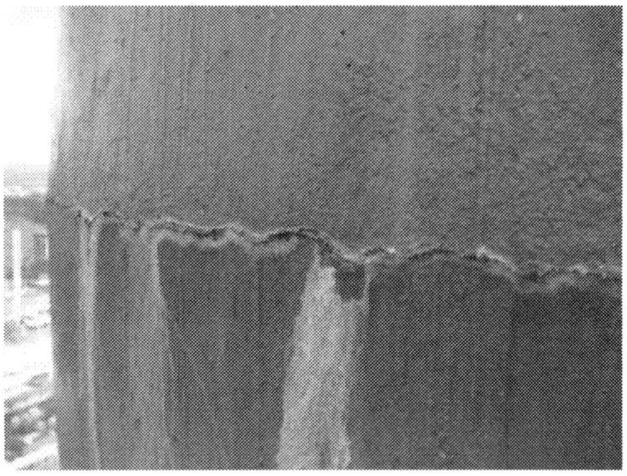

Fig. 2 The condensate flowing form horizontal cracked joints.

7 Some examples of repairing of silos

After many years of exploitation some silos construction need to be renovated and strengthened. Strengthening of silo construction (silo walls, bottom, foundations or overlap) can be carried out generally in two ways, according to formula (1) [3]. Decreasing of value of left side (loads) of this equation and increasing value of right side (strength) of this equation separately or simultaneously makes silo construction more safety. The first simplest solution used for silo is reducing the level of filling of the silo of material stored in silo or using perfect control of kinematic of discharging material and change the kinematic of flowing to chimney flow. It is possible using the discharging device decreasing value of material pressure. Second solution is increasing strength of the silo wall, using additional materials of higher strength as steel hoops, prestressed tendons (Fig. 5) [4].

Fig. 3. Destroyed upper reinforced concrete crown.

Typical concrete silos for rape built 40 years ago in many places in Poland were renovated and most of them were strengthened. On Fig. 6 it is shown the rechange of cover and corroded reinforcement.

Fig. 4. Renovated cooling tower.

Fig. 5. Most of concrete cover and corroded steel bars are replaced.

Fig. 6. Silo battery for rape partially strengthened by steel hoops and prestressed tendons.

8 The methods used to restore the durability of reinforced concrete structures

The typical methods used to restore the durability and strength of reinforced concrete silos, chimneys and cooling towers are: removing the shattered parts of carbonated concrete and outer coatings in joints and cracks in the wall, reprofiling walls and cleaning reinforcement steel bars and supplement missing bars, repair corroded reinforcement, replacement of damaged parts of the concrete wall filling cracks and joints by injection materials, location of additional reinforcement, hoops or prestressed, or carbon fiber tapes, location of the additional layers of concrete, restoration of the thermal insulation and moisture of the concrete wall (chimneys), additional calculations strength of the new repaired structure [5]

The strength calculations took into account the many layers renovated walls. So in this case new object was created. New concrete was used with another strength parameters as adhesion, plasticity, cohesion, moisture etc. The surface of the repaired structure is subjected to shear, delaminated stresses etc. The most important for new materials is fact that they were not rejected by the parent material. New parts of the structure are not as strained as those that already exist, don't undergone the process of rheology, introducing additional self stresses in the joint place between old and new material. These physical phenomena must be taken into account in the calculation of strength of new construction.

9 Conclusions

All designed constructions are subject to aging. You cannot design a structure where all elements at the same time lose reliability. All constructions are subject to periodic repairs in order to restore the designed resistance. However economic and technological aspect decides about the repair. It is very important to choose the method of formwork and it seems that the formwork requires a lot of attention. It seems that in the initial design of construction process should be included restoration of the durability of the object whole or part of it. During the initiate construction design process should be consider and indicate the places with maintaining the durability of materials. Such designed repairs should be considered in terms of the need strengthening, possibility of technology, economics, health of workers, the impact on the ecology of the environment, etc.Renovated structure is not the same as original designed object. Should be reconsidered basic static schemes and equations applied equations strength.

References

1. A. Herbut, J. Rybak, Guidelines and recom-mendations for vibration control in the case of rapid impulse compaction, *Advances and trends in engineering sciences and technologies II*, 761-766 (CRC Press, Taylor & Francis Group, 2017)
2. J. Rybak, A.G. Tamrazyan, Calibration of rapid impulse compaction on the basis of vibration velocity control, *16th International Multidiscip. Scientific GeoConference SGEM 2016, Albena, Bulgaria, 30 June-6 July, 2016* vol. **1**, 715-722, (2016)
3. H. O. Madsen, *Methods of structural safety* (Prenice Hall, 1986)
4. Murzewski J., *Niezawodność konstrukcji inzynierskich* (Arkady, Warszawa, 1989)
5. Kaminski, M. Maj, M., Ubysz A., Chimney cracked reinforced concrete walls as a problem of durability exploitation, SEMC 2013, *The 5th International Conference on Structural Engineering, Mechanics and Computation*, , 2-4 (Cape Town, South Africa, 2013)

Evaluation of the carrying capacity of the building structures in the industrial heritage for its adaptation

Sergej Kudrjavtsev[1,*]

[1]Ural Federal University, Department of civil and industrial engineering, 620002 Yekaterinburg, Russian Federation

Abstract. The article presents the results of verification calculations of carrying capacity of building structures of "White Tower". This former water tower is a constructivist monument of state-wide significance and one of the oldest reinforced concrete structures in Yekaterinburg. Currently the design of adaptation of the object as a museum or an exhibition hall is being developed. The purpose of the study is to answer the question: Will the structures of the 1930s withstand the loads regulated by modern building codes? The results of carried out field inspections and a finite-element model of the LIRA-SAPR software are also described. The data of the weakest sections and elements of the structure are presented. The result of the work will allow carrying out a complex of efforts for the restoration of building structures of the object and will help in deciding whether to develop one or another alternative for its adaptation.

1 Introduction – Brief history of the construction

There are more than 15 thousand monuments of federal significance in Russian Federation. One of these is the former water tower, better known as the "White Tower" in Yekaterinburg [1]. This facility is a remarkable example of the industrial architecture of the Soviet avant-garde and one of the first reinforced concrete structures in Sverdlovsk (the former name of Yekaterinburg).

The initial outline design of the tower was developed by the architect M.V. Reisher in 1929 [2]. The draft project provided for installation of a reinforced concrete water tank with a volume of 540 m^3, an observation deck on the roof of the structure and a bus stop pavilion at the base of the tower. The support of the water tank was provided on two reinforced concrete columns and on the internal stairwell structure. For technological reasons a steel tank was provided instead of a reinforced concrete water tank, with its volume increased to 700 m^3. In addition, two extra external columns were added in order to support the structure of the tank. Constructive drawings of the tower were developed by the design office "Techbeton", Moscow [3]. The design drawings for water tank of Otto Intze system were carried out by the design department of "Uralmash" factory. It was the first construction of this sort made by welding and without rivets in USSR. The construction of tower was

* Corresponding author: s.v.kudryavtsev@urfu.ru

carried out in 1930-1931. During hydraulic tests taken on June 5, 1931, the bottom of the tank collapsed. After that it was decided to make a new tank bottom from reinforced concrete, retaining the steel walls of the tank. The tower was commissioned in 1931. In the 1960s the tower ceased to perform its basic functions in water supply of residential houses and the question arose of its further conversion to other needs. In 1971 an architect M.V. Reisher developed a project to house cafe in the tower, but it was never implemented, thus since the 1970s the tower is not functioning as a water tower.

Fig. 1. Sketch of the water tower, designed by the architect M.V. Reisher in 1929 and structure completed by construction in 1931. [2]

2 Characteristics of building structures

Nowadays the structure represents two interrelated spatial volumes: a cylindrical part with a diameter of 13.5 m, supported by four reinforced concrete external columns and a prismatic part of a width of 3.2 m and a length of 5.75 m.

In the cylindrical volume of the structure there was a water tank which is partially dismantled nowadays and only the reinforced concrete tub bottom has survived. The preserved structures of the water tank are supported over a circular reinforced concrete beam with a cross-section of 900 x 1800 mm which has rigid fittings by the four outer columns of the tower and the two internal columns of the stairwell. The diameter of the circular beam is 7.8 m. The frame tent with reinforced concrete columns with two rows of reinforced concrete crossbars is made around the water tank to serve as a shield from the atmospheric precipitation. The wall enclosure of the tent is made from bentonite stones [4] 200 mm thick with reed insulation. The coating of the tank tent is made in the form of a reinforced concrete dome.

In the prismatic part of the tower there is a reinforced concrete staircase for moving between the construction levels and premises for laying water supply pipelines from the tank to the basement of the structure. Floor slabs at different levels of the tower are made from a reinforced concrete ribbed plate, consisting of the main and secondary beams and slab of 80 mm and 100 mm thick.

Fig. 2. The current look of the facility.

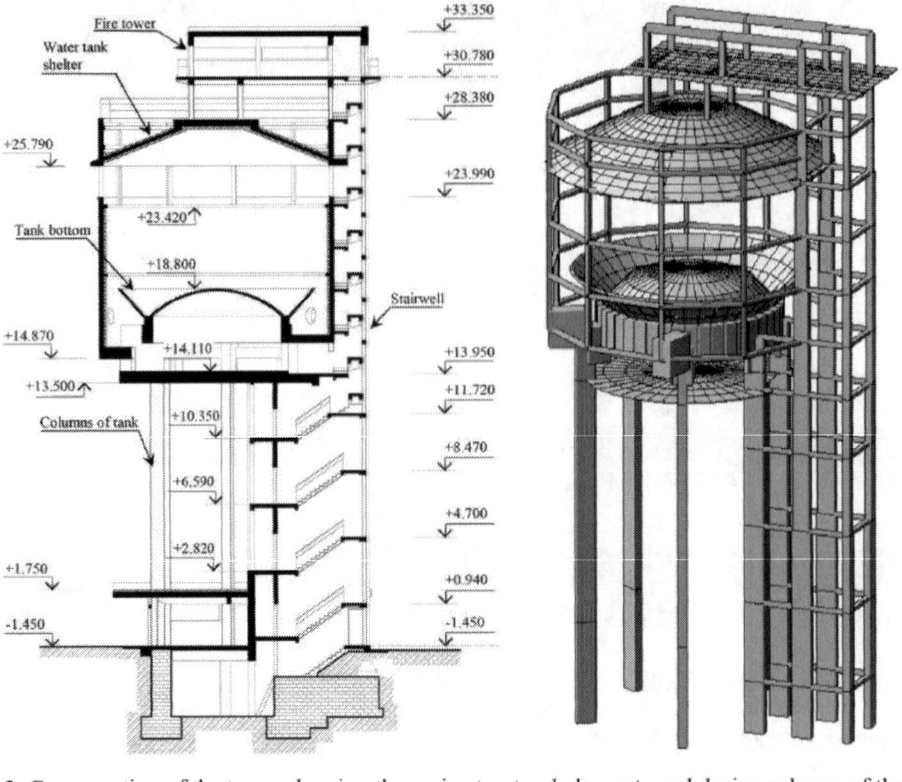

Fig. 3. Cross-section of the tower showing the main structural elements and design scheme of the structure.

Reinforcement of bearing structures is made of hot-rolled bars of steel grade St0 with circular and square cross-sections. In order to reinforce the structures the bars were used executed both in metric and American assortment, for example, 3/8" and 1/2".

3 The results of structural and technical survey of the tower

In August 2013 students of the Brandenburg Technical University, Cottbus, Germany under the supervision of Prof. W. Lorenz carried out a survey of the tower [5]. In this work the following major defects and structural damages were noted: damage caused by soaking atmospheric precipitation from the roof and through open window openings in the walls of the structure; insufficient or completely absent protective layer of concrete; defects of concreting shaped as shells and voids that appeared due to insufficient compaction of the concrete mixture.

In May 2014 the Research Institute "INTER/TEK" carried out a survey of the bearing structures of the tower [6]. At this survey, numerous defects and structural damages of the tower constructions were noted, caused by two main factors. First of all, the destruction of the waterproofing layer of the coating and the complete destruction of the window glazing structures. Secondly, for a long time the construction was available for unauthorized access by unauthorized persons.

Special attention was paid to the future functional assignment of the construction at the "International Forum of High-Rise and Unique Construction 100+" held in Yekaterinburg on September 23-25th, 2015 [7]. The opinion of the expert community tends to use the facility as an exhibition site devoted to the themes of architectural avant-garde and industrialization in the Urals, but the question of the actual bearing capacity of the structure remains unresolved.

Fig. 4. Typical damages of the structures: cracks in floor slabs, exposure and corrosion of working reinforcement, breaks in the floors.

4 The bearing capacity of the tower's structures

To perform the static calculation and determine the design forces in the elements of the bearing structures of the facility, computer modelling of the tower bearing skeleton and its finite element analysis in program LIRA-SAPR 2013 are performed.

Columns, girders and circular beams of the facility were modelled by rod elements. Lamellar finite elements with a size of 0,5 x 0,5 m are used for modelling slabs at different levels of structure, the dome of the tank shelter and the preserved bottom of the tank. The grid of lamellar finite elements was compacted in the joints to adjacent structures. The stiffness of the elements of the design scheme was adopted according to the results of the survey and according to the data of the remaining original design documentation.

The calculation is made for two cases of load distribution:

- case 1 corresponds to the current state of the structure;
- case 2 corresponds to the design state of the structure and its operation as a museum.

Dead loads on the structure were taken based on the results of the structural survey. The estimated value of the snow load s_g = 180 kg/m^2 and the specified value of the wind pressure w_0 = 23 kg/m^2 are accepted for the area of the city of Yekaterinburg according to the set of rules SP 20.13330.2011 [8]. The specified live loads on the floors and the staircase structure are taken from SP 20.13330.2011 [8]: for case 1 – p_n = 150 kg/m^2 as for the maintenance and repair areas in industrial premises; for case 2 – p_n = 400 kg/m^2 as for visual and concert halls and meeting rooms.

Table 1. The results of estimating the bearing capacity of structures.

Element of structure	**Cross-section *h* x *b* and the area of the stretched reinforcement**	**M_x, kNm**	**$M_{ult,x}$, kNm**	**M_y, kNm**	**$M_{ult,y}$, kNm**	**Use of bearing capacity, %**
Water tank columns	900mm x 600mm A_{sx} = 1356 mm^2 A_{sy} = 1808 mm^2	803 877	1101 1204	440 517	844 948	72 73
Tank shelter columns	300mm x 300mm A_{sx} = 850 mm^2 A_{sy} = 850 mm^2	74 86	124 133	90 110	108 119	83 92
Stairwell columns	900mm x 300mm A_{sx} = 850 mm^2 A_{sy} = 1133 mm^2	424 471	588 654	158 184	264 288	72 73
Stair stringer	200mm x 150mm A_{sx} = 380 mm^2	4,5 7,7	11 11,5	–	–	41 67
"Fire tower" columns	250mm x 250mm A_{sx} = 253 mm^2 A_{sy} = 253 mm^2	26,1 30,7	51,5 56,2	33,9 38,6	46,5 52,0	72 74
"Fire tower" beams	400mm x 200mm A_{sx} = 820 mm^2	88,4 136,6	124,4 147,3	–	–	71 93
"Fire tower" floor slab	1000mm x 80mm A_{sx} = 427 mm^2	2,48 3,90	4,17 4,17	–	–	59 94

In the above table, the values above the line refer to the first case for calculating the structure – under the action of existing loads; the values below the line refer to the second case for calculating – under the action of the estimated loads.

Values for the strength of concrete for compression R_b = 85 kg/cm^2 and the strength of reinforcing steel for tension R_s = 1600 kg/cm^2 are taken from structural survey in accordance with the recommendations of SP 13-102-2003 [9].

The bearing capacity of the building structures of the facility was evaluated using the critical force approach, laid down in SP 63.13330.2012 [10] from expression

$$M \leq M_{ult}\,, \tag{1}$$

where M – is the bending moment from the external load;

M_{ult} – is the ultimate bending moment, that can be perceived by cross-section of the element according to SP 63.13330.2012 [10].

The bending moment for elements subject to bending was taken from the static analysis. The bending moment for elements subject to eccentric compression was determined from the formula

$$M = N \cdot e\,, \tag{2}$$

where N – is a longitudinal force from external load, determined from static analysis;

e – is a distance from the point of application of the longitudinal force to the centre of gravity of the section of the stretched or least compressed reinforcement.

Due to the structural features of the facility, most of the tower columns are under conditions of a complex stress-strain state and are subject to oblique eccentric compression [11], therefore the following condition was additionally checked

$$(M_x / M_{ult,x})^k + (M_y / M_{ult,y})^k \leq 1 , \quad (3)$$

where M_x and M_y – are the bending moments from the external load relative to the centre of gravity of the cross section of the element in two mutually perpendicular planes of symmetry of the cross section;
$M_{ult,x}$ and $M_{ult,y}$ – are ultimate bending moments, that the element cross section can be perceived in the planes of symmetry;
$k \leq 1{,}6$ – numerical factor, depending on the boundary height of the compressed area of concrete and steel ratio of concrete.

5 Conclusions

It is established that the main constructions of the structure have sufficient bearing capacity to place the museum in the tower. It is worth noting that the planned live loads that will affect on the tower's construction structures are not the main ones, i.e. the loads that are causing the main loads onto the elements. The main impact, which makes the greatest contribution to the exhaustion of the bearing capacity of the elements of the structure, is the weight of the structures themselves. Live loads that will arise in the construction when operating it as a museum are the main ones for the floor slab of "fire tower", the structures of the stairs and the columns of the tank shelter. The last one is the weakest part of the entire structure.

The author hopes that this work will allow carrying out a complex of activities for the restoration of the building structures of the facility and will help in deciding whether to develop one or another alternative for its adaptation.

References

1. White tower. Available on http://www.tower1929.ru/ (2017)
2. M.V. Reisher. *Memories of the construction of the tower.* Archive of the Museum of the History of UZTM. Yekaterinburg (1971)
3. I.A. Kazus. *Soviet architecture of the 1920s. Organization of design* (2009)
4. S.L. Prokhorov. Recent achievements in the construction of bentonite stones. Modern Architecture, **2**, 49-60 (1928)
5. O. Arkhipkina, G. Albrecht. *Untersuchung, Bemessung und Bewertung historischer Bausubstanz – Der "Weiße Turm" in Jekaterinburg (Russland): Aufgabenstellung für eine Bachelorarbeit im Studiengang Bauingenieurwesen.* Cottbus (2013)
6. S.V. Kudryavtsev, Y.V. Gurjanov. *Technical report 05-2014-OBS. Visual inspection and assessment of the technical state of the water tower at: Yekaterinburg, ul. Bakinskih Commissarov, 2a/1 for the development of the project of conservation of facility.* Research Institute "INTER/TEK". Yekaterinburg (2014)
7. International Forum of High-Rise and Unique Construction 100+. Available on http://www.forum-100.ru (2015)
8. *SP 20.13330.2011. Set of rules. Loads and effects* (2011)
9. *SP 13-102-2003. Set of rules. Rules for inspection of bearing constructions of buildings and structures* (2003)
10. *SP 63.13330.2012. Set of rules. Concrete and reinforced concrete structures. Principal rules* (2012)
11. S.V. Kudryavtsev. Evaluation of bearing capacity of concrete thickeners with carbon plastic reinforcement. Akademicheskij Vestnik UralNIIproekt RAASN, **2**, 91-96 (2013)

Comparison of engineered cementitious composites and concrete for strengthening of concrete structural details using RBSM

Yuliia Khmurovska[1] , and *Petr* Štemberk[1,*]

[1]Czech Technical University, Thákurova 7, 166 29 Prague 6, Czech Republic

Abstract. This paper presents a numerical simulation and subsequent comparison of strengthening performance of an ordinary concrete overlay and an overlay made of engineered cementitious composite (ECC) with polyvinyl alcohol fibers (PVA). The comparison is performed on an L-shaped joint when the overlay is placed on the outer surface so that the applied bending moment causes tension in the overlay. The nonlinear numerical analysis is based on the three-dimensional rigid-body-spring model (RBSM). The results show the beneficial effect of the PVA fibers within the ECC matrix when the damage is distributed evenly so that only thin microcracks open. The observation is easy to obtain when the RBSM is employed. On the contrary, the overlay made of ordinary concrete fails due to localization of the damage into a single crack. The applicability of the RBSM is discussed.

1 Introduction

Durability, which is one of the basic requirements for concrete structures, is to a large extent ensured by a cover layer. Since the cover layer of concrete is subjected to direct exposure to aggressive environment, its degradation is a frequent cause of structural repair. Carbonization can be taken as the typical example of concrete cover degradation (Figure 1), which can be divided into 3 stages. Firstly, the cover layer is damaged by the imperfection of material and volumetric changes that are caused by shrinkage, temperature difference etc. Secondly, the concrete cover layer loses its protective properties so much that the reinforcement begins to corrode, which causes further cracking of concrete. Thirdly, rebars corrode up to the load bearing limit and destruction of cover layer [1]. Due to this process, the damaged part of concrete has to be removed and replaced with suitable material.

In this work, the engineered cementitious composite (ECC) is considered as an alternative repair material. 2% by volume of polyvinyl alcohol fibers (PVA), [3], are added to the mixture to reduce shrinkage effect and to increase adhesion properties of ECC. The main advantage of PVA-ECC is its tensile strain capacity in the range of 3-7% [4], and strain hardening behavior that is achieved by cement matrix multiple cracking [5]. The performance of PVA-ECC in tension is characterized by appearance of large number of

* Corresponding author: stemberk@fsv.cvut.cz

microcracks that provide slower transport of environmental products as compared with ordinary concrete, where the crack localization take place from the very beginning.

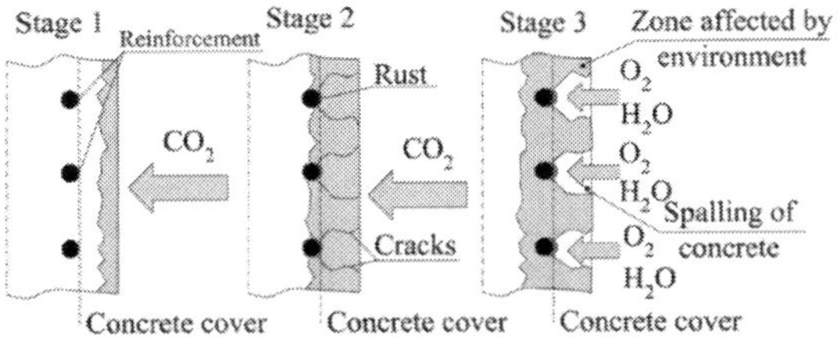

Fig. 1. Deterioration of reinforced concrete [1].

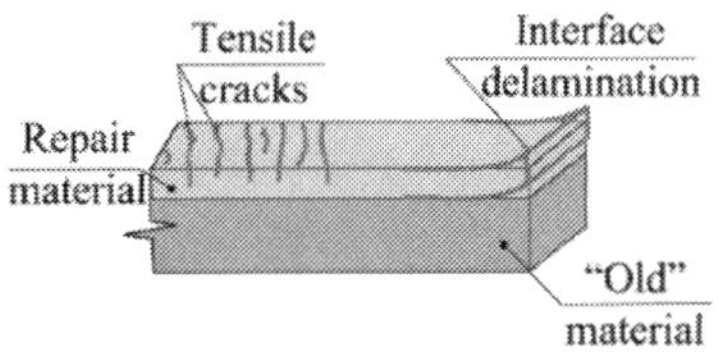

Fig. 2. Cover layer failure modes [2].

The repair material layer of structural elements subjected to bending has to withstand tensile stresses that develop in a tensile part of bended structure and shear stresses that appear along the interface between "old" and "new" materials [2]. Tensile stresses may provoke the repair material cracking, while shear stresses may be a cause of repair layer delamination. Possible failure modes of repair system are shown in Figure 2.

3D Rigid-Body-Spring Model (RBSM) is used to perform the numerical simulation in this work in order to analyze both the crack patterns and the repair layer delamination.

2 Material model

The Rigid-Body-Spring Model (RBSM) is a type of a discrete method model which was invented by Kawai [6]. The main benefit of the RBSM is its relatively simple constitute formulation that can be validated easily by classical benchmark tests. The random geometry of the particles provided by Voronoi tessellation [7], helps to simulate crack patterns and their propagation and to reduce mesh bias. The material is modeled as a set of undeformable rigid bodies interconnected by sets of springs on their surfaces. In the presented model, each rigid body has six degrees of freedom (3 translational and 3 rotational) defined at the nuclei of the Voronoi cells. In the modified formulation of RBSM, each split triangular surface consists of a set of one normal and two shear springs, which account for the effect of bending and torsion without the use of rotational springs [8].

2.1 Concrete material model

The material model within the 3D RBSM for the concrete substrate (original concrete part of the structure) in tension, compression and shear is defined in Figure 3 for easy understanding. All material parameters used later in the analyses are provided below. The tensile stress-strain relation for the normal springs is linear elastic with the slope corresponding to the elastic modulus, E, up to tensile-strength dependent value σ_t with the bilinear softening behavior after cracking as shown in Figure 3a) [8, 9]. In Figure 3a), f_t is the tensile strength, E_{exp} is the experimentally obtained elastic modulus, G_f is the fracture energy and h is the distance between the nuclei of the neighboring rigid-bodies. The compression model for the normal springs is modeled as a combination of two quadratic functions with the resultant S-shape curve (Figure 3b)). In Figure 3 b), f_c is the compressive strength. The model considers only the failure in tension or shear, as compressive failure does not occur [8, 9]. The compression model is given by the following formulas

$$\sigma = \begin{cases} a_0\varepsilon^2 + b_0\varepsilon & \varepsilon > \varepsilon_{c1} \\ a_1\varepsilon^2 + b_1\varepsilon + c_1 & \varepsilon \leq \varepsilon_{c1} \end{cases} \tag{1}$$

where $a_0 = -\frac{E(1-\alpha_{c1})}{2\varepsilon_{c1}}$, $b_0 = E$, $a_1 = -\frac{E(\alpha_{c2}-\alpha_{c1})}{2(\varepsilon_{c2}-\varepsilon_{c1})}$, $b_1 = -\frac{E(\alpha_{c1}\varepsilon_{c2}-\alpha_{c2}\varepsilon_{c1})}{\varepsilon_{c2}-\varepsilon_{c1}}$,

$c_1 = -a_1\varepsilon_{c1}^2 + b_1\varepsilon_{c1} + \sigma_c$.

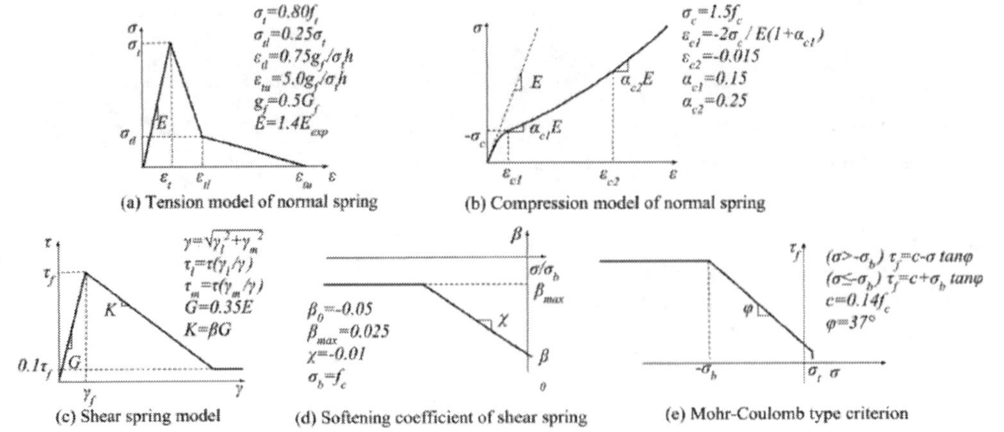

Fig. 3. Concrete material model.

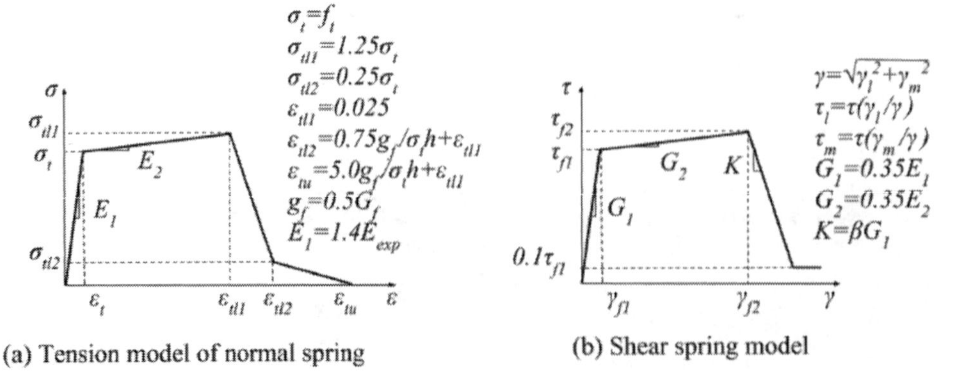

Fig. 4. PVA-ECC material model.

The shear stress-strain relation is defined as a combination of the stresses and strains of two shear springs (γ_l, γ_m), where l and m denote two perpendicular directions. The shear stress increases elastically with the slope of the shear modulus, G, up to the shear strength, τ_f. Then, the softening behavior is assumed (Figure 3c)). The slope of softening, K, depends on the normal spring stresses through the softening ratio, β, that increases up to the maximum value with increase of the tensile stresses (Figure 3d)) [8, 9]. The shear behavior is defined by the following formulas

$$\tau = \begin{cases} G\gamma & \gamma < \gamma_f \\ \max(\tau_f + K(\gamma-\gamma_f), 0.1\tau_f) & \gamma \geq \gamma_f \end{cases} \quad \text{and} \quad \beta = \min(\beta_0 + \chi(\sigma/\sigma_b), \beta_{max}) \tag{2}$$

The Mohr-Coulomb type criterion is assumed as the shear spring failure criterion (Figure 3e)) [8, 9]. The decrease of the shear stress with the increase of the crack width is taken into consideration as follows [8, 9],

$$\tau = \begin{cases} \beta_{cr} G\gamma & \gamma < \gamma_{ft} \\ \beta_{cr} \max(\tau_{ft} + K(\gamma_{\max} - \gamma_{ft}), 0.1\tau_f) & \gamma \geq \gamma_{ft} \end{cases} \tag{3}$$

where $\beta_{cr} = \frac{\varepsilon_t}{\varepsilon} \exp\left\{\frac{k}{\varepsilon_{tu}}(\varepsilon - \varepsilon_t)\right\}$, $\tau_{ft} = c - \sigma_t \tan\varphi$.

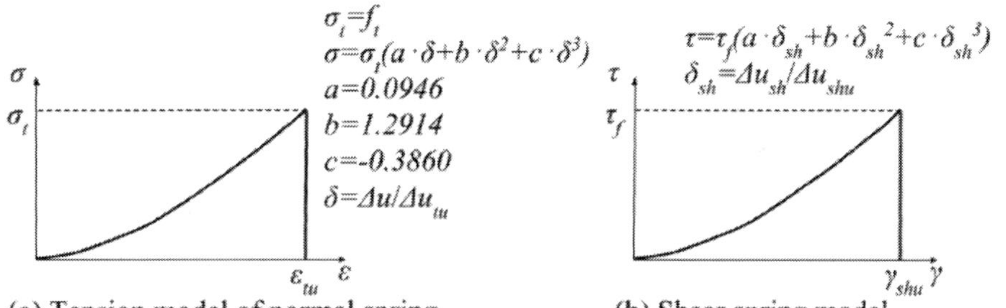

(a) Tension model of normal spring (b) Shear spring model

Fig. 5. PVA-ECC-to-concrete bond material model.

2.2 PVA-ECC material model

The material model for the PVA-ECC in tension was taken from the experimental investigation of the PVA-ECC with 2% fiber content [10], and is shown in Figure 4a). The normal spring in tension is modeled as linear elastic up to the tensile strength σ_t with a linear hardening branch after cracking and a bilinear softening behavior after crack localization. The strain characteristics of the PVA-ECC derived from the fracture energy of concrete with the assumption that PVA-ECC behavior before cracking and after crack localization is similar to concrete behavior [10]. Since the compression behavior of the PVA-ECC is close to concrete (see Figure 3 b)) [10], the compression model of the PVA-ECC is assumed to be the same as in the material model for concrete and is given by Eq. 1.

The shear stress-strain diagram is shown in Figure 4b) where the slope of softening K also depends on the normal spring stresses through the softening ratio β (see Figure 3d)) and thus is also given by Eqs. 2 and 3. The Mohr-Coulomb type failure criterion is also assumed (see Figure 3e)), where

$$\tau_{f1} = \tau_f \quad \text{and} \quad \tau_{f2} = \begin{cases} c - 0.8\sigma\tan\varphi & \sigma > -\sigma_b \\ c - 1.25\sigma\tan\varphi & \sigma \leq -\sigma_b \end{cases} \tag{4}$$

2.3 Bond between concrete and PVA-ECC material model

The trend of the stress-strain curve of the bond is similar to the stress-strain curve of the PVA fibers in tension [11], which means that the bond is provided mostly by the PVA fibers. The tensile bond behavior was presented in previous numerical and experimental study [5], where the tensile stress-strain relation is expressed by a cubic parabola up to the failure without any softening brunch (Figure 5a)), where Δu is the relative normal displacement of the spring and Δu_{tu} is its ultimate relative normal displacement.

The bond compression strength is accepted as the average of the strengths of the PVA-ECC and concrete. The stress-strain relation is shown in Figure 3b).

The shear behavior is similar to that in tension and is shown in Figure 4b), where Δu_{sh} is the relative shear displacement of the spring and Δu_{shu} is its ultimate relative shear displacement. The shear spring model has no softening branch, but the Mohr-Coulomb type failure criterion shown in Figure 3e) is assumed.

3 Numerical analysis

A concrete frame corner with and without a PVA-ECC repair layer subjected to the bending moment (Figure 6) is considered in this study. This structural detail is chosen due to concave corner presence, which may produce serious problems because of non-uniform stiffness distribution and restriction of deformation [12]. Numerical analysis is performed in Matlab using non-linear 3D RBSM. The material parameters considered in the analyses were as follows. For concrete: E_{exp} = 35GPa, f_t = 3.5MPa, f_c = 40MPa and G_f = 100N/m. For the PVA-ECC: E_{exp} = 20GPa, f_t = 5MPa, f_c = 50MPa. For bond between concrete and PVA-ECC: f_t = 1.5MPa, $f_c = (f_{c,conc}+f_{c,ECC})/2$ = 45MPa, Δu_{tu} = 0.473mm and Δu_{shu}, = 1mm.

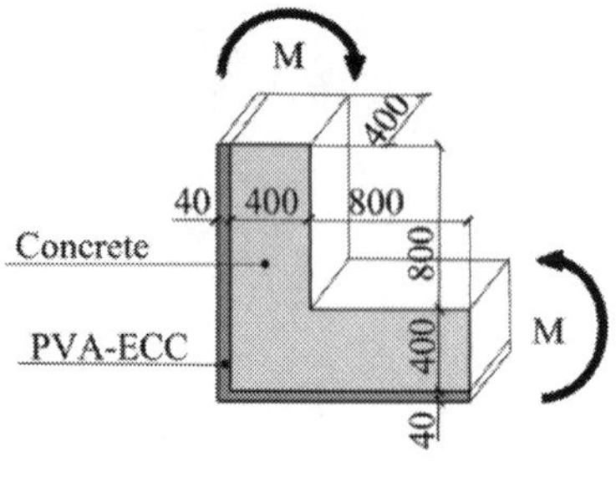

Fig. 6. Analyzed frame corner.

The analyzed frame corner was loaded with displacement as shown in Figures 7 and 8. The load was divided into 30 load steps of 8 μm each. Displacement-controlled modified Newton-Raphson method was used as the convergence algorithm. Convergence of the model is attained when the Σ(Reaction of elements in the model)^2 becomes less than 10^{-6}. When the model did not converge at a given maximum iterations (maximum number of iterations = 400), the analysis proceeded to the next step [13].

4 Results and discussion

The crack patterns obtained from the numerical analyses are shown in Figures 7 and 8. The results show that the sample destruction starts in concrete part. PVA-ECC layer affects weakly the crack pattern and cracks width in concrete, but it provides thinner cracks on the surface of concrete frame corner, which slows the chemical particles transfer, which is verified by experimental investigation [14].

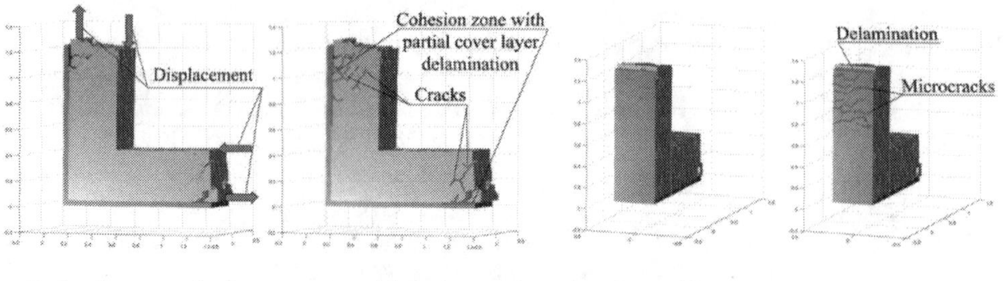

Fig. 7. Crack pattern in frame corner with PVA-ECC overlay. (magnification=150).

The PVA-ECC layer is partially delaminated and creates a cohesion zone between the two materials (Figure 7) because of high ductility and lower strength of bond in tension, so the damaged concrete parts are slightly pulled out of PVA-ECC layer, while the PVA-ECC layer is not broken due its high ductility and strain hardening in tension.

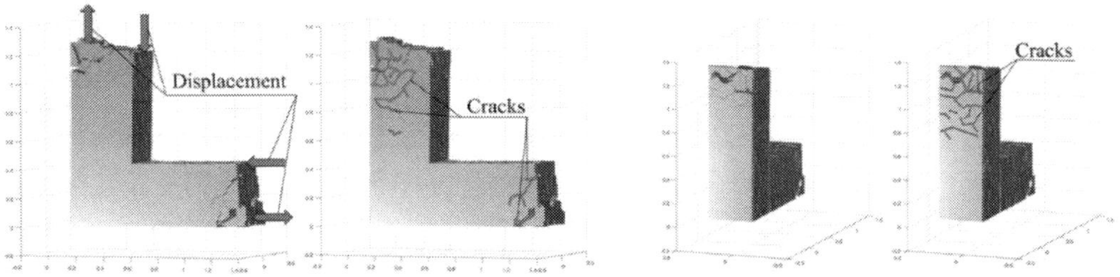

Fig. 8. Crack pattern in frame corner without PVA-ECC overlay. (magnification=150).

5 Conclusions

According to the results of the numerical analyses, the PVA-ECC overlay helps to reduce crack width on a surface of the modelled structural detail so that the rate of penetration of aggressive chemicals in concrete can be reduced, which is in accordance with the experimental investigation reported in [14]. The PVA-ECC provided a relatively strong bond with concrete with occurrence of insignificant delamination. Thus, the PVA-ECC becomes a suitable material for concrete cover repairs.

The Rigid-Body-Spring Model (RBSM) method with random Voronoi tessellation proved suitable for analyses of repair layers, since it is able to determine the crack pattern, crack width and possible cover layer delamination, while using simple material model.

This work was financially supported by the Czech Science Foundation, project 15-11753S, and the Czech Technical University in Prague (CTU), project SGS17/048/OHK1/1T/11, which are gratefully acknowledged.

References

1. J. Ślusarek, A. Kostrzanowska, ACEE. **1**, 85-92 (2010)
2. M. Li, V.C. Li, Restoration of Buildings and Monuments. **12(2)**, 143-160 (2006)
3. V.C. Li, Concrete: material science to application. **SP 206-23**, 373-400 (2002)
4. V.C. Li, S. Wang, C. Wu, ACI Mater. J. **98(6)**, 483-492 (2001)
5. G. Fischer, V.C. Li, ACI Struct. J. **99(1)**, 104-111 (2002)
6. T. Kawai, Nucl. Eng. Des. **48**, 207-229 (1978)
7. J.E. Bolander, S. Saito, Eng. Fract. Mech. **61**, 569-591 (1998)
8. Y. Yamamoto, H. Nalamura, I. Kuroda, N. Furuya, Eur. J. of Environ. Civ. En. **18**, 780-792 (2014)
9. K. Nagai, Y. Sato, T. Ueda, J. Adv. Concr. Technol., **3**, 385-402 (2005)
10. I. Paegle, G. Fischer, Proc. Of the Int. Conf. "Innovative Materials, Structures and Technologies", 122-128 (2013)
11. M.S. Magalhaes, R.D. Toledo Filho, E.M. Rego Fairbairn, Matéria (Rio de Janeiro), **18**, 1587-1595 (2013)
12. Y. Yang, Y. Wang, F. Fu, J. Liu, Thin Wall Struct. **95**, 674-388 (2015)
13. Y. Ishikawa, H. Nakamura, T. Tanabe, Proc. of JCI meeting, **25**, 467-472 (2003) (In Japanese)
14. M. Sahmaran, V.C. Li, Cement Concrete Comp., **33**, 72-81 (2008)

Interaction of collar roof with the bottom support structure of a building

Josef Musilek[1,*]

[1]Institute of Technology and Business in České Budějovice, Department of Civil Engineering, Okružní 517/10, 370 01 České Budějovice, Czech Republic

Abstract. Collar roofs belong to the traditional structures used for roofing buildings. In present, the collar roofs are widely used for their ability to omit internal vertical support structures (columns, support walls), which might be sometimes a complication during a design of the layout of the building. Collar roof acts on the bottom support structure of the building by quite big lateral horizontal forces. These forces are being often neglected in common practice. This fact leads to failures of the bottom support structures. Reinforced concrete rim made on the top of the brick wall is one of the methods, how to deal with these horizontal forces. The article deals with the interaction. The article deals with the interaction between the structure of the collar roof and the reinforced concrete rim.

1 Introduction

Collar roofs are historically one of the oldest systems that have been used throughout all periods and since their expansion in the Middle Ages they are one of the most widespread construction systems. Their significance is noticeable at present as they are very often used for roofing of simple roofs of smaller objects. Collar roofs allow to free up ground floor layout because they do not require internal vertical support elements. However, the collar roofs cause a significant lateral horizontal force on the lower support structure in the area of anchoring in the lace, which, unfortunately, is often underestimated and not considered in practice. The effects of these forces are even greater than in the case of plain or purple systems. This often leads to disturbances of the lower bearing structure of the walls, because the walls are made from common bricks, and are not able to pass these horizontal forces.

One way to capture these transverse horizontal forces is to design a reinforced concrete rim to be carried out at the top of the wall. This reinforced concrete rim then acts as a horizontal continuous beam that transmits horizontal forces from the roof. Horizontal support of this beam is made of shielded walls, where the rim is sufficiently covered or otherwise anchored and then internal support. The inner supports can be formed either by pulling the rim into the inner transverse load-bearing wall or, for example, by using a steel rod anchored to the supporting structure of the ceiling, which may optionally be hidden, for example, in a non-load-bearing wall. This reinforced concrete rim has to be dimensioned to

* Corresponding author: musilek@mail.vstecb.cz

transfer horizontal load from hornbeam truss. However, the question of how far the rigidity of this reinforced concrete rim affects the behavior of the individual truss connections, it means the internal forces in the individual bonds and the individual horizontal forces, can arise. On the example of a concrete roof, a calculation will be made which will try to take into account the compliance of the reinforced concrete rim.

2 Specification of the roof for calculation

Layout and the section of the roof is show on the Fig. 1.

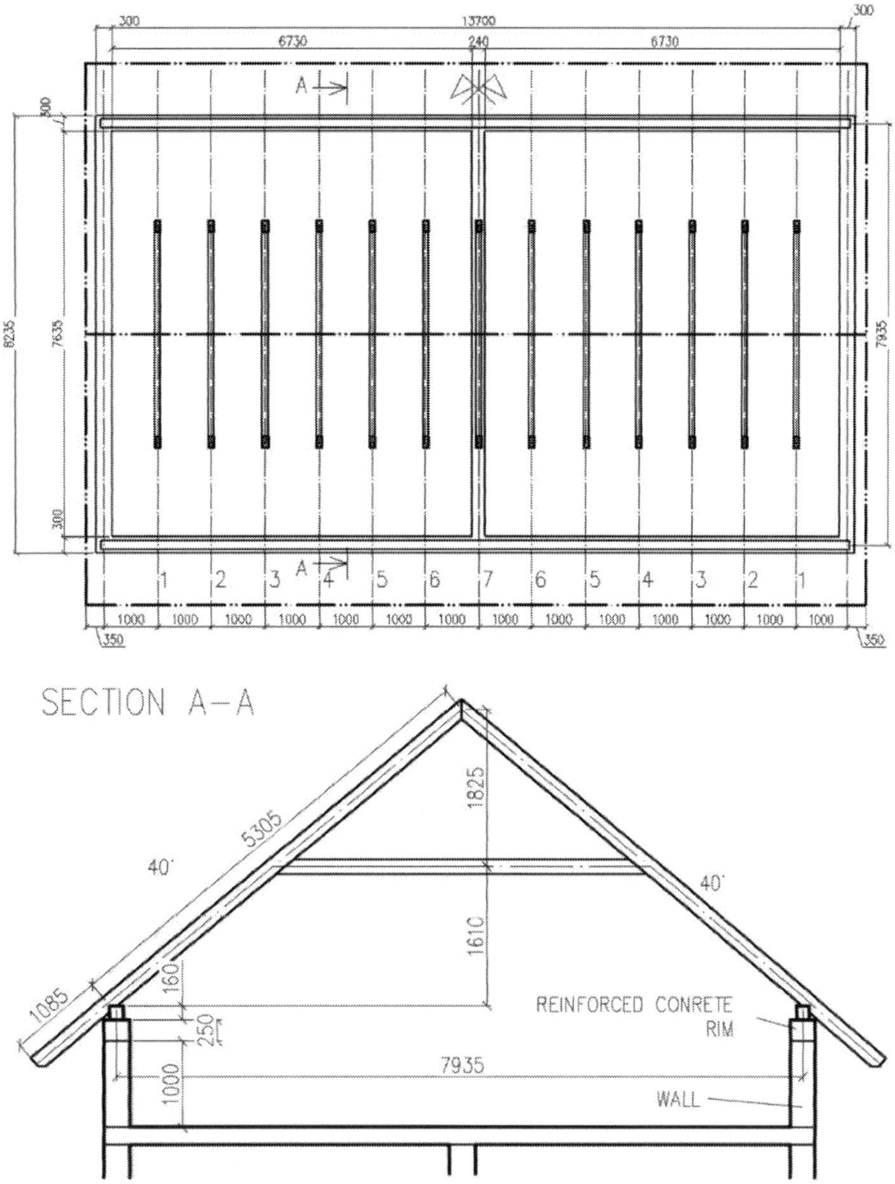

Fig. 1 Layout and the section of the roof.

It is a roof of a typical family house of a rectangular ground plan. In the middle of the house there is one transverse wall into which a reinforced concrete rim is inserted. The rim is further introduced into the gable walls. Concrete C20 / 25 concrete material. B500B reinforcement material. Diameter of the reinforcement bars is 14mm.. The reinforcement scheme is in Fig. 2.

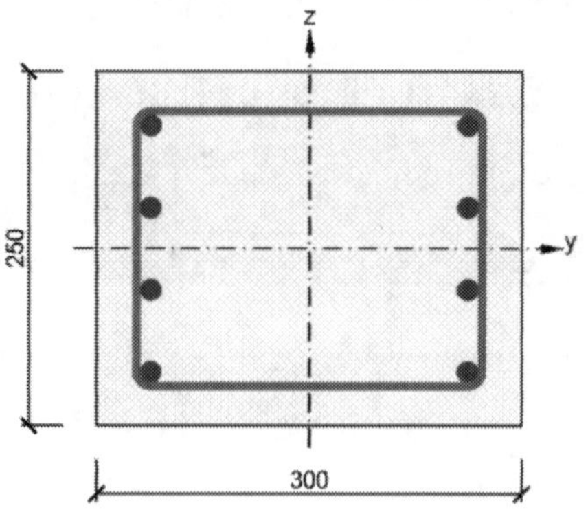

Fig. 2 Reinforcement of the rim.

It is assumed that the reinforced concrete rim is designed according to the present valid standards, for example [1], or [2] and the timber structure is designed for example according to [3] or [4].

3 Computational model

The computational model considers a reinforced concrete rim, which is simulated as a beam of C20/25 isotropic concrete. The reinforcement of the rim cross section is not taken into account in this calculation. This computational model also does not consider the fact that concrete can´t transfer the tension (practically is transmitted by the tension reinforcement) and also does not consider reducing the bending stiffness of the rim due to cracks. Spatial computing model is at Fig. 3

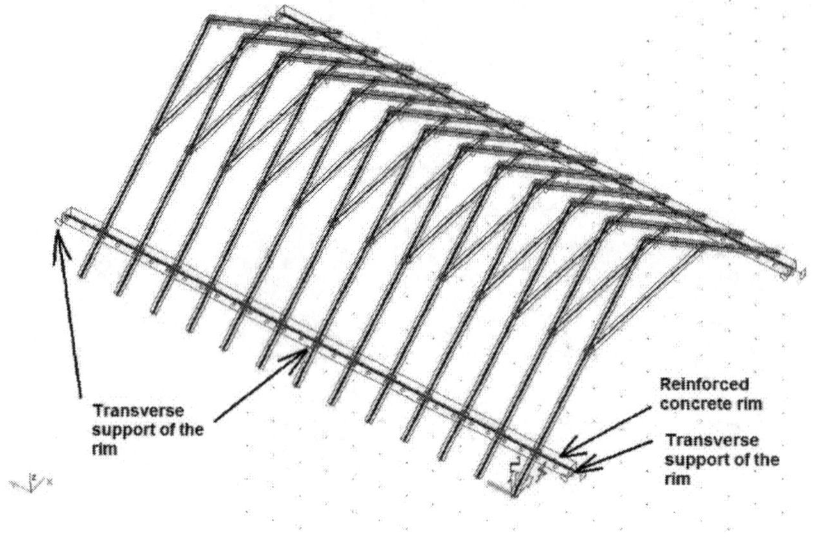

Fig. 3 Computational model.

The rim is modeled as a continuous beam, which is vertically supported continuously in the vertical direction. It is assumed that cracks will not occur. In the transverse (horizontal) direction it is supported at its ends at the gable walls and in the center of the building at the site of the considered internal cross-sectional wall. Fig. 4 shows the shape and size of the bending moment from the considered load combination on the rafters in row no.7. Row no. 7 is located at the transverse load bearing wall which supply the rigid transverse support for the rim. The horizontal displacement at this point is zero, as is commonly contemplated in the ideal collar roof system. The Fig. 5 shows the bending moment on the rafters for the row no. 3. It can be seen that the lateral horizontal deformation of the rim caused a change in the internal forces in rafters compared to the row no. 7. In Table 1 you can see other selected results of internal forces and deformations for the third and seventh row.
The Fig. 6 shows the bending moment on the rim, which is assumed as ideally rigid. The Fig. 7 shows the bending moment on the rim assumed made from material C20/25.

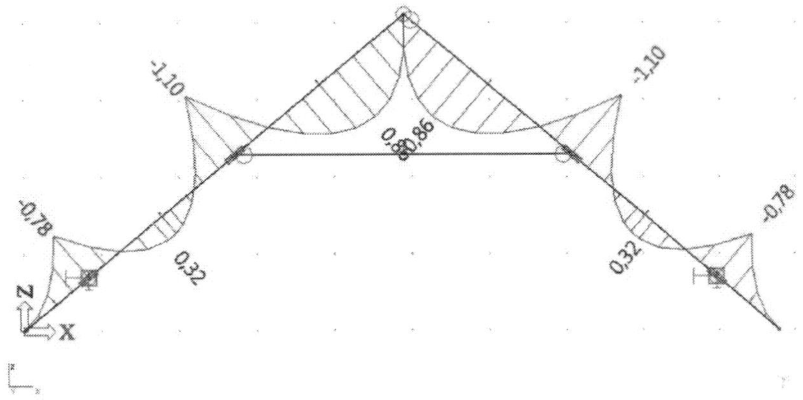

Fig. 4 Shape and size of the bending moment on the rafters in the row no. 7 [kNm].

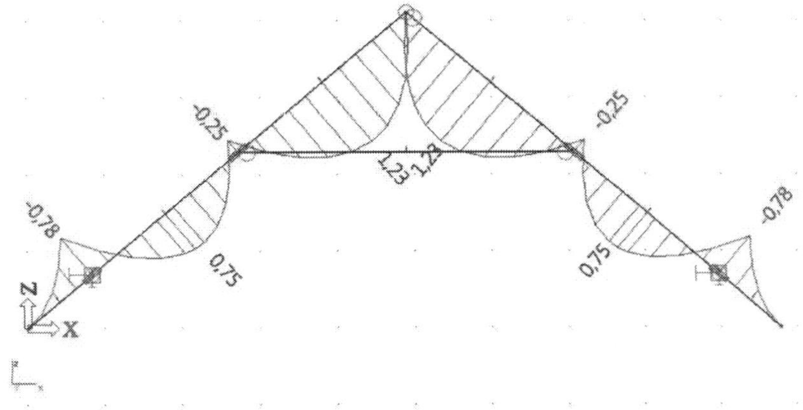

Fig. 5 Shape and size of the bending moment on the rafters in the row no. 3 [kNm].

Table 1. Selected results.

Row	Normal force in collar beam (compression) [kN]	Transverse horizontal force into the rim [kN]	Transverse displacement of the rim [mm]	Max transverse displacement of the rim [mm]
7	8.41	10.74	0.0	8.0
3	7.41	10.22	8.0	

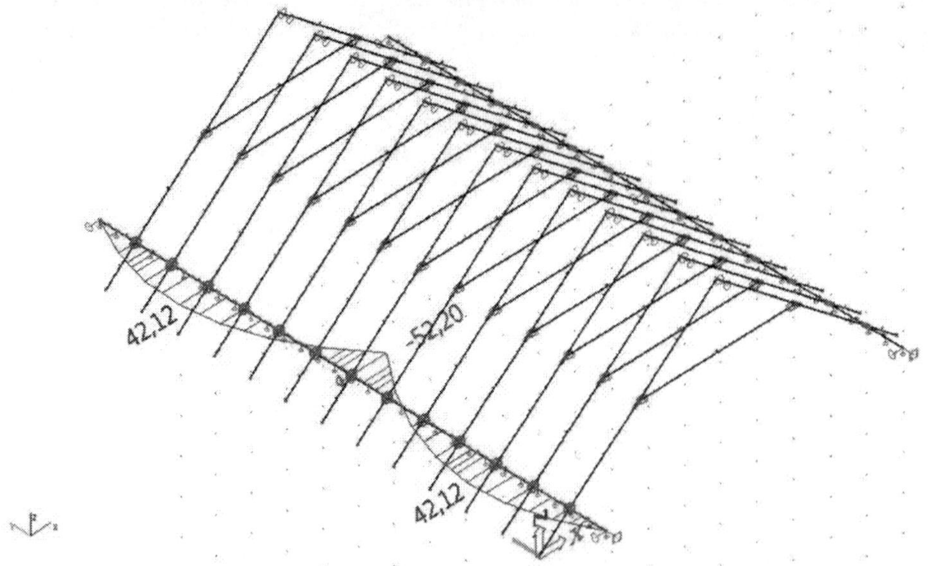

Fig. 6 Shape and size of the bending moment on the rim, which is assumed as ideally rigid [kNm].

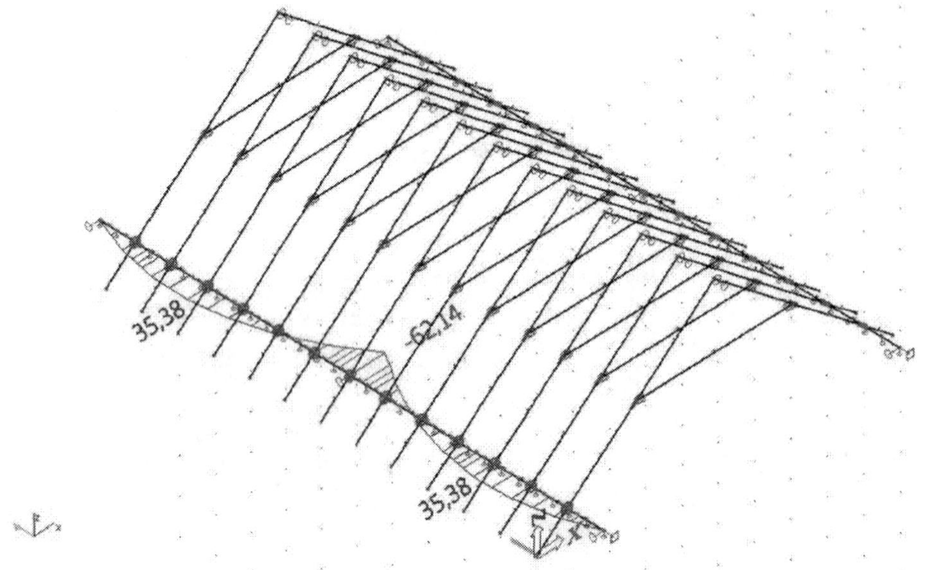

Fig. 7 Shape and size of the bending moment on the rim, which is assumed as concrete C20/25 without cracks [kNm].

4 Conclusion

This paper presents how to integrate the horizontal rigidity of a reinforced reinforcing wreath into a global hull truss analysis. This procedure assumes the rim as an isotropic flexible bead of plain concrete that also acts under pressure and tension. The rows found in the lateral support of the rim are possible to consider as an ideal collar roof behaviour.

It is possible to see, that the rows near the middle of span of the rim show differences in internal forces compared to rows near the transverse supports. The rows near the middle of span of the rim act on the rim by smaller transverse force than the rows near the transverse rim supports. This fact corresponds to the change of force in collar beam.

However, the computational model presented in this paper doesn´t take into account the influence of the reinforcement of the rim and also the influence of the cracks in the rim which will surely occur. It is highly probable that the cracks in the rim will have big impact on the behaviour of the structure and the big influence on the internal forces in the elements in the structure. This influence of these cracks on the rim is the subject of the further research, as well as finding a suitable computational methodology to include these cracks in a global analysis of the structure.

References

1. *EN 1992-1-1 Design of concrete structures - Part 1-1: General rules and rules for buildings. European Standard* (2004)
2. J. Prochazka, *Design of concrete structures* (Czech concrete society, Prague, 2005)
3. *EN 1995-1-1 Design of timber structures - Part 1-1: General - Common rules and rules for buildings* (2006)
4. P. Kuklik, *Timber structures 10* (Czech Technical University in Prague, Prague, 2002)

Limitations and risk related to static capacity testing of piles – "unfortunate case" studies

Jarosław Rybak[1,*], and *Maciej* Król[2]

[1] Wrocław University of Science and Technology, Faculty of Civil Engineering, 50-370 Wrocław, Wyb. Wyspiańskiego 27, Poland
[2] KELLER POLSKA Sp. z o.o., Specjalistyczne Techniki Fundamentowania, 05-850 Ożarów Mazowiecki, ul. Poznańska 172, Poland

Abstract. Static load testing of foundation piles is still considered the main capacity check technology for almost any kind of piling works (bored, displacement, driven). The rates of load, time increments or imposed displacement are the key factors for various testing methods. The load-displacement relationship, which can be transformed into a coordinate system, is always the point of departure for the analysis of the results. For axis (vertical) loading, both, applied load and measured pile head settlement are examined during the entire test. During the lateral capacity tests, the horizontal displacement and inclination of free pile head is measured. In either case, all recorded values must be thoroughly examined in order to avoid systematic errors which could flaw the results of the analysis. In this paper, the authors gather and discuss their experience from field tests in which equipment malfunctions or external circumstances had the impact on the range of load test and, thus, their results. The possible ways of test extrapolation with regard to load range are also the subject of further observations presented in this publication.

1 Introduction – the need of pile testing

Static load tests are used for direct determination of capacity, verification of static calculations or for calibration of other tests (e.g. dynamic testing) at similar soil conditions. In accordance with Eurocode 7 [1], pile foundations should preferably be designed on the basis of results from load capacity tests [2]. During these tests, increasing loads are applied stepwise specific time intervals; after each step the settlement is measured. The obtained load settlement curve is further interpreted [3]. In this technique, the load should increase until the pile reaches the limit state (i.e. the situation in which the increment in settlements is unhindered without the need of increasing the load applied), which leads to the identification of the ultimate capacity. In practice, one never reaches this moment for various reasons, like limited strength of a testing station, insufficient capacity of anchoring piles or too small a kentledge (see Fig. 1), therefore it is often impossible to achieve the loads which would actually cause the unhindered pile penetration in the plastic soil. In the end, the scope of the test does not enable one to get relevant information about the pile

* Corresponding author: j.rybak@pwr.edu.pl

ultimate capacity and would not allow to calibrate the calculation model, even if the force applied during the test is larger than the design capacity and still, the structure safety is confirmed. Without knowing the ultimate capacity, we are not able to take advantage of the margin that would make it possible to decrease the number of piles in the foundation or to reduce their length.

Fig. 1. Static tests with anchoring pile group and kentledge (water tanks).

In designing practice, we often assume the maximum load effected in field testing as the ultimate capacity and accept that limitation due to uncertainty of extrapolation methods, as recommended by Fellenius [4]. In nowadays' trends of giving primary importance to economic factors, the fact that static tests are costly and time-consuming is especially brought to the fore, and because of that it is emphasized that they should be used to get as much information about the pile capacity as possible. This paper outlines how the number of measurements (for the range of test load effected) affects the accuracy of determining the ultimate capacity- Q_u. The review was based on data from 30 site logs of test loads applied to prefabricated driven piles. Examinations were made within supervision of Wrocław University of Science and Technology on construction and modernization of road infrastructure in 2010-2012.

2 Traditional methods of load test extrapolation

There are numerous methods of determining the ultimate capacity with extrapolating the results gained before pile reaches the state of ultimate load bearing capacity. Polish Code of Practice PN-83/B-02482 [5] makes it possible to determine Q_u from the auxiliary waveform dQ/ds. The condition $dQ/ds=0$ means unhindered increment in settlement without increase of load. The most popular extrapolation methods worldwide for foundation piles, described in detail by Fellenius [4], were tested for their accuracy (possible error of estimating Q_u). Each extrapolation method employed in these analyses may be described as fitting the results of settlement measurements in successive pile loading steps into the assumed functional form. The co-ordinate system is re-scaled (ordinates and abscissas) in order to facilitate the fitting of extrapolating function parameters; because of that, the settling-strain relation is roughly linear in the final stage of pile operation. This allows to fit the function extrapolating the test load curve with the use of the simplest linear regression in the spreadsheet. The slope of a straight line and the free term need to be fitted to estimate the ultimate capacity.

Of key importance are the last points on settlement-strain curve, which represent elastic/plastic operation of pile prior to reaching the ultimate capacity. The method presented in Fig. 2 proposed by De Beer [6] to evaluate the pile capacity may be helpful as the criterion of selecting the range of point fitting. The settlement-loading relation is given in logarithmic scale; hence it is roughly linear in intervals. The operating ranges of pile are presented in Figure 2. The intersection point of curves drawn on the diagram represents approximate value of load above of which the pile operates in plastic range. It should be emphasized that the concept of De Beer [6] was not aimed at determining the value of Q_u but merely the load above which permanent pile displacement occurred. Similar nature (however for another system of co-ordinates) had the interpretation of test loads according to Polish Code of Practice [5]. In fact, the ultimate capacity can be determined by means of points on the straight line representing the plastic range of pile operation. The more points are recorded in this range, the higher precision of ultimate capacity estimation is guaranteed. If the number of measurements is too small, errors occur and make the extrapolation method useless. Errors generated by a selected method are compared and the effect of the number of measurements in elastic/plastic range and in plastic range of pile operation on extrapolation accuracy is verified.

3 Decourt's, Brinch-Hansen's 80% and Chin-Kondner's method

According to Decourt's method [7], the settlement-loading relation (s–Q) is transformed into the coordinate system: Q - abscissa and Q/s - ordinates (Fig. 3). For the last points of the performed test at which the settlement reached stabilization the following linear dependence is found (by way of approximation): $Q/s = A \cdot Q + B$.

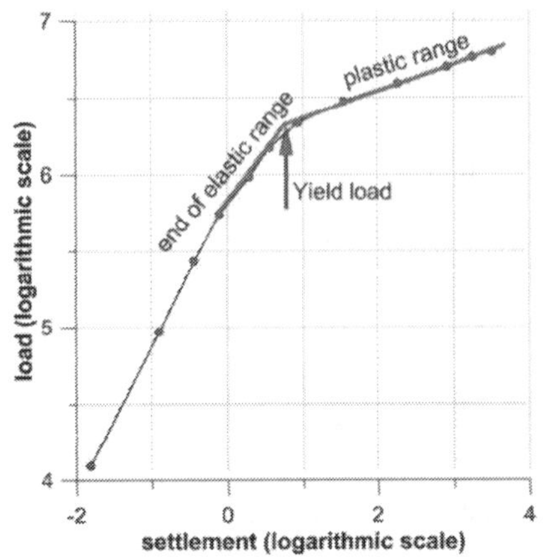

Fig. 2. De Beer double logarithmic method [6].

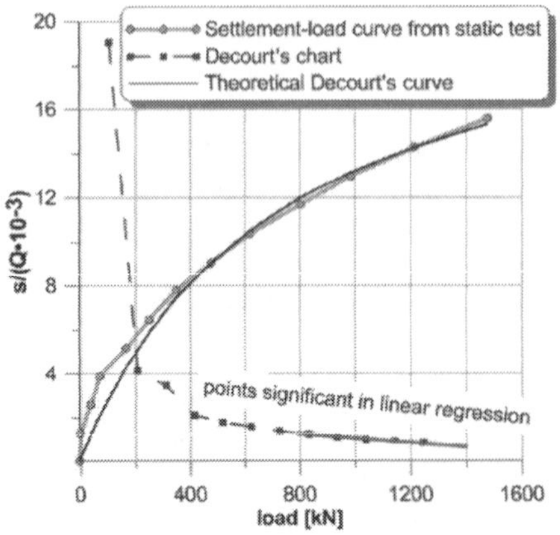

Fig. 3. Decourt's method [7].

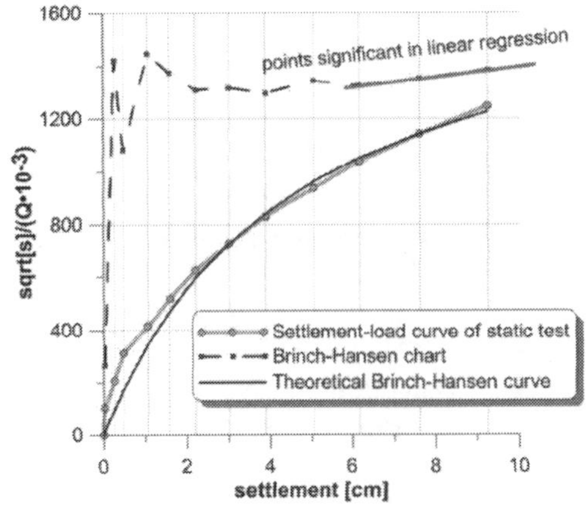

Fig. 4. Brinch-Hansen 80% method [8].

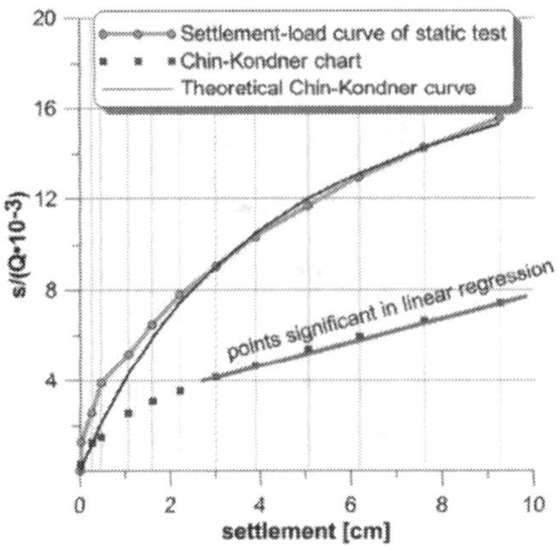

Fig. 5. Chin-Kondner method [9].

According to Brinch-Hansen, in a so-called 80% method [8], the ultimate capacity Q_u is understood as such loading for which the settlement s_u exceeded four times the value s measured after 80% of the load Q_u has been applied. The settlement-loading relation (s–Q) is then transformed into the coordinate system: s - abscissa and $\sqrt{s}/Q$ - ordinates (Fig. 4). For the last points of the performed test at which the settlement reached stabilization the following linear dependence is found (by way of approximation): $\sqrt{s}/Q = A \cdot s + B$.

According to Chin-Kondner method [9], the settlement-loading relation (s–Q) is transformed into the coordinate system: s - abscissa and s/Q - ordinates (Fig. 5). For the last points of the performed test at which the settlement reached stabilization, the following linear dependence is found (by way of approximation): $s/Q = A \cdot s + B$.

The ultimate capacity and settlement-load relation the for all the methods can be then derived from formulae juxtaposed in Table 1.

Table 1. The ultimate capacity and settlement-load relation the for all the methods.

method	Decourt's	Brinch-Hansen 80%	Chin-Kondner
ultimate capacity	$Q_u = -B/A$	$Q_u = \dfrac{1}{2\sqrt{A \cdot B}}$	$Q_u = 1/A$.
settlement-load relation	$Q = \dfrac{B \cdot s}{1 - A \cdot s}$	$Q = \dfrac{\sqrt{s}}{A \cdot s + B}$	$Q = \dfrac{s}{A \cdot s + B}$

where: A – slope of regression straight line;
B – intersection point between regression straight line and 0Y axis.

4 Example of evaluation of ultimate capacity inaccuracy

It must be noticed that all the extrapolation methods are strongly dependent on the range of the performed test in relation to the computed ultimate capacity. In this study, the authors chose the Chin-Kondner method for its simplicity; similar calculations, however, for Decourt's and Brinch-Hansen 80% criteria gave very similar results.

A series of calculations were performed to check to what an extent we can be mistaken when the test range reaches a certain percentage of ultimate capacity. For load-settlement charts from failed load tests – an estimation of (known) capacity was done on the basis of insufficient data. In some of the cases the result was quite satisfactory. In other cases (see Fig 6) extrapolation led to serious under- or overestimation of ultimate capacity Q_u.

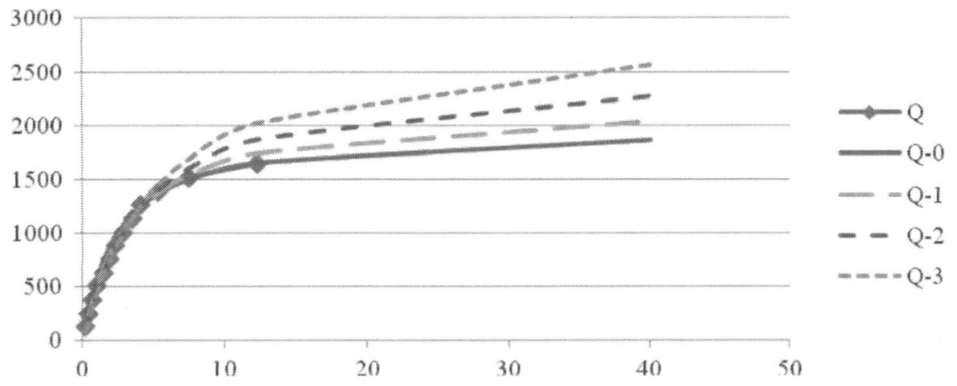

Fig. 6. Chin-Kondner extrapolation for reduced test information (very risky).

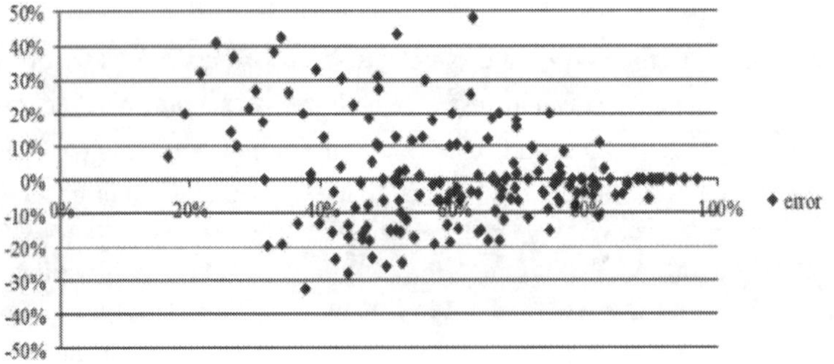

Fig. 7. Error in ultimate capacity estimation for Chin-Kondner method.

The error shown in Fig. 7 is the distance between the ultimate capacity estimated on the basis of only some part of the obtained data (i.e. the values preceding the limit state) and the actual limit state values measured during the test. In most of the cases the result is positive, which points to an overestimation of capacity (as in Fig 6).

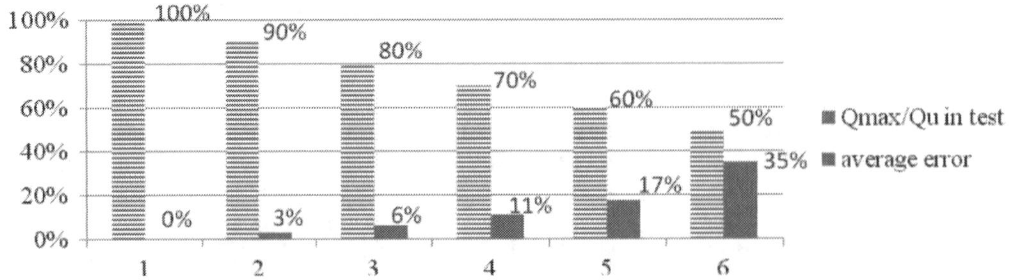

Fig. 8. Average error in ultimate capacity estimation for short series.

It can be noticed (Figure 8) that if the range of test does not exceed 70% of Q_u, then the average absolute error reaches almost 20%. Such result will be unacceptable for design.

5 Summary and conclusions

The authors evaluated each method described in this paper with respect to its accuracy of estimating the ultimate capacity. The review of 30 site logs of pile static testing proves that the Brinch-Hansen 80% method is the most accurate of all extrapolation methods. If we want to be able to apply it, however, we need a relatively large number of load steps. When used for extrapolating the points from the elastic range of pile operation (i.e. after the limit capacity has been reached), the method turns out overtly futile. Calculations made by way of steps proposed by Chin-Kondner and Decourt show that the estimated capacities differ significantly from those determined in static tests. However, the features of applied approximating functions make it possible to use them for loads performed in a narrower range. If former authors' works [10] it was already proved that the loss in accuracy in determining the capacity is the consequence of the having the advantage of possible extrapolation for points from elastic/plastic interval of pile operation. Fellenius [4] expressed doubts about the sense of using the Chin-Kondner and Decourt method, labelling them as insufficiently conservative. It must be underlined that same doubts must be emphasised when dealing with numerical modelling of pile behaviour under test [11], especially in the last stages of static load test [12].

Table 2. Method comparison – derived errors.

	Brinch-Hansen 80%		Chin-Kondner		Decourt	
Points used	Mean	Standard deviation	Mean	Standard deviation	Mean	Standard deviation
2	24.86%	9.76%	50.71%	10.67%	45.13%	10.48%
3	16.84%	10.12%	42.93%	9.99%	37.42%	11.43%
4	11.09%	9.47%	36.20%	8.03%	31.18%	10.39%
5	10.18%	7.04%	29.09%	7.78%	25.24%	8.66%
6	6.46%	4.75%	20.83%	7.14%	18.93%	8.24%
7	4.43%	2.92%	9.59%	4.73%	14.60%	7.84%

It is also an essential information that measurements from the first steps of pile loading have no effect on the result. Attempts to extrapolate the ultimate capacity from the initial measuring points are completely useless, as the result would not be even slightly similar to real value. Theoretical curves fit well in plastic range, while large discrepancies are observed in the range of pile operation. Results of three methods of estimating the ultimate capacity are presented in Table 2 on the basis of data from 30 site logs of static tests.

References

1. EN-1997-1:2008 Geotechnical design. Part 1: General rules.
2. D. Sobala, G. Tkaczyński, Interesting Developments in Testing Methods Applied To Foundation Piles, IOP Conf. Series: Materials Science and Engineering, **245,** 022074, (2017)
3. M. Drusa, F. Gago, J. Vlček, Contribution to Estimating Bearing Capacity of Pile in Clayey Soils, J. of Civil. and Environ. Eng., De Gruyter Open, **12 (2)**, 128-136, (2016)
4. B.H. Fellenius, What capacity value to choose from the results of a static loading test. We have determined the capacity, then what? Deep Foundation Institute, Fulcrum Winter, pp. 19 – 22 and Fall, 23 – 26, (2001)
5. PN-83/B-02482 (1983): Foundations. The capacity of piles and pile foundations.
6. E.E. DeBeer, Proefondervindlijke bijdrage tot de studie van het grensdraag vermogen van zand onder funderingen op staal. *Tijdshift der Openbar Verken van Belgie*, No. **6**, (1967) and No. **4**, **5**, and **6**, (1968)
7. L. Decourt, Behavior of foundations under working load conditions. *Proc. of the 11th Pan-American Conf. on Soil Mechanics and Geotechnical Engineering*, FozDoIguassu, Brazil, August 1999, Vol. **4**, 453-488, (1999)
8. J. Brinch Hansen, Discussion, Hyperbolic Stress-Strain response, Cohesive soil., J. of Soil Mechanics and Foundation Engineering Division, ASCE, **89**, 241-242, (1963)
9. F.K. Chin, Estimation of the Ultimate Load of Piles Not Carried to Failure, *Proc. 2nd Southeast Asia Conference on soil Engineering*, 81-90, (1970)
10. M. Król, J. Rybak, Accuracy of extrapolation methods for non-failed static load tests *Effect of water on geotechnical structures – Proc. of the 11th Slovak Geotechnical Conference*, Bratislava, Slovak University of Technology, 285-294, (2013)
11. M. Baca Michał, W. Brząkała Włodzimierz: Numerical modeling of pile installation influence on surrounding soil, *17th Int. Multidisciplinary Scientific GeoConference, SGEM 2017*, **17** (Issue 12), Sofia : STEF92 Technology, 619-626, (2017)
12. M. Drusa, J. Vlček, Numerical approach to pile load test using 3D finite element method, *Komunikácie* **16 (4)**, 94-97, (2014)

Dependence between the width of strip overlap and the strength of joints of one-layer bitumen sheets

Tomáš Petříček[1,*], and *Petr* Kacálek[1]

[1]Brno University of Technology, Faculty of Civil Engineering, Institute of Building Structures, Veveří 95, 602 00 Brno, Czech Republic

Abstract. This paper focuses on the examination of mutual dependence of overlap width of bituminous sheets and the strength characteristics of their mechanically fastened joints. Bituminous sheets with two different widths of overlap and different degree of modification were selected as samples for experimental measurements. They are supposed to be used as single-layer mechanically fastened waterproofing membranes. The aforementioned dependence has not been explicitly documented yet. The aim of this paper is to confirm or to refute a hypothesis that a relationship exists between a different width of overlap of bituminous sheets and the strength of their joints. The formulation of the results was based on the statistical evaluation of laboratory measurement data.

1 Introduction

The use of one-layer mechanically fastened systems is mostly based on meeting the basic requirements and technological regulations specified by manufacturer.

High quality bitumen sheets with the total width of min. 5.0 mm are used for one-layer systems from bitumen sheets. Bitumen coating layer must be of sufficient quality to allow good welding properties and long-term flexibility. The sheets are most commonly reinforced with a combination of PES fleece strengthened with glass fibres laid in one or both directions. The sheet overlaps are usually set to min. 120 mm, although there are also sheets for which manufacturer specifies the overlap to just 100 mm.

The critical joint spot is the presence of fastening element as a spot where force effects caused by wind sucking are captured. Full scale fusion of sufficient width of the joint is a basic requirement to provide waterproofing system ability and to transfer force effects caused by wind sucking.

2 Overlaps of bitumen sheets

The existing standards in force related to waterproofing membranes fail to specify the size of joints. Joints are only specified in ČSN EN 13707+A2. Flexible sheets for waterproofing

* Corresponding author: petricek.t@fce.vutbr.cz

– Reinforced bitumen sheets for roof waterproofing – Definitions and characteristics [2], however, minimum joint length is missing.

Therefore, the minimum overlap width is fully in competence of bitumen sheet manufacturers. The overlap width particularly depends on the application method of the bitumen sheet (roof cladding, substructure, one-layer or multi-layer system), on the presence of a fastening element in the joint, on surface treatment of bitumen sheet, and on joint type – whether longitudinal or transversal overlap is concerned. Regarding longitudinal overlap of one-layer mechanically fastened bitumen sheets, the minimum width is usually set by manufacturers to 120 mm, exceptionally 100 mm. The width in transversal direction (front joint) ranges between 100 mm and 150 mm.

3 Experimental measurement

A pressure chamber was used in which the composition of a roof membrane from a real joint segment of mechanically fastened waterproofing could be subjected to defined stresses. This wind load test is defined in the Approval Guideline for systems of mechanically fastened waterproofing membranes ETAG 006 [6].

Experimental measurements tests were carried out on a standard tensile testing machine. The samples were roof membrane segments, not just the waterproofing membrane. Two metal fastening strips in the arms of the upper grip of the machine held the waterproofing membrane in place. These arms transferred the force of the applied stress to the tested material. A pad, to which the waterproofing membrane was fastened by thermal insulation, was clamped into the lower grip. As the lower grip of the tensile testing machine receded, it applied stress on the joint and subsequently its rupture.

The test samples were prepared according to composition variant S1, i.e. analogically for a roof without thermal insulation with the waterproofing membrane mechanically fastened to OSB board of 22 mm thickness. The fastening element to the hard base was a screw with a metal oval washer. The edge of the pressure washer of the fastening element was located 10 mm from the edge of the lower bitumen sheet. The use of the oval pressure washer was intentional because it provides better pressure on the lower sheet thanks to its larger surface area. As a result, the rupture of the test samples depended on the parameters of the bitumen sheet rather than on the failure of the fastening element.

Three bitumen sheet variants were chosen as the waterproofing membranes:

1. **Test sample F** - SBS modified bitumen sheet 5.2 mm thick, carrier: PES 180 g/m^2, reinforced on both sides with glass fibres, flexibility at low temperatures: -25 °C, joint length 100 mm;
2. **Test sample G** - SBS modified bitumen sheet 5.2 mm thick, carrier: PES 180 g/m^2, reinforced on both sides with glass fibres, flexibility at low temperatures: -25 °C, joint length 120 mm;
3. **Test sample A** - SBS modified bitumen sheet 5.0 mm thick, carrier: PES 180 g/m^2, reinforced longitudinally with glass fibres, flexibility at low temperatures: -20 °C, joint length 100 mm;
4. **Test sample D** - SBS modified bitumen sheet 5.2 mm thick, carrier: PES 180 g/m^2 reinforced on both sides with glass fibres, flexibility at low temperatures: -25 °C, joint length 120 mm.

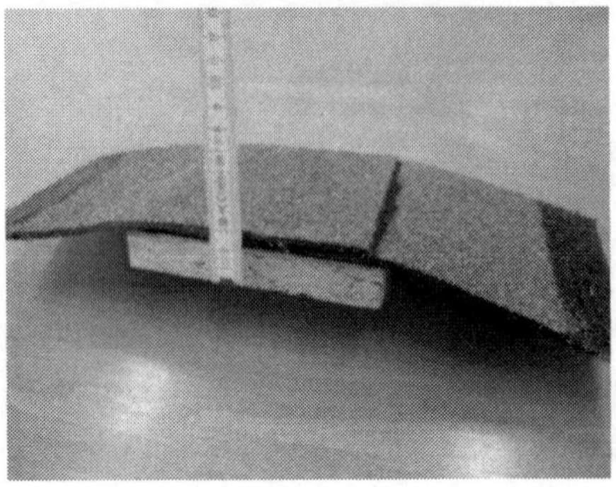

Fig. 1. Testing sample - composition variant S1 [1].

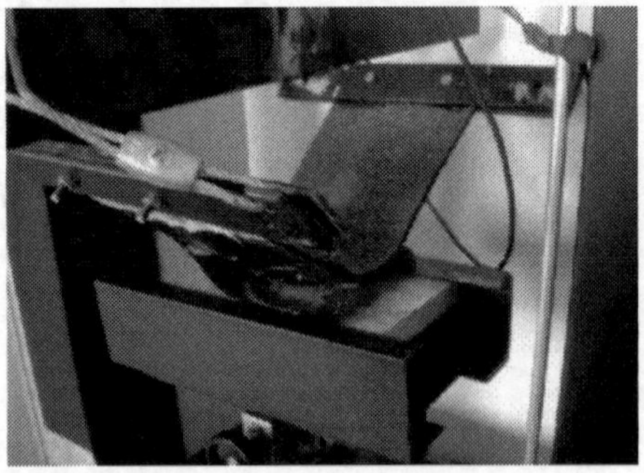

Fig. 2. Testing sample in tensile testing machine during test [1].

4 Evaluation

Approximately 10 samples were used for each testing sample. The resulting force is in the form $\bar{x} \pm \bar{\vartheta}(\bar{x})$, where $\bar{x}$ is the arithmetic average and $\bar{\vartheta}(\bar{x})$ is the probable measurement error. The latter was determined on the basis of the error theory for deviations of arithmetic average using the formula:

$$\bar{\vartheta}(\bar{x}) = \frac{2}{3}\sqrt{\frac{1}{n(n-1)}\sum_{i-1}^{n}(x_i - \bar{x})^2} \quad (1)$$

5 Results of experimental measurement and discussion

Comparing the measurement results of samples F and G (Fig. 5), there is just a small difference in reached strengths of mechanically fastened joints. Sample F with joint length of 100 mm only showed 5% lower values than sample G (joint 120 mm).

This difference would probably be higher (Fig. 6) for bitumen sheets with lower degree of modification (flexibility at low temperatures -20°C). This comparison is not fully representative, since the parameters of samples A and D differed by carrier type and total sheet width. Nevertheless, the strength of testing sample A (joint 100 mm) was approx. 22% lower than the strength of testing sample D (joint 120 mm).

Detailed results and the comparison of results of all samples are shown in Table 1.

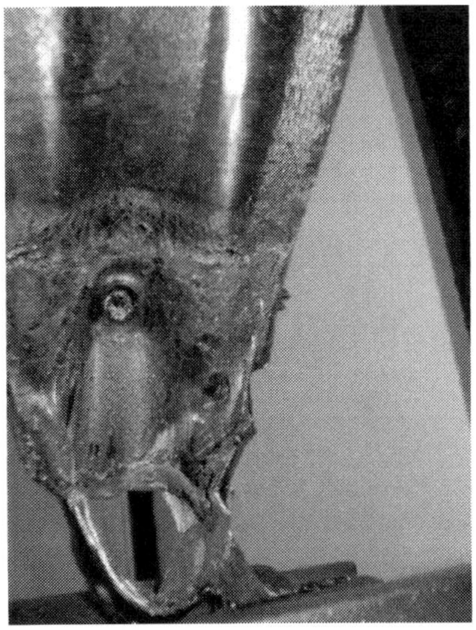

Fig. 3. Unsealed joint near a fastening element and beginning of its deformation – sample G [1].

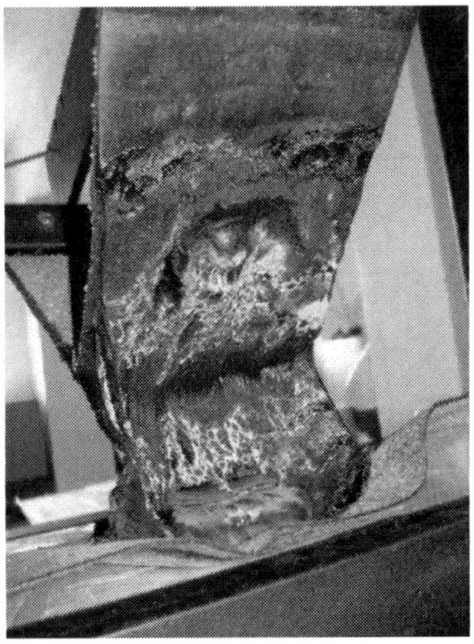

Fig. 4. Distinct unsealing of joint of testing sample A [1].

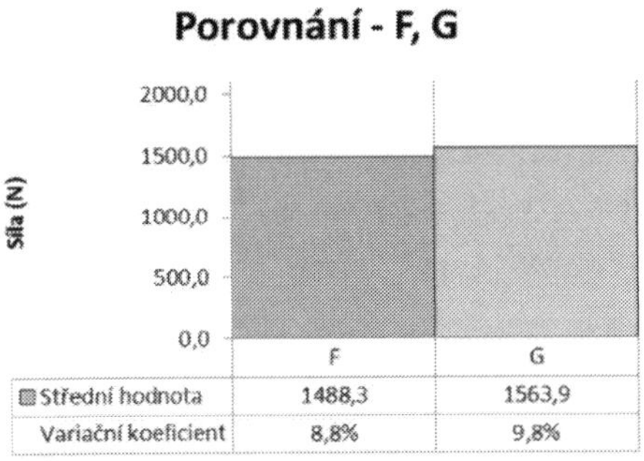

Fig. 5. Comparison of samples F and G (SBS modified bitumen, -25°C) [1].

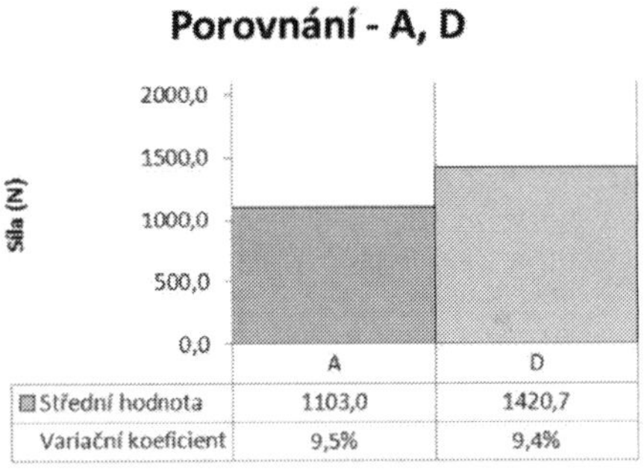

Fig. 6. Comparison of samples A and D (SBS modified bitumen, -20°C) [1].

Table 1. Comparison of testing samples.

Test sample:	F	G	A	D
Bitumen substance:	SBS modified			
Flexibility at low temperatures:	-25 °C		-20 °C	
Sheet thickness:	5.2 mm		5.0 mm	5.2 mm
Carrier:	PES reinforced on both sides, 180 g/m^2		PES reinforced longitudinally, 180 g/m^2	PES reinforced on both sides, 180 g/m^2
Overlap width:	100 mm	120 mm	100 mm	120 mm
The highest tensile strength:	longitudinally: 900 (± 200) N/50mm, transversally: 650 (± 150) N/50mm		longitudinally: 900 (± 200) N/50mm, transversally: 700 (± 150) N/50mm	
Ductility:	longitudinally and transversally: 45 (± 10) %		longitudinally and transversally:	longitudinally and

			50 (± 10) %	transversally: 45 (± 10) %
Mean value:	**1 488.3 N**	**1 563.9 N**	**1 103.0 N**	**1 420.7 N**
Standard deviation:	131.6 N	153.2 N	104.8 N	134.0 N
Variation coefficient:	8.8%	9.8%	9.5%	9.4%

6 Conclusion

Based on the experiment results, it can be assumed that the length of bitumen sheets with higher degree of modification has no significant effect on the resulting strength of a mechanically fastened joint. Regarding sheets with lower degree of modification, it is assumed that the resulting strength depends on the length much more.

This paper has been worked out under the project No. LO1408 "AdMaS UP - Advanced Materials, Structures and Technologies", supported by Ministry of Education, Youth and Sports under the „National Sustainability Programme I".

This work has been financially supported by a research projects FAST-S-17-4765 from the Internal Grant Agency, Brno University of Technology, Czech Republic.

References

1. T. Petříček , *Adhesion of joints of mechanically fastened bitumen sheets* (Doctoral thesis, University of Technology, Faculty of Civil Engineering, Brno, 2013)
2. *ČSN EN 13707+A2 Waterproofing sheets and foils – Reinforced roof waterproofing bitumen sheets – Definitions and characteristics* (The Institute of Technical Standardization, Metrology and State Testing, Prague, 12/2009)
3. J. Plachý. MATEC Web of Conf. **93**, 02004 (2017)
4. J. Plachý. MATEC Web of Conf. **93**, 02006 (2017)
5. R. Smolka, J. Sobotka, J. Šulistová, MATEC Web of Conf. **93**, 01009 (2017)
6. J. Sobotka, R. Smolka, MATEC Web of Conf. **93**, 01008 (2017)
7. P. Kacálek, T. Petříček, MATEC Web of Conf. **93**, 03006 (2017)
8. *ETAG 006. European Technical Approval Guideline – Systems of Mechanically Fastened Flexible Roof Waterproofing Membranes. Brussels: European Organisation for Technical Assessment - EOTA, 03/2000.*

Failure of pillar of sports and entertainment hall structure

Tomasz Trapko[1], *Michał* Musiał[1,*], and *Andrzej* Ubysz[1]

[1]Wroclaw University of Science and Technology, Faculty of Civil Engineering, Wybrzeże Wyspiańskiego 27, 50-370 Wrocław, Poland

Abstract. This paper deals with the failure of one of the sports and entertainment halls in Poland. The facility was built in the years 1977-85. Its main load-bearing members were three-hinged double-sloped frames with a total span of 45 m. The main girders of the frames were made of glued-laminated timber. The girders rest on reinforced concrete pillars. Because of the considerable expansion forces the facility was founded on concrete padstones. In the summer of 2016 one of the reinforced concrete pillars was found to have failed. It was established that the failure condition was mainly due to construction errors. A way of strengthening all the pillars with additional RC members was proposed.

1 Introduction and general description of facility

The sports and entertainment hall was built in Poland in the mid-1980s. Its west and south elevations are shown in fig. 1. The hall houses a full-size (20 m × 40 m) handball pitch, an amenity area and stands with 800 seats. The construction of the hall began in 1977, but the facility was put in operation as late as in 1985. The facility had been designed by Professor Wojciech Zabłocki. It may be of interest to know that he is not only an outstanding architect, but also a well-known fencer, the winner (as a team) of three Olympic medals (Melbourne'56, Rome'60, Tokyo'64) and the many times world fencing championships medallist (both as an individual and a team).

Fig. 1. South and west elevations (with main entrance) [1].

* Corresponding author: michal.musial@pwr.edu.pl

The facility's principal load-bearing structure consists of 11 three-hinged frames spaced at every 6.00 m. A typical cross section of the frame is shown in fig. 2. On plan the structure occupies a 45 m × 60 m rectangular area. The roof framework is wooden. The main roof girts are made of glued-laminated timber and their dimensions are: 2 × 14 cm × 124 cm (no. 1 in fig. 2) and 28 cm × 124 cm (no. 2). They are joined together with a steel hinge.

X-type bracings, made of beams 12 cm × 30 cm in cross section, were used in all the fields of the roof planes. Nine pairs of bracings were made in the longer roof plane between the main girders, and five pairs in the shorter roof plane.

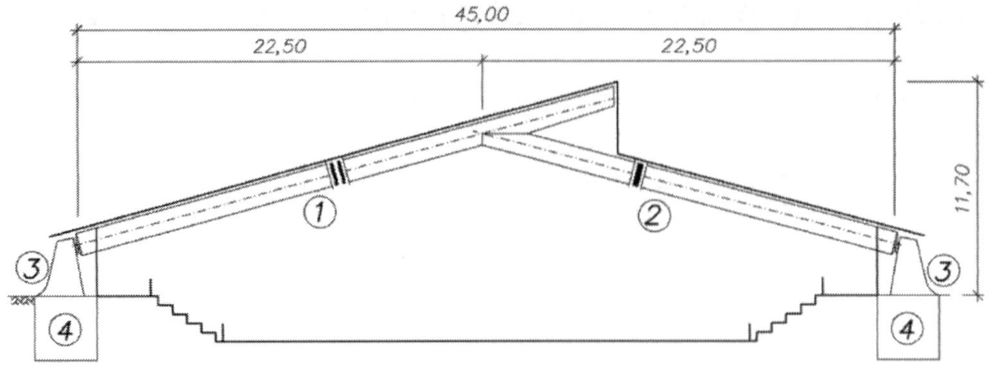

Fig. 2. Typical cross section of hall (dimensions in m, description in text) [1].

The wooden girders rest, via steel bearings, on reinforced concrete columns (no. 3) with a tapered cross section with dimensions $b \times (h_{min} \div h_{max}) = 60$ cm × (70÷210) cm. The columns were to be anchored (the anchoring proved to be ineffective) in the 180 cm × 300 cm and 360 cm deep concrete padstones (no. 4).

2 Failure condition. Structural defects

In the summer of 2016, during clean-up work, one of the supporting columns on the facility's south side was found to be damaged. In the zone of the joint with the concrete padstone the cover of concrete debonded extensively along the length of the lap of the starting foundation rebars and the tension reinforcement of the column (fig. 3). At the same time the column turned relative to its compression zone and a gap appeared between the column and the foundation. In its widest part the gap was 3.5 wide. Numerous cracks (most with allowable width) were found on the column's sides. A macroscopic visual examination of the other structural members did not reveal any other negative effects of the failure. This could be owing to the static system (the three-hinged frame), the interaction (via the bracings) between the individual frames in carrying loads and the adaptation properties of the wood.

As the column was damaged, a direct survey of its reinforcement could be carried out. Eight ribbed bars 16 mm in diameter, made of steel with the characteristic yield strength of 410 MPa, extended from the concrete padstone. The bars were joined by a 59-65 cm long lap splicing with the column's reinforcement of the same density (8 steel ribbed bars 16 mm in diameter), but made of steel with the characteristic yield strength of 355 MPa. The reinforcement starting from the foundation and the column's reinforcement suffered from corrosion, with the corrosive losses estimated at 10%. No lateral reinforcement was found in the exposed area. Also above this area examinations carried out by means of a reinforcement detector did not show any lateral rebars. In the gap which had appeared at the joint there were no vertical rebars along the column's longer side. The concrete cover of the exposed rebars amounted to 5-15 mm. All the other columns were examined using the reinforcement detector. The same faulty reinforcement layout was found.

Fig. 3. Damaged column (A – overview, B – place of damage initiation) [1].

Destructive tests, consisting in taking core samples from the columns and the concrete padstones, were carried out [1]. Concrete grades: C30/37 and C20/25, respectively, were found. Comparing the identified concrete strength classes with the current recommendations concerning structural durability [2], the class of the concrete in the pillars can be regarded as satisfactory. Whereas the class of the concrete in the padstones should be higher. However, considering the overall dimensions and the absence of reinforcing rods near the surfaces, the too low class of the foundation concrete did not require an intervention.

3 Design check calculations

The reactions originating from the roof framework are of key importance for the stress intensity level in the reinforced concrete pillars. In order to determine most accurately their values a spatial bar structure model (fig. 4) was used and static linear analyses were carried out [3].

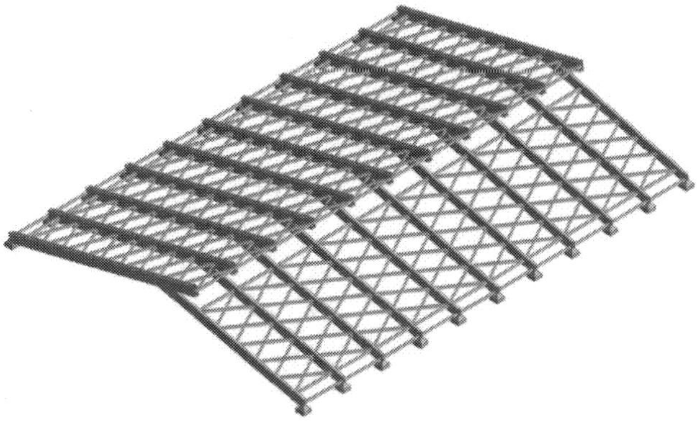

Fig. 4. Numerical model of structure [1].

The roof bracings consisted of two types of bars: full (girder-to-girder) bars and cut (girder-to-bracing) bars. This fact was taken into account in the computational model. The interconnections between the girders and the support were modelled as hinged.

Static load calculations were performed in accordance with [4]. The following loads were taken into account: the deadweight of the structure, the deadweight of the roof cover, the snow load and the wind load. Weightless panels were adopted in the computational model to determine the distribution of the roof plane loads among the main load-bearing members.

The resultant maximum design vertical reactions (denoted as **3D-V**) and horizontal reactions (denoted as **3D-H**) for the individual supports are presented in the chart (fig. 5). Also the reactions computed in the planar system (denoted as **2D**) for the load originating from the band corresponding to the spacing of the frames is included. The differences between the values yielded by the 3D model and the 2D model can be regarded as negligible for the 3-9 supports.

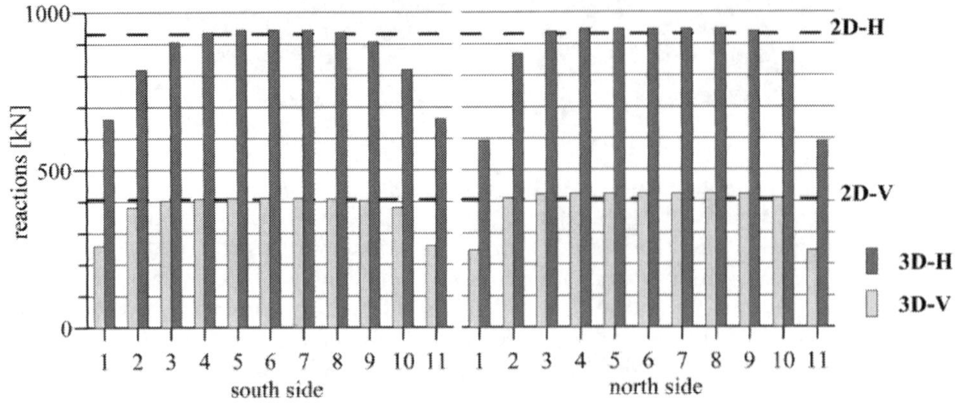

Fig. 5. Maximum design reactions [1].

An interaction curve [5], which can be equated with the load-carrying effects, was plotted for the analysed columns. The bending moment and the axial force, resulting from the interactions, were calculated on the basis of the maximum reactions acting on the damaged column. The results are compared in the diagram shown in fig. 6.

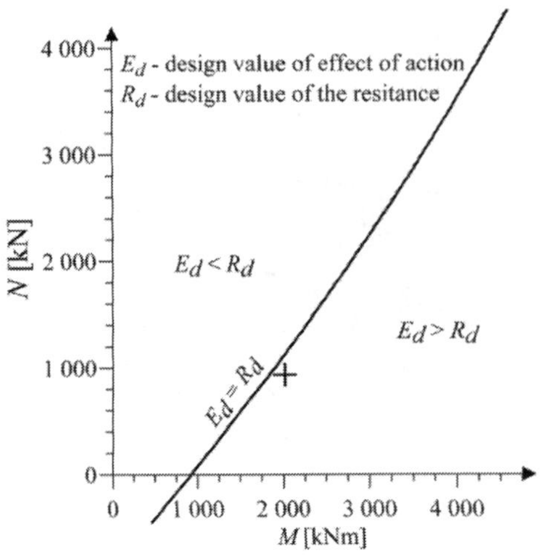

Fig. 6. Interaction force and cross sectional forces for damaged column.

The design check calculations showed a relatively small exceedance of the load-bearing capacity, which cannot be regarded as the single direct cause of the failure, considering that the damage was noticed in the summer (there was no snowload). The snow load amounts to about 25% of the total reaction. The couple of internal forces M, N produced solely by the permanent loads would be within the safe range. The possible overloading of the structure in the winter period can be regarded as a factor which speeded up the failure, but did not initiate it. Nevertheless, it should be stated that the pillar needs to be strengthened because of the exceedance of the strength limit state.

4 Causes of failure. Repair proposals

It is usually difficult to point to a single cause of a failure or a structural collapse. In most cases it is the coexistence of several adverse circumstances, which only together reveal the structural defects. This was so also in this case.

In the authors' opinion the primary cause of the failure was the faultily constructed reinforcement, particularly:

- the absence of transverse reinforcement which should surround the longitudinal reinforcement and which would hinder the separation of the rebars from the concrete, the disruption of the interaction between the rebars, and the pulling out of the longitudinal rebars along the length of the lap;
- the absence of vertical reinforcement at the lateral edge of the column; this reinforcement could prevent the intensive propagation of cracks between the column and the foundation (in the years when the facility was built, the recommended maximum spacing of longitudinal rebars in a column was 40 cm [6]);
- the too small length of the lap of the reinforcement extending from the foundation and that of the primary reinforcement, which did not exceed 65 cm (according to the then guidelines, the required length of anchorage was 56 cm; moreover, it should have been increased even as much as two times since all the tension rods were connected together in one cross section [2]).

The next cause of the failure was the insufficient rebar cover (5-15 mm). This adversely affected the durability (concrete carbonation, reinforcement corrosion) and adhesion (the layer of concrete surrounding the rebars was too thin). Finally, another cause could be the overloading of the structure during periods of intensive snowfalls.

Taking into account the above circumstances, it was proposed to strengthen all the columns [7]. The strengthening was to consist in erecting 20 cm thick concrete walls on both sides of each of the columns (fig. 7). The walls were to take over the whole load from the roof. A transfer slab (a kind of cap) was to introduce uniform stresses into the padstone. The existing members and the strengthening ones were to be joined together by means of epoxy resin bonded rebars. The new structure was to transfer the entirety of the roof load.

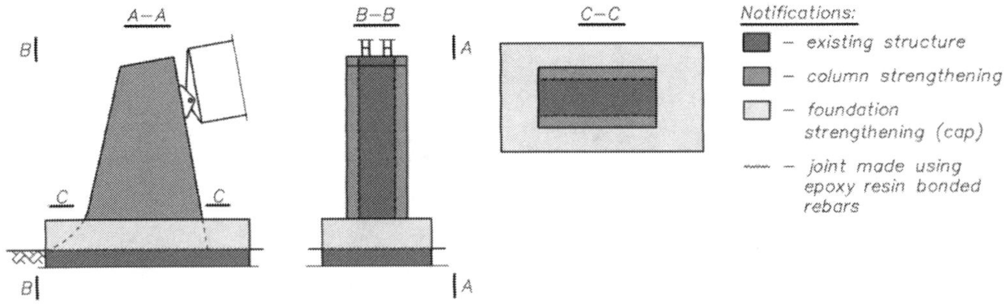

Fig. 7. Proposed way of strengthening [1].

The proposed way of strengthening was implemented in the autumn of 2016 (fig. 8).

Fig. 8. Strengthened columns of south side.

5 Conclusion

The failure of the hall structure is an example of the construction negligence typical of the forty years after World War II in Poland. The low professional standards of the contractors and the shortage of materials were the source of many errors, and sometimes ended in tragedy. In the considered case, it was only a failure, which was a serious warning for the users of the facility. It was also a compelling reason for an inspection of all the key structural members of this facility.

The presented example illustrates well the construction practices of the bygone era. A characteristic feature of a large population of buildings from this period are very carefully prepared design documents and detailed specifications and drawings for construction. As opposed to this, the workmanship was poor and construction projects tended to be carried out recklessly, which would often end in structural collapses.

An important message for today is to develop a system of monitoring buildings, especially the ones characterized by a high level of risk (e.g. prestressed structures), and documenting their condition.

References

1. A. Ubysz, T. Trapko, M. Musiał, Emergency conditions of pillar supporting roof girder of sports and entertainment hall, *Awarie Budowlane 2017. Wybrane zagadnienia diagnostyki, napraw i rekonstrukcji*, 837-846 (Publishing House of West Pomeranian University of Technology, Szczecin, 2017) – in Polish
2. *EN 1992-1-1. Eurocode 2: Design of concrete structures - Part 1-1: General rules and rules for buildings*
3. K. Marsh, *Autodesk Robot Structural Analysis Professional 2015: Essentials*, (Marsh API LLC, United States, 2014)
4. *EN 1990. Eurocode 0: Basis of structural design*
5. B. Mosley, J. Bungey, R. Hulse, *Reinforced Concrete Design to Eurocode 2. Sixth edition*, (Palgrave Macmillan, United States, 2007)
6. J. Kobiak, W. Stachurski, *Reinforced Concrete Structures*, Vol. I, (Publishing House Arkady, Warsaw, 1984) – in Polish
7. A. Ubysz, T. Trapko, M. Musiał, Materiały Budowlane, **538,** 44-45 (2017) – in Polish

Application of recycled plastic in flat roofs

Radim Smolka[1,*], and *Jindřich* Sobotka[1]

[1]Brno University of Technology, Faculty of Civil Engineering, Institute of Building Structures, Veveří 95, 602 00 Brno, Czech Republic

Abstract. The article describes and verifies the possibility of applying secondary plastic materials to construction details such as flat roofs. The created element is a placement considered both in the area and in the finishing detail. The secure connection of the embedded element with the waterproofing layer is an integral part of the detail functionality. This is verified by a peeling resistance test at joints. Peel resistance is the tensile force required to completely separate the prepared joints. During testing, the test piece is pulled at constant speed until complete separation.

1 Introduction

The article describes and verifies a possibility of applying secondary plastic materials to construction details, such as flat roofs.

In terms of its position, an element can be placed in a flat part as well as in finishing details, e.g. in the attic. The secured connection of the embedded element with the waterproofing layer is an integral part of the detail functionality.

2 Application of plastic in flat roofs

Secondary materials, mostly of polymeric composition, can be a suitable element to substructures for technological systems in roof structures. Nowadays, insulators of higher strength are inserted, or spreading layers are created on the roof, where they may prevent smooth water drainage. Therefore, application of polymeric materials, potentially with additives, which guarantee all necessary parameters, seems to be the best solution.

2.1 Production and testing of testing samples

2.1.1 Production of testing samples

Testing boards were produced with dimensions of 150 x 150 x 15 mm from waste HDPE [1]. In order to improve heat, technical, and mechanical properties, other boards were

* Corresponding author: smolka.r@fce.vutbr.cz

produced with sawdust and blowing agents as additives. After measuring the coefficient of heat penetration, the boards were tested for peeling at connections [2-3].

Fig. 1. Testing board – HDPE with added sawdust.

Four elements of bitumen sheets were applied on the testing board by fusing. Two on each side.

2.1.2 Resistance from peeling at connections

The test was performed in a laboratory of company Dehtochema bitumat-tn a.s. The testing method is designed for testing connections in mechanically fixed one-layer bitumen roofing.

Resistance to peeling is a tensile force necessary to complete separation of prepared connections. During testing, the testing body is pulled by a constant speed up to its complete separation.

The test was performed with the use of a testing tearing device LABORTECH 2.050, which is shown in Fig. 2. The device is equipped with continuous recording of force which corresponds with the distance of grips and it is able to keep balanced speed of separating the grips. The grip width may not be lower than 50 mm. The speed of separating is defined as 100 mm ± 10 mm per minute. The testing body is secured against displacement by an inserted stopper. Therefore, the displacement of the sample did not exceed 2 mm.

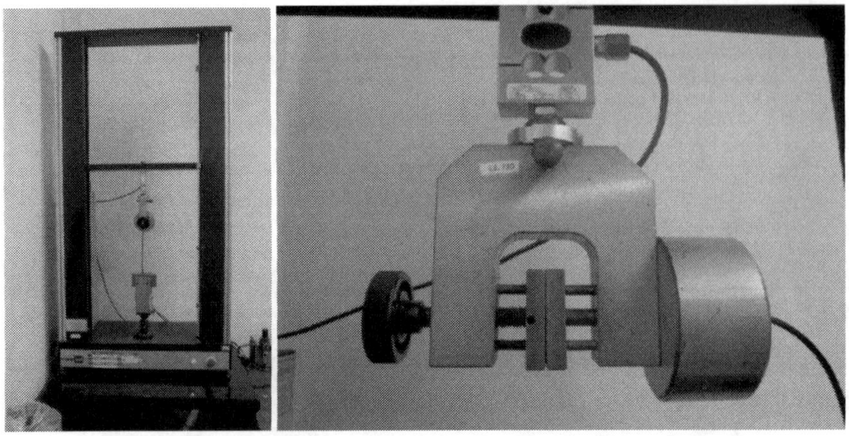

Fig. 2. Testing tearing device Labortech 2.050 and a detail of its upper grips.

The testing samples for connections were tempered (left) for 48 hours at the temperature of 22°C – 24°C under relative humidity of 60 %. Therefore, the tempering conditions complied with the standard conditions [3], i.e. 20 hours at the temperature (23 ± 2) °C and relative humidity of (50 ± 20) %.

2.2 Used samples of bitumen sheets

The following types of bitumen sheets were used for the measurements:
a) modified sheet – SBS – Skloelast extra, thickness 4 mm, insertion – fiberglass, flexible at low temperature -25°C,
b) oxidised sheet – Extrasklobit PE, thickness 4 mm, insertion – fiberglass, flexible at low temperature 0 °C.

12 testing waterproofing stripes were prepared from two different bitumen sheets for the experiment. The bitumen sheets were cut into testing stripes of corresponding lengths, based on the type of test and prescribed width of 50 mm ± 1 mm. Each stripe was measured by a digital calipers before the test. The application of bitumen sheets was performed by flame fusing according to manufacturer's technical sheets. Subsequently, the sample was modified by rolling up to the phase when glut is made. The testing plastic bodies were not treated by any type of coating in order to improve adhesion.

Based on the test type, the lengths of 210 mm (± 1 mm) and 270 mm (± 1 mm) of individual bitumen stripes were selected; the lengths comply with the requirements of standards for the given test, i.e. 50 mm of the stripe is fixed to tearing grips of the device, 100 mm is the initial distance of tearing grips from the testing sample (connections), the remaining lengths (60 mm and 120 mm) are fused to the testing body.

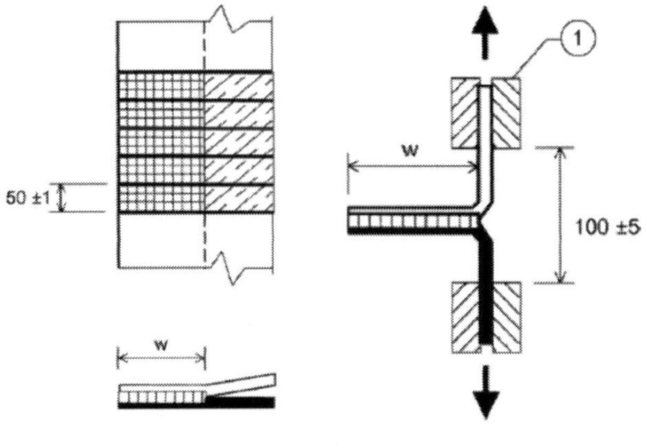

Fig. 3. Preparation of testing body according to standard; course of peeling test [3], 1 – tearing grips, W – connection length.

When calculating average resistance to peeling in connections in [N] for 50 mm, the standard [3] states that the first and the last quarter of recorded data cannot be taken into account. For illustration, Table below shows all recorded values. Therefore, the testing evaluation omitted the data (first and last quarters) in accordance with the standard.

The following Table 1 describes individual testing samples and summarizes testing results (maximum tensile force).

Table 1. Overview of tested samples.

sample	plastic (type)	sheet type	application method	fusion length [mm]	maximum force [N]	evaluation
sample a01-60	HDPE with sawdust	modified	flame fused	60.00	**152.509**	**NO***
sample a02-60	HDPE with sawdust	modified	flame fused	60.00	**234.303**	**YES**
sample a03-60	HDPE with sawdust	modified	flame fused	60.00	**376.710**	**YES**
sample b01-60	HDPE with sawdust	oxidised	flame fused	60.00	**485.616**	**YES**
sample b02-60	HDPE with sawdust	oxidised	flame fused	60.00	**367.259**	**YES**
sample b03-60	HDPE with sawdust	oxidised	flame fused	60.00	**421.094**	**YES**
sample a01-120	HDPE with sawdust	modified	flame fused	120.00	**398.282**	**YES**
sample a02-120	HDPE with sawdust	modified	flame fused	120.00	**375.341**	**YES**
sample a03-120	HDPE with sawdust	modified	flame fused	120.00	**482.553**	**YES**
sample b01-120	HDPE with sawdust	oxidised	flame fused	120.00	**484.900**	**YES**
sample b02-120	HDPE with sawdust	oxidised	flame fused	120.00	**405.908**	**YES**
sample b03-120	HDPE with sawdust	oxidised	flame fused	120.00	**407.472**	**YES**

The unsatisfactory sample in Tab. 1 marked NO* - Sample a01-60 was not activated by a hot flame before the fusing of the bitumen sheet. The fused sheet was just mechanically attached and rolled. Due to the imperfect application, the mentioned sample shows considerable range of values. A specific feature of fusing waterproofing materials is the presence of human factor. Uneven heating of elements, uneven fusing, poor rolling, and other technological faults may occur, which leads to imperfect application of an element.

The following Figure 4 of force courses shows the selected reference sample. The courses of measured values are crossed by a 4th order polynomic trendline, and its equation is shown.

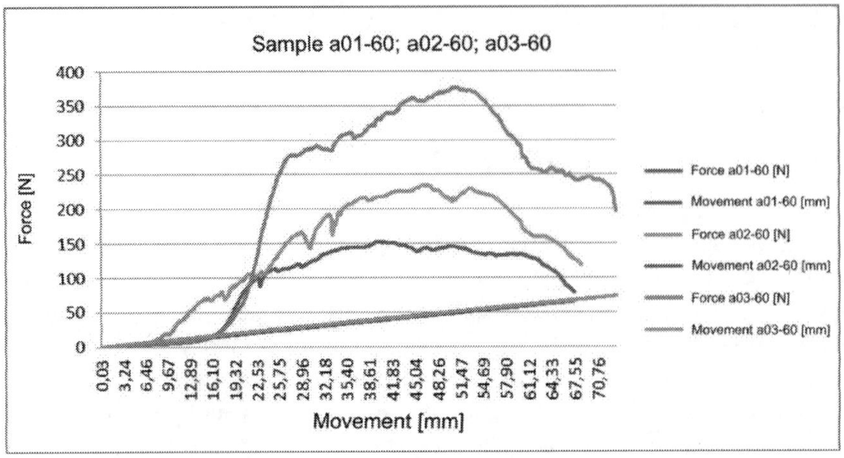

Fig. 4. Total of samples a01-60; a02-60; a03-60 – courses of forces.

Commonly reached values for tested waterproofing bitumen sheets during peeling tests in connections generally range between 150 N and 400 N.

During the testing the sheet either got damaged in the area around the insertion, or the lower part of the bitumen sheet from the bearing insertion was separated (Fig. 5). Therefore, the connection of the bitumen sheet and the secondary plastic material (HDPE) is stronger than the bitumen sheet material itself [4-6].

Fig. 5. Sample detail a03-60 – separation of bitumen material from the bearing insertion.

The test aimed to prove the feasibility of an idea to apply secondary plastic materials to roof structures under the waterproofing layers while still maintaining the required safety. It is necessary to perform the test with the final product. The test is only applied to a single type of secondary plastic material.

3 Conclusion

The main points of a new product for civil engineering were set. While maintaining the required flat roof slope gradient, the product should transfer even loading to heat insulating sublayers or be used as a substitute for a common heat insulator. In addition, it should allow simple and safe installation of waterproofing layers without changing the technological instructions. This product could be installed in new buildings (standard, passive, and energy-saving buildings) as well as in the existing buildings during reconstructions or modifications, mostly during modernisation works.

The following basic properties which the element should meet are selected:
- long life span,
- volume stability,
- tolerance to other materials (material corrosion),
- strength – resistance to impacts and stable loading,
- fast and all-year installation without technology change,
- meeting requirements for fire resistance,
- harmless to health,
- sound insulation.

The above mentioned properties are required for correct function and long-term life span of the designed product [4-6].

Materials from recycled plastic have a great potential for application in a wider scope of civil engineering. Their physical, mechanical, and chemical properties puts them in compositions where traditional heat insulation materials are used. Using a product from secondary materials results in a lower thickness of a structure, since no other waterproofing or spreading layers are necessary, while the requirements of technological standards, regulations, and legislation acts are met. A considerable advantage is lower financial costs for the construction.

This paper has been worked out under the project No. LO1408 "AdMaS UP - Advanced Materials, Structures and Technologies", supported by Ministry of Education, Youth and Sports under the „National Sustainability Programme I"

References

1. L. Matejka, R. Smolka, *patent EP 2159364 B1 Support structures for a doors threshold* (Brno University of Technology, Munchen, 2012)
2. V. Duchacek, *Polymers – production, properties, processing, applications* (Institute of Chemical Technology in Prague, Prague, 2006)
3. *ČSN EN 12316-1 (727638) Flexible sheets for waterproofing - Part 1: Bitumen sheets for roof waterproofing - Determination of peel resistance of joints* (Prague, 2000)
4. J. Sobotka, AMR **1122**, 65-69, 2015
5. R. Smolka; T. Petříček; P. Kacálek, AMM **824**, 188-195 (2016)
6. P. Kacálek; R. Smolka; T. Petříček, AMR **1041**, 288-292 (2014)
7. T. Petříček; P. Kacálek; R. Smolka, AMR **1041**, 71-74 (2014)

Monitoring of underground subsurface structures in Ekaterinburg

Anna B. Galieva [1,*], *Vladimir N.* Alekhin [1], and *Liliya G.* Pastukhova[1]

[1]Ural Federal University, Institute of Civil Engineering, Ekaterinburg, Russia

Abstract. This paper presents the issues of technical state of subsurface structures in operation and the erection of buildings under available development. It is stated that detailed monitoring of technical state of underground tunnels is essential, since the structure is unique and technologically complex. The aspects of geotechnical monitoring of underground tunnels are considered. The principles of geodesic monitoring of underground rings, tension increase in tube lining and pressure from the building under construction on the ground are described. Recommendations for underground tunnel monitoring are presented.

1 Introduction

The problem of safety is one of the urgent problems in the building sector. At present, infill development is actively underway including the development near existing subsurface structures in operation. Meanwhile, subsurface structures such as underground tunnels are unique, technically complex and strategically essential for a megapolis as safety of the public depends on the reliability of subsurface structures. This calls for the monitoring of their technical state in order to take timely measures to prevent from negative consequences deteriorating this state [1, 2]. The experts from UrFU began their work in autumn 2014 to inspect and monitor underground tube lining in line sections between two stations under the building being constructed above the ground. The functional purpose of the object under construction is a multistoried block of flats with nonresidential premises.

2 The description of the object under construction and the specific features of the site

The building under construction has a trefoil form in plan consisting of three sections. The load-bearing framework of the building is designed from in-situ reinforced concrete. The joining of walls, pillars, columns with floor structures is considered to be rigid. The construction arrangement of each temperature block of the dwelling provides the stability of its geometric shape, the overall stability and taking up both lateral and vertical loads.

From hydrogeological perspective, the territory under consideration is located in the East-Ural hydrogeological area of crustal water basins. The considerable groundwater level

* Corresponding author: a.b.galieva@gmail.com

lowering in the given site at the present stage is recorded on basis of the analysis of hydrogeological situation in the region. The change in the groundwater level position is influenced by the location of the site in and around the depressed effect zone of the existing underground route. The water drawdown work is stopped, the level is gradually recovered. Referring to the results of the research one can see that the level of ground waters on the investigated site gradually grows and eventually reaches the position of natural water table formed during the track of many years.

3 The research of the object and the identification of geometry parameters

The inspection of the object is executed before the excavation for a structure stripping. The lining structures of both right and left main line tunnels are rings from ribbed, cast-iron tubes. The soil mass, where the main lime tunnels are located, is composed of filled soil, drift and eluvial sandy clays. According to the customer requirement specification, one should examine the sections of two main line tunnels entering the area of the construction influence and define geometrics of the object. Then, these data were compared with design variables. One should define the ellipticity of each ring in tunnel lining on four diameters: horizontal, vertical and two other diameters at an angle of 450 (Fig. 1). Structural deformation of lining should not affect clearness gauge [3 – 7].

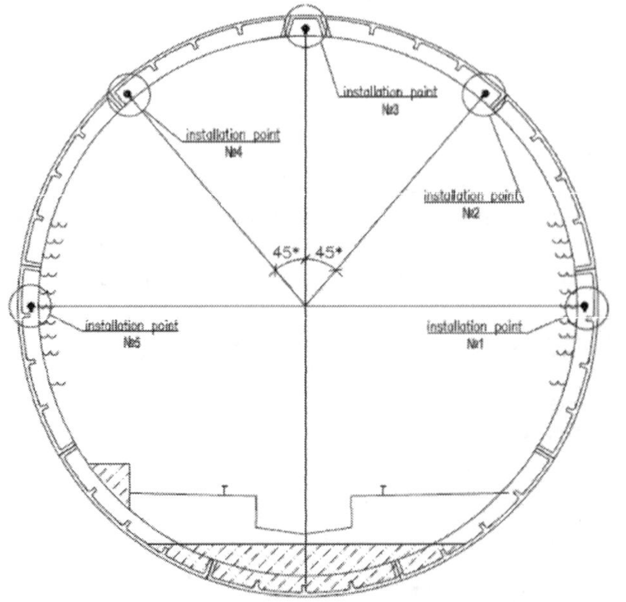

Fig. 1. Installation diagram of deformation control benchmarks to define the ellipticity of tunnellining

When performing visual inspection of tunnels it became evident that tacheometric survey based on five lining points does not provide a complete picture of the strain state of tunnel rings. One may observe local defects on tubing stiffeners that will generally lead to the misperception of clearness gauge (Fig. 2) when entering them. In addition, some essential points of tubing linings are inaccessible for observation because of cables, cantilever brackets and other maintenance-engineering members of tunnel operation system laid along tunnels.

Fig. 2. The example of local reduction in the cross-section area of an underground tubing stiffener.

The experts from UrFU decided to scan the object using new-generation three-dimensional laser tacheometer Leica MS50.

Taking into account the object secrecy, supervised and time limited access, the scanning mesh was chosen so that it was optimal in terms of obtaining fully detailed high quality and time-effective picture. The tunnel scanning from all five basic observation points of the tacheometer was carried out for 25 minutes, and the scanning from all four auxiliary intermediate observation points was carried out for 12 minutes.

Figure 3 gives the example of the image of a separate tunnel ring according to the results of the work.

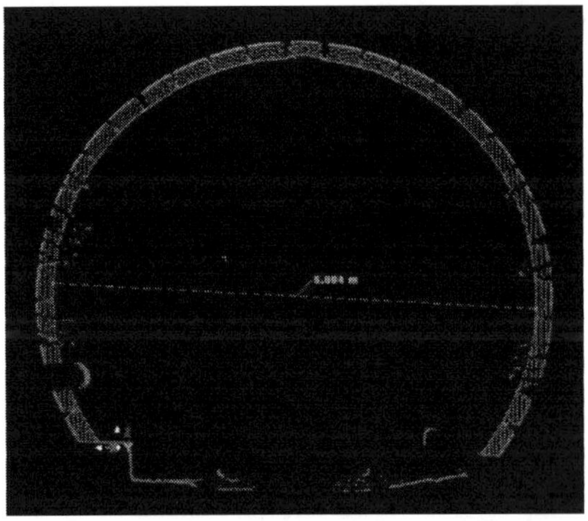

Fig. 3. The example of the scanned lining ring in the underground tunnel.

The inspection of tunnels by means of laser scanning technique allowed investigating the deformation behaviour of the object in detail and detecting weaknesses [8 – 12]. The quality of the images received enables to scan the members of the tunnel structure including bolts, plug fittings, and structural reinforcement members (Fig. 4). The licensed software packages Leica Cyclone and Leica Infinity were used to process data.

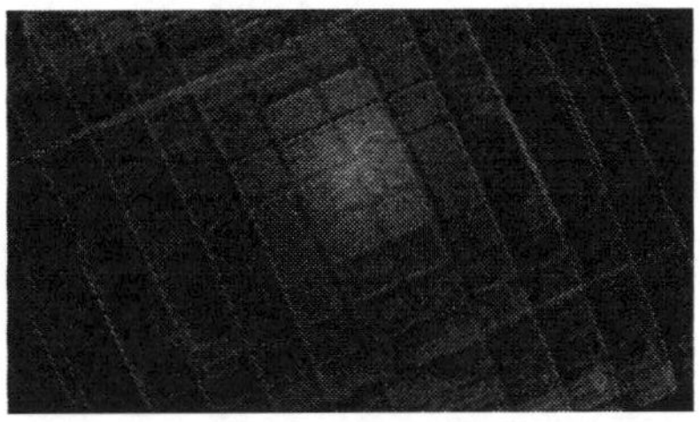

Fig. 4. Overall view of the tunnel structure in the software Leica Infinity.

4 Geotechnical monitoring of underground tunnels

The next stage of the work was to perform geotechnical monitoring of the object including geodesic monitoring (tacheometric survey with the tachometer Leica TM30), tensometry (monitoring system of tension increase in tunnel lining) and registration system of the pressure from the building under construction on the ground. For geodesic monitoring of the tubing stiffeners, five prism deflectors were installed on each ring under test.

At least once a month the inspection is conducted to control the strain state of the structures. The geodesic measurement processing is performed by means of the licensed software packages Leica Geomos Monitor and Leica Geomos Analyzer [13, 15, 16]. To realize the tensometric monitoring, strain gauges were installed on the tubing stiffeners to observe the tension changing in the cast-iron lining of the tubing during the building construction above the tunnels under test [14]. Stringed strain gauges are high-precision gauges and their measured quantity is relative deformation. The strain gauge is a hollow tube with a strain stretched inside. One of the tube ends is stationary; the other is able to move lengthway causing tension or relaxation of a string [17].

Underformed strain gauges were installed into the strain gauges already deformed under the tunnel external load. So, the task of defining the "initial" deformation prior to the building erection as well as tensions acting in lining is technically impractical. Thereby, the technique to record the changes of the relative deformation of the tunnel lining in the points of the gauge installation during the building erection and the tension increase calculation was employed. Tunnel tubes operate under complex multiaxis stress state. When calculating tunnel lining by means of the classical approach, the tubing is assumed to operate mainly under compression with bending, the tunnel cross-sections operate under plain deformation. There is tension system in the tubing net section: radial, tangential and tangent. Tangential tension obtains greater values (in absolute term). Gauges installed on the tunnel lining are oriented in tangential direction (the axis of gauges is pointed tangentially to the lining contour). Strain gauges are installed on the points nearby the edges of longitudinal ribs since the distribution of tangential tensions in tunnel lining thickness is nonuniform, and greater tension values are obtained on the internal and external surfaces of the lining. Thus, the strain gauges are installed in pairs on three points, corresponding to the maxima of bending moments – the point of the largest tangential tensions.

In spring 2015, pressure gauges were installed under the base slab of a dwelling under construction in the areas with maximum tension in order to register the pressure of the building erected on the ground. The gauge consists of two stainless round perimeter-welded plates. The hollow between the plates is filled with deaerated liquid and connected with the sensor by means of stainless steel tube. Any change in pressure on the plate surface results in the change in the hydraulic fluid between the plates, which, in its turn, influences the sensor and causes either the tension or the compression of a string inside the sensor. To record the changes in the string tension, the vibration frequency proper, which is the tension function, is measured. Thus, the changes of pressure on the sensor plates are converted into electric signal that is sent towards the registration point via the cable. The signal cable of each gauge in the corrugated cable is led outside the slab and marked for the further identification when taking the readings.

When the gauge depth level is lower than that of ground waters, pressure gauges should be accompanied with threshold pressure gauges to make provisions for the separation of effective ground pressure from hydrostatic one [18 – 23]. Figure 5 shows the overall view of the construction site and the gauge installation position together with laid signal cables.

Fig. 5. Overall view of the installation site of pressure gauges.

The strain and pressure gauge readout is performed by means of a portable reader (datalogger) able to transmit the readings via Bluetooth to any android device. The Department of CAD in Civil Engineering, UrFU designed the application for android-platform-based devices capable of processing gauge readings. The mobile application allows to add photos, drawings, object layouts, to process the readings of a portable datalogger or to enter data in the program manually as well as to construct a table or a graph and to send them to draw an annual report. The overall view of the application work window is shown in Figure 6.

Fig. 6. Overall view of the application work window. The arrangement of strain gauges.

5 Conclusion

Monitoring of the object is currently under way; statistical information on the underground tube behaviour is gathered. The behavioural analysis of the subsurface structure performed on schedule allows to size up the impact of new construction on the structure and to develop solutions preventing negative process development.

Measured changes to the shape of the lining are to inform about the overloading of the structure or the fall of the subsoil under the objects above the underground structure. Modern laser measurement currently replaces old methods of monitoring using strain gauges used in the past due to the speed of its implementation over the original methods.

Implementation of computer-based structure monitoring systems, determination of real structure geometrics by means of three-dimensional laser scanning are recommended when monitoring the existing underground tunnels.

After commissioning the object or stabilizing the design parameter changing, monitoring is permitted quarterly. In the absence of stabilizing of design parameter measuring, geotechnical monitoring should be continued. By the absence of stabilizing of design parameter changing, we consider the excess of parameter values by more than a half of precision of measurement compared with previous cycles.

References

1. Y. Y. Su, Hashash Y. M. A., Liu L. Y. Journal of construction engineering and management **132**, 12, 1234-1241 (2012)
2. Y. M. A. Hashash, J. J. Hook, B. Schmidt, J. I-Chiang Y., Tunnelling and Underground Space Technology **16(4)**, 247–293 (2001)
3. GOST 23961-80. Subways. Clearance obstructions of work equipment and rolling stock. (Introduced 12.29.1979)
4. SP 120.13330.2012. Subways. The updated edition of SNiP 32-02-2003.(Introduced 01.01.2013)
5. B.E. Slavin. *Fundamentals of scientific research of underground structures: Textbook* (2004)
6. V.N. Suchenko. *Mine surveying support for the construction of tunnels subway: Tutorial* (2001)
7. V. A. Garber, Transport construction **5**, 12-14 (2012)
8. E.I. Gorokhova, I.V. Aleshina, E.V. Romanovich and others. *VIII Intern. Sci. Congress «Interexpo GEO-Sibir-2012»* (Novosibirsk, 10 - 20 April, 2012)
9. A.V. Komissarov. *Method of research of metric characteristics of scans [Text]: the author's abstract.* (Novosibirsk, 2007)
10. E.I. Gorokhova, A.V. Ivanov, *Collected materials of scientific. Congress. "GEO-Siberia-2009"* (Novosibirsk, April 20-24, 2009)
11. H. Park, H. Lee, H. Adeli, I. Lee, Computer-Aided Civil and Infrastructure Engineering **22**, 19-30 (2007)
12. R. Van Gosliga, R. Lindenbergh, N. Pfeifer, International Archives of Photogrammetry, Remote Sensing and Spatial Information Sciences **36** (Part 5), 25-27 (2006)
13. Bernd Hiller. *Open urban planning forum, session "Safety of construction and operation of objects"* (Novosibirsk, 2010)
14. B. He, B. Wang, C. Li, T. Y. Yang, Z. C. Cai, F. Xiao, AMR **718-720**, 837-841 (2013)
15. H.H. Bogomolov, Proceedings of the Mining Institute **204**, 40-45 (2013)
16. D.A. Afonin, H.H. Bogomolov, M.Ya. Bryn, Geodesy and cartography **1**, 7-11 (2014)
17. F. Wang, D.M. Zhang, H.H. Zhu, et al. CMC: Computers, Materials & Continua **34 (1)**, 63-81 (2013)
18. K. P. Bezrodny [et al.], Metro and tunnels **1**, 30-32 (2012)
19. Boldyrev G.G., Zhivaev A.A. *Proceed. Of the 8th Internat. Workshop on Structural Health Monitoring* (Fu-Kuo, Stanford University, USA, 2011)
20. V.M. Gorpinchenko, M.I. Egorov, Industrial and civil construction **10**, 39-41 (2004)
21. K.A. Aime, Construction materials, equipment, technologies of the century **11**, 37-39 (2005)
22. E.V. Potapkin, Industrial and Civil Engineering **12** (2006)
23. V.K. Tarakanovskiy, N.K. Kapustin, A.N. Klimov, Geoecology. Engineering geology. Hydrogeology. Geocryology **6**, 551-562 (2010)

MATEC Web of Conferences **146**, 02011 (2018) https://doi.org/10.1051/matecconf/201814602011
Building Defects 2017

Influence of the frontal joint of the bitumen sheet on the final strength of the joint

Jan Plachý[1,*], *Jana* Vysoká[1] , and *Radek* Vejmelka[1]

[1]Vysoká škola technická a ekonomická v Českých Budějovicích, Okružní 10, 37001 České Budějovice

Abstract. This article focuses on the frequently discussed topic how to make the frontal joints of bitumen sheets with coarse-grained surface coating. Bitumen sheets have a spill in the place of these joints. Insufficient strength of the frontal joint is one of the possible faults in waterproofing using bitumen sheets. The paper compares different ways of performing these joints in terms of force stress based on surface treatment. Surface treatment is possible by immersing the sprinkle, scraping it into the bitumen sheet mass, or simply heating the surface and leaving it unpainted.

1 Introduction

One of the frequent failures that arise on the roof cladding is the release of bitumen sheets in joints. Bitumen sheets joints can be divided into longitudinal (lateral) or transverse (front). See Figure 1.

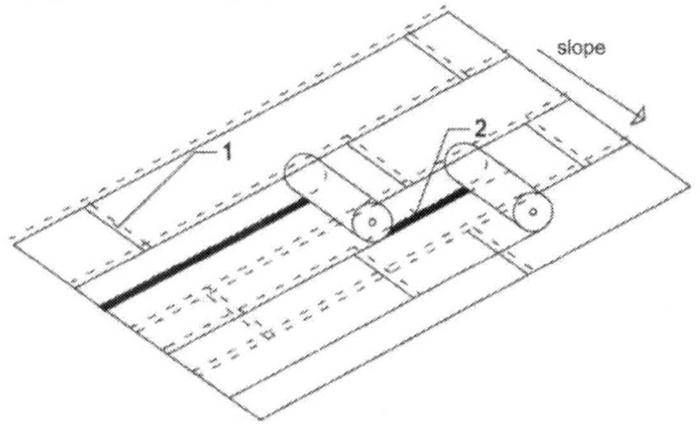

Fig. 1. Types and nomenclature of joints of bitumen sheets (1 - frontal, transverse, 2 - lateral, longitudinal) [1].

In the case of single-liner systems, bitumen sheets with coarse-grained spreading are used as the final liner of the waterproofing liner. For better execution of the joints, bitumen sheets with coarse grain on the upper surface are connected at the lateral joint site with polymeric tape connections. Sprinkle is omitted here. At the site of the frontal joint of these bitumen sheets sprinkle is left, there is no strip without surface treatment. See Figure 2.

* Corresponding author: plachy@mail.vstecb.cz

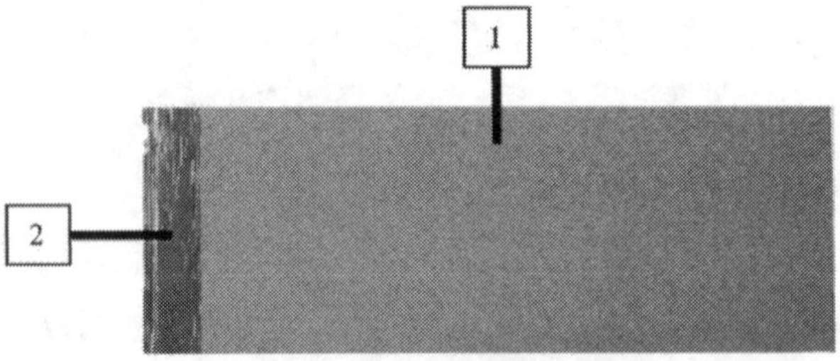

Fig. 2. Bitumen sheet (1 - frontal, transverse, 2 - side, longitudinal). Source: author.

One of the problems involved in the practical implementation of coarse-grained bitumen sheets joints is the question of how the respective surface treatment of the sheet will affect the resulting joint strength. The problem of bitumen sheets has been dealt with [1-5], the topic of adhesion to bitumen cover is described in [6]. The surface treatment of the coarse-grained bitumen sheets prior to the actual construction of the transverse joint, only [2] dealt with it. The author states that when coarse-grained sprinkle is spattered, it occurs a 5-10% increase of shear strength [7], comparing to non-spattered, depending on the type of bitumen sheet.

In practice, it is possible to meet a total of three possible adaptations of the surface of the bitumen sheet and therefore this experiment was carried out. The first method is to heat the lower cheek of the upper sheet and the upper cheek of the lower sheet. This is a classic, commonly performed method. The second option is to spatter the lower sheet at the joining point and the third option is to immerse the bottom sheet into the mass of the bitumen sheet, see Figure 3

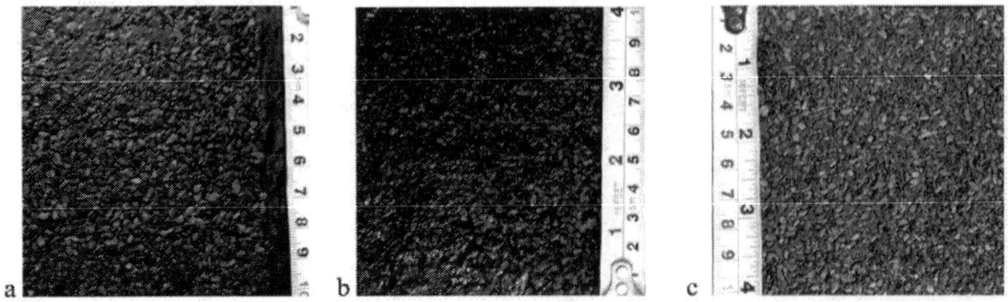

Fig. 3. Modification of the frontal (transverse) connection of the bitumen sheet. a – immersion of the lower sheet, b – sprinkle is spattered, c – warming of the lower cheek of the upper sheet and the upper cheek of the lower sheet. Source: author.

A – Immersion of the lower sheet sprinkle into the mass of the bitumen sheet. When this treatment is carried out, the asphalt mixture, upon burning the bitumen sheet, exits to the surface and forms more than 50% of the surface of the sheet. In the case of bitumen sheet with a polyester carrier liner (PES), this carrier liner is damaged by the increased temperature. See Figure 3a.

B – Spattering the bitumen sheet by means of an insulating spoon. The adjustment consists in heating the surface of the heat with a burner and the sealing, using specially treated spoon, of sprinkle into asphalt. There is a risk of the liner being exposed, which can cause problems in smelting the upper sheet. See Figure 3b.

C – Fomenting the lower cheek of the upper bitumen sheet and the upper cheek of the lower bitumen sheet (the classic method). With this option, the asphalt mixes up to about 20% after heating. See Figure 3c.

The strength of the joint [7] affects the type and thickness of the bituminous cover material of the bitumen sheet, the adhesion between the support liner and the bitumen cover, the adhesion between the coarse-grained sprinkle and the bitumen cover, and the type of the bitumen sheet carrier liner.

It is possible to say the hypothesis that, according to theoretical assumptions, the highest strength should have a connection where the immersion in the bitumen mass is immersed. In this variant, A is the largest contact surface in combination with bitumen mass – bitumen mass. In the case of variant B – the lower bitumen sheet sprinkle spattering and variant C – fomenting of the lower cheek of the upper bitumen sheet and the upper cheek of the lower bitumen sheet as a result of the smaller contact surface in the bitumen mix – bitumen mix, the resulting joint strength should be less than that of variant A. Another important factor that affects the properties of the joint is the skill of the worker who performs the joint. The surface treatment of the bitumen sheet at the joint site by immersing and spattering was compared with the conventional embodiment.

2 Material and methods of testing

2.1 Materials

For the tests, a 5.2 mm thick bitumen sheet was selected with cover mass of modified copolymer styrene - butadiene - styrene (SBS), carrier liner of polyester fleece (PY) weighing 230 g/m^2 and the slate slivers sprinkle. The lengths of the transverse joint (W) were determined on the basis of [1] in two variants of 120 mm and 150 mm. In total, six samples were prepared.

2.2 Methods of testing

For each variant of the joint surface treatment, nine test bodies were always prepared. An increase in the number of test bodies out of five, which is prescribed [7], has been done for statistical evaluation. Calculations were performed with precision to two decimal places. Tests were performed at (23 ± 2) °C. The test procedure was performed according to [7]. For the test scheme, see Figure 4.

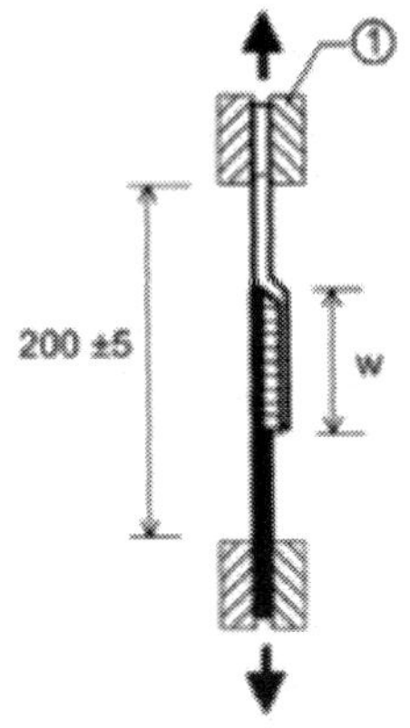

Fig. 4. Diagram of the joint shearing resistance test in accordance with ČSN EN 12317-1. W – joint size. Source: author.

2.3 Evalution

Nine test samples of the determined combinations were used for testing. The final force is presented in the form $\bar{x} \pm \bar{\vartheta}(\bar{x})$, where $\bar{x}$ is the arithmetic average and $\bar{\vartheta}(\bar{x})$ is the probable measurement error. The latter was determined on the basis of the error theory for deviations in the arithmetic average using the formula:

$$\bar{\vartheta}(\bar{x}) = \frac{2}{3}\sqrt{\frac{1}{n(n-1)}\sum_{i=1}^{n}(x_i - \bar{x})^2}. \qquad (1)$$

3 Results

The measurement results and estimates of standard deviations are nearly the same for each frontal joint size. The 150 mm front link strength was higher than 120 mm in length. A summary of the individual measurement results is summarized in Table 1.

Table 1. Shear strength of the frontal joint of the bitumen sheet with various sheet surface treatments. Source: author.

Samples	Shear strength of the frontal joint [kN/m]					
	Joint width 120 mm			Joint width 150 mm		
	Sprinkle immersed	Classic treatment	Sprinkle spattered	Sprinkle immersed	Classic treatment	Sprinkle spattered
The mean value [kN/m]	18.64	19.11	18.96	19.48	19.64	19.26
Probable error [kN/m]	0.69	0.67	0.88	0.65	1.22	1.40
Standard deviation [kN/m]	1.54	1.59	1.52	1.13	1.49	1.71
Coefficient of variation [%]	8.25	8.30	8.03	5.82	7.60	8.89

4 Discussion

The original hypothesis of increasing the strength of the frontal joint in the immersed sprinkle was not confirmed. The effect of the higher percentage of bitumen mass on the surface of the bitumen sheet is therefore not a decisive factor. The effect of the higher percentage of bitumen mass on the surface of the bitumen sheet is therefore not a decisive factor. In the results, it is interesting to note that for both sizes of joints, greater strength was achieved in the classic design of the front joint of the sheet.

On the basis of these measurements, it can be stated that the impact of spattering (sprinkle push off) and immersing of the sheet has no significant influence on the shear joint strength in terms of the lacerated load of the joint. Individual measurements and estimates of standard deviations were comparable. The question of the implementation of the frontal joints remains open and may be the subject of further testing and research

5 Conclusion

The results of the experimental measurement showed that with the correct implementation technology the difference in shear strength is very small. It is not necessary to perform the surface treatment at the location of the frontal joints by spattering and immersing the sprinkle. These adjustments are, on the contrary, risky in terms of possible damage to the carrier liner. In the case of bitumen sheets, which are used as the final one in the waterproofing strata, and in the case of single-liner applications, polyester is used.

References

1. J. Plachý, *Bitumen sheet joints under the influence of temperature and time in the composition of single-membrane roofs* (Brno University of Technology, Faculty of civil engineering, Brno, 2005)
2. R. Šmehyl, *Stability of joints of bitumen membranes* (University of Technology in Bratislava, Faculty of civil engineering, Bratislava, 2007)
3. T. Petříček, P. Kacálek, R. Smolka, AMR **1041**, 71-74 (2014)
4. T. Petříček, P. Kacálek, T. Hlavačka, MATEC Web of Conf. **93**, 03007 (2017)
5. J. Plachý, MATEC Web of Conf. **93**, 02006 (2017)
6. J.A. Marques, J. G. Lopes, J. R. Correia, Construction and Building Materials **25(**1), 138-144 (2011)
7. *ČSN EN12317-1 Flexible sheets for waterproofing - Part 1: Bitumen sheets for roof waterproofing - Determination of shear resistance of joints* (Prague, 2000)

Deep excavation in urban areas – defects of surrounding buildings at various stages of construction

Jarosław Rybak[1,*], *Alexander* Ivannikov[2], *Elena* Kulikova[2], and *Tomasz* Żyrek[3]

[1]Wroclaw University of Science and Technology, Faculty of Civil Engineering, 50-370 Wrocław, Wyb Wyspianskiego 27, Poland
[2]MISiS National University of Science and Technology, Department of Underground Construction, Leninsky Ave, 4, Moskva, Russia, 119049
[3]Silesian University of Technology, Faculty of Civil Engineering, 44-100 Gliwice, Akademicka 2A, Poland

Abstract. Deep excavation and tunnelling works in city centres always bring some risks to surrounding structures, especially in the case of old town centres, where the technical condition and structural stiffness of historical buildings is rather doubtful. When the new desired excavation depth goes deeper than the foundation of the surrounding buildings or when tunnelling works are conducted directly under them, the existing objects are subject to stress, vibrations and displacements imposed at almost every stage of building the new construction. The presented paper outlines, on the basis of the authors' experience, the typical damages appearing during the supporting wall construction (sheet pile driving, piling and formation of diaphragm walls) and tunnelling works. Other damages appear due to soil mass unloading (caused by excavation stages) and horizontal loading during pre-stressing of struts or ground anchors. The selected case studies of steel sheet pile wall installation is given with regard to typical failures caused by an unplanned excavation and its impact on neighbouring structures.

1 Introduction – the need for development of infrastructure

Constant development of town centres, accompanied by earthworks and tunnelling, brings about many problems that need resolving. Some interesting experiences from Warsaw (Poland) were presented in recent works by Siemińska-Lewandowska [1, 2]. Both, tunnelling and deep excavation works, create some risks to the surrounding area [3]. In the old city centres, investors are required to ensure respective number of parking bays, which in turn demands 2-3 storeys of underground car parks. Apart from engineering problems, very stringent requirements of heritage conservator supervision are also raised. The problems with ensuring stability of excavation walls must be solved by examining foundations of neighbouring structures and possible strengthening them at all the stages of

* Corresponding author: j.rybak@pwr.edu.pl

earthworks, i.e. installation of the excavation protection walls, the excavation works themselves, and constructing basement storeys. Another problem refers to constructing underground storeys below the level of groundwater. Brief summery of those problems was given by Kulikova [4]. This requires efficient cutting-off or long-term lowering of water table inside the excavation with possibly limited intervention in hydrological regime beyond the project in progress. In the typical case of old cities in river valleys such "hoarding off" the excavation and cutting off groundwater leads to temporary or permanent disturbances of groundwater flow and possible local swellings. Contemporary technologies make it possible to protect vertical fault and simultaneously to cut-off groundwater inflow by means of steel sheet pilings, diaphragm walls or secant pile walls. Using them, we are able to construct monolithic reinforced concrete structures of underground tunnels or storeys thus ensuring both their tightness and high rigidity of foundation in the neighbourhood. Many examples of case studies concerning the geological risks related to tunnelling works can be found also in the references.

2 Design and execution problems at subsequent stages

Every project starts from an outline proposal and preliminary design. Usually, the amount of field data at this stage is limited, so the structural engineer must rely on parametric studies. Interesting examples of such procedure were given by Mitew-Czajewska in work [5] and recently [6] where very sophisticated hypoplastic constitutive model of soil body was implemented to Finite Element Model computations in clays. Once the construction starts, it is possible to verify primary assumptions by means of continuous monitoring of displacements. That leads to a so-called "observational method" of design, where the design model may be currently updated on the basis of displacement monitoring. An example of such procedure was given by Gorska in articles [7, 8], on the basis of steel sheet pile wall design. Data for model updating may be provided by various monitoring systems, usually based on standard surveying techniques like geometrical precise levelling for vertical, and structural monitoring for horizontal displacement. Structural monitoring makes it possible to record automatically the state of the structure, cyclically at predefined time intervals. The obtained values of displacement are compared with the permissible values, visualized and sent remotely to relevant supervisory bodies. The main component of such a system is the motorized electronic total stations, which offer the precise measurement of the angles and distances. These total stations have the function of the automatic targets detection (indicated by means of geodetic prisms). Recent developments can be found in works of Drusa [9], Bednarski [10] and Muszyński [11], where the suitability of terrestrial laser scanning for the control of deep excavation support was examined. Recommendations concerning inclinometric measurements of soil mass displacements (slope stability) can be found in references given by Segalini [12] and Drusa [13].

Another group of problems is related to the vertical soil support installation. Most of the methods of sheet pile embedment (driving, vibratory driving) as well as other technologies impose some dynamic impact on the surrounding area by means of single blows or continuous vibrations. Wide studies on this impact and the formal limitations given by various codes of practice were gathered in works of Rybak [14], Brząkała [15, 16] and Oliveira [17]. In the current study however, only the static effects of deep excavations will be presented and examined.

Various factors may affect the safety of deep excavation: improper soil parameters taken for the design, unexpected surcharge load, changing water levels, and last but not least, the static scheme taken for a structural analysis. This last factor is often neglected, which brings serious threats, especially for cantilever structures that relay only on embedment.

3 Case study – the consequences of an unplanned overdig

The presented case study is taken from the building site in southern Poland. The excavation was necessary for the planned underground storey (car-park). Due to the limited depth of the excavation (4.0m to 4.5m) and for the simplicity of works, the designed scheme of the sheet pile wall was just the cantilever system with embedment providing stability and necessary stiffness of the support.

Geotechnical conditions presented on geological cross-section (Figure 1) were constituted by anthropogenic fill layer to the depth of around 1.5m. The underlying sand layers consisted of medium sands (MSa) with a free water table and deeper layers of hard plastic and stiff clays. The designers had sufficient data for proper static computation of required length and profile of sheet piles. Depending on the local geological profiles, the designed lengths of the sheet piles varied from 8.0m to 10.0m.

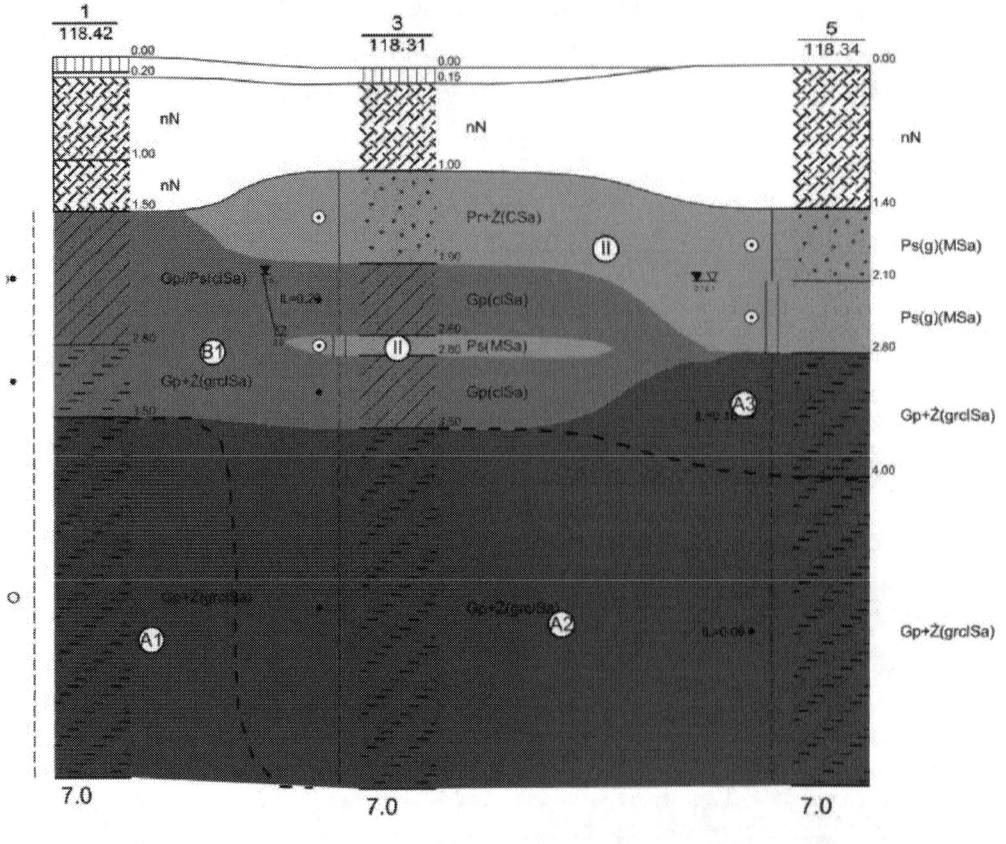

Fig. 1. Geological cross-section with anthropogenic deposits (Nn), sands (green) and clays (brown).

Just after the proper completion of earthworks (excavation) to the designed depth of 4.5m, the contractor made partially an additional excavation for the foundation plate of a crane lift. Unfortunately, the unplanned overdig (4.0m wide and 0.8m deep inside the excavation) was located adjacent to the wall which simply lost its designed embedment. Due to this unfortunate decision the measured horizontal displacement of the sheet pile wall reached in this region almost 0.3m (immediately – Figure 2) and went up to 0.5m due to swelling in the next days. Such large deformations caused problems on both sides of the wall. First – for the stability of the structures in the vicinity of the excavation. Second, even more difficult to solve, was related to the insufficient space for the designed basement walls.

The scheme of the measured horizontal displacements around the excavation is given on Figure 2. One may notice that values measured in the situations corresponding to design assumptions (ROB1, ROB3, ROB4) fitted very well the preliminary assumptions (corresponding with the proper depth of 4.5m) presented on Figure 3a. Only in measuring point ROB2, the current displacement reached almost 0.3m.

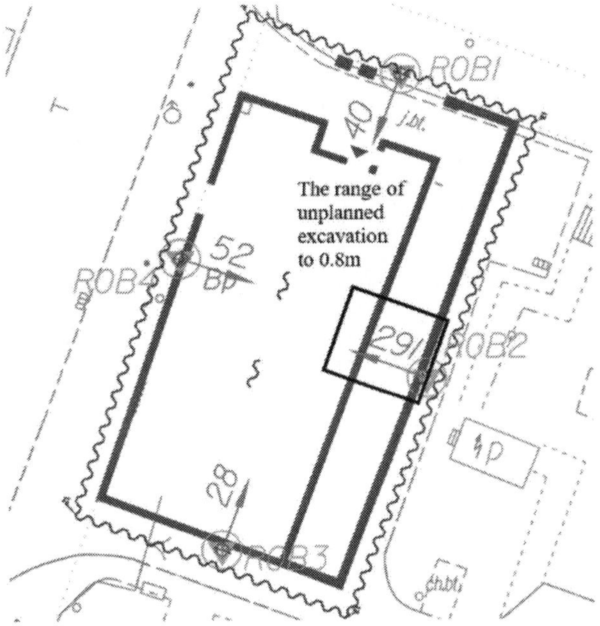

Fig. 2. Monitoring of horizontal displacements of sheet pile heads at ground level.

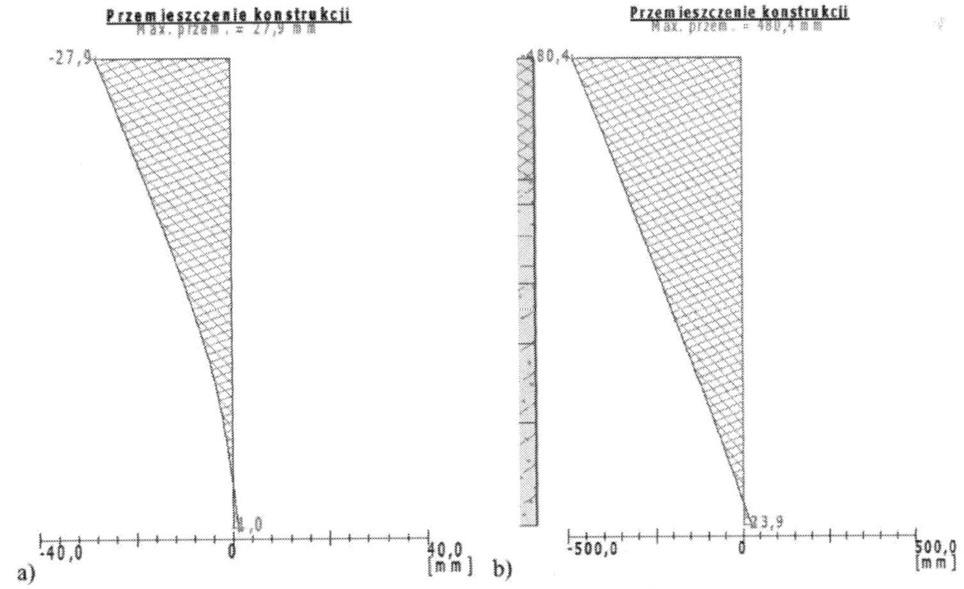

Fig. 3. Computed horizontal displacements of wall at ground level a) for H=4.5m, b) for H=5.3m.

The computed schemes of horizontal displacements prove again that the change of the static scheme results in large displacements. Since the presented case was qualified as defect on the building site and the deformations (horizontal displacements) increased during subsequent days, a decision was made to construct an additional temporary support by means of steel pipe struts acting against combined piles driven inside the excavation (Figure 4). Such a solution made it possible to continue the works on the foundation slab, however seriously impeded them (delivery of materials, reinforcement assembly, work breaks).

Fig. 4. Temporary support of deflected sheet pile wall.

4 Conclusions

Design and dimensioning of the retaining walls embedded in soil and their anchorages needs to adopt adequate calculation model and proper estimation of data used in calculations. The values of earth pressures on a retaining structure depend especially on the physical and strength properties of soil, such as: internal friction angle, cohesion, soil bulk density or possible over consolidation ratio (OCR). When the wall crosses the natural flow of the ground water, the possible rise of water level should also be considered. The presented case study shows that the decisive factor which conditions the retaining structure safety can be also the change of static scheme, unpredicted in calculations and caused, for example, by excavation below the formation level or by washing away the bottom. The earth pressures are also dependent on the value and way of loading applied to ground level (considered, e.g. as the increased height of excavation). Taking into account the geometry of the excavation, wall deformation and terrain sinking are always highest in the middle of each wall section, between the retention corners. This is caused by the reduced soil pressure against the retaining wall around the retention corners and often referred to as 'the corner effect'. This effect is dealt with in the works of Ou & Shiau [18] and Wu & Ou & Tung [19] where the result of FEM modeling is referred to the real-life (on-site) observations made during the cutting of deep trenches.

References

1. A. Siemińska-Lewandowska, M. Mitew-Czajewska, The effect of deep excavation on surrounding ground and nearby structures, *Geotechnical Aspects of Underground Construction in Soft Ground - Proceedings of the 6th International Symposium, IS-SHANGHAI 2008*, 201-206 (2009)
2. A. Siemińska-Lewandowska, M. Mitew-Czajewska, The second metro line in Warsaw - Lessons learnt, *ITA-AITES World Tunnel Congress 2016*, **4**, 3148-3157 (2016)
3. L.A. Molokov, Interaction of engineering structures with geological environment, [Взаимодействие инженерных сооружений с геологической средой], 222 (1988)

4. E.Y. Kulikova, Filtration reliability of structures of urban underground structures, [Фильтрационная надежность конструкций городских подземных сооружений], Издательство «Мир горной книги», 316 (2007)
5. M. Mitew-Czajewska, Numerical analysis of a 36m deep diaphragm wall - Parametric study, *Geotechnical Aspects of Underground Construction in Soft Ground - Proceedings of the 8th Int. Symposium on Geotechnical Aspects of Underground Construction in Soft Ground*, TC204 ISSMGE - IS-SEOUL 2014, 257-262 (2014)
6. M. Mitew-Czajewska, FEM modelling of deep excavation - Parametric study, Hypoplastic Clay model verification, MATEC Web of Conferences **117**, 00121
7. K. Gorska, Z. Muszyński, J. Rybak, Displacement monitoring and sensitivity analysis in the observational method, Studia Geotechnica et Mechanica **35**, No 3, 25-43 (2013)
8. K. Gorska, M. Wyjadłowski, An analysis of excavation support safety based on experimental studies, Studia Geotechnica et Mechanica **37** (3), 19-29 (2015)
9. M. Drusa, V. Chebeň, R. Bulko, New technologies implemented in geotechnical monitoring on transport constructions, *14th International Multidisciplinary Scientific GeoConference Surveying Geology and Mining Ecology Management, SGEM*, **2** (1), 651-656 (2014)
10. Bednarski Ł., Sieńko R., Howiacki T.: Supporting historical structures technical condition assessment by monitoring of selected physical quantities, Procedia Engineering **195**, 32-39 (2017)
11. Z. Muszyński, Assessment of suitability of terrestrial laser scanning for determining horizontal displacements of cofferdam during modernization works on the Rędzin sluice. *14th International Multidisciplinary Scientific GeoConference, SGEM 2014: GeoConference on informatics, geoinformatics and remote sensing*, **2**, 81-88 (2014).
12. A. Segalini, L. Chiapponi, M. Drusa, B. Pastarini, New inclinometer device for monitoring of underground displacements and landslide activity. Komunikácie **16 (4)**, 58-62 (2014)
13. M. Drusa, V. Chebeň, P. Prohovníková, Functionality of TDR piezometers and inclinometers for monitoring of slope deformations. *SGEM GeoConference on science and technologies in geology, exploration and mining: 13th int. multidisciplinary scientific geoconference*, **Vol. II**, Sofia: STEF92 Technology, 157-164 (2013)
14. W. Brząkała, A. Herbut, J. Rybak, Recommendations for ground vibrations survey in course of geotechnical works. *14th International Multidisciplinary Scientific GeoConference SGEM 2014*: **2**, Hydrogeology, engineering geology and geotechnics. Sofia : STEF92 Technology, 747-754 (2014)
15. W. Brząkała, M. Baca, The measurement and control of building vibrations in course of sheet pile wall and Franki pile driving, *17th Int. Multidisciplinary Scientific GeoConference, SGEM 2017* **12**, Science and technologies in geology, exploration and mining, 929-936 (2017)
16. F. Oliveira, I. Fernandes, Influence of geotechnical works on neighboring structures, *17th Int. Multidisciplinary Scientific GeoConference, SGEM 2017* **12**, Science and technologies in geology, exploration and mining, 993-1001 (2017)
17. J. Rybak, K. Schabowicz, Survey of vibrations generated in course of geotechnical works. *NDE for Safety: 40th int. conf. and NDT exhibition: proceedings*, Czech Republic, Brno University of Technology, 237-246 (2010)
18. Chang-Yu Ou, Bor-Yuan Shiau, Analysis of the corner effect on excavation behaviors. Canadian Geotechnical Journal; Jun 1998; **35**, 3 ProQuest Central, 532 – 540 (1998)
19. Chao-Hui Wu, Chang-Yu Ou, N. Tung, Corner effects in deep excavations – establisment of a forecast model for Taibei Basin T2 Zone. Journal of Marine Science and Technology, 18 (1), 1-11 (2010)

Complex method of defects diagnostics in underground structures

Alexey Beskopylny[1,*], *Alexander* Lyapin[1], *Maxim* Kadomtsev[1], and *Andrey* Veremeenko[1]

[1]Don State Technical University, pl. Gagarin, 1, Rostov-on-Don, 344010, Russia

Abstract. The article presents a comprehensive method for diagnosing underground structures using the example of an underground pedestrian crossing located in Rostov-on-Don. The problem of assessing the condition of buildings and structures is very relevant at all stages of the life cycle. There is a special need for continuous monitoring of bearing structures for many buildings with critical applications especially if reliability of which determines the life and health of people. This work considered complex method included geodetic research, geological study of the state of sub-soil, analysis of state of steel and reinforced concrete structures, vibro diagnostics of load-bearing structures under dynamic technogenic impacts and an assessment of the mechanical characteristics of steel and reinforced concrete structures in order to develop a conclusion on the overall state of the transition and the possibility of its further operation. Analysis of the response from arbitrary non-stationary effects on structural elements is carried out on the base of the calculated spectrum of reference impacts.

1 Introduction

Recently the problem of assessing the condition and working capacity of buildings and structures becomes increasingly important. At the same time, the need for technical survey not depends on the phase of construction and building purpose. Modern trends in construction, such as the increase in the number of storeys of buildings, the compaction of urban buildings, the tightness of construction sites, the development of underground space, and the saturation with engineering communications invariably lead to the emergence and subsequent increase in the negative techno genic impact of the construction carried out on already constructed facilities located in adjacent areas. So the special importance of the problem of monitoring the technical condition of buildings and structures is become in order to prevent the emergence. It is obvious that the assessment of the technical state of load-bearing structures should be systematic and allow for the evaluation of the changes taking place on the basis of quantitative criteria, based on procedures to identify the compliance of the actual strength, rigidity and stability of structural elements to regulatory requirements.

There are many methods of mechanical properties assessment that are useful for monitoring. Depending on the building purpose the methods to determine the structures

*Corresponding author: besk-an@ya.ru

properties are different. These are non-destructive tests based on indentation methods that used for metal structures [1,2], determination of the rheological parameters for the creep models [3], identification of the structures properties [4].

Complex and multidisciplinary approach to the diagnosis of historical buildings is shown in [5]. Items related to various issues as structural and thermal aspects, acoustic requirements and efficiency of active energy systems are took into account.

Damages and defects happening throughout structure execution or transport and consignment on bridge structure are considered in [6]. The research represents an automated laser remote sensing data with Bridge Information Model for a constructed bridge.

A case study of different approaches to planning and building control focus on physical and technical issues and on design and construction and in a social and environmental context at the other hand is in [7]. This paper suggests thinking and acting in terms of complex ecosystems and addressing the issues from an anthropological perspective.

The defects, faults and accidents of contemporary constructions are considered in [8]. The article focuses on the defects and disruptions of buildings, which caused death or serious injuries (human or animal), or major property damage.

All cases show that a complex assessment of the state of a building is divided into a study of the state of building elements, their strength and other physical characteristics.

2 Problem statement and equipment

The objectives of the survey included:
- Preliminary (visual inspection) with the registration of damages and drawing up of the defective sheet;
- Carrying out geodetic measuring works, deviations from axes, deplanation;
- Non-destructive testing of the strength of structural elements;
- Non-destructive testing of strength of steel structures, reinforcement bars;
- Laboratory evaluation of strength of elements of stone structures on the basis of destructive methods;
- Geological analysis of soil in which there are buildings;
- Study of the effect of vibration exposure on the bearing capacity transitions from passing over them transport;
- Analysis of the causes of defects in structures;
- Development of conclusions about the technical condition of building structures, pedestrian subways.

Tests of monolithic reinforced concrete structures made by the method of surface ultrasonic testing according to standard methods. The strength of concrete in monolithic reinforced concrete structures is determined by an ultrasonic method for the calibration dependence, built in accordance with the requirements of MDS 62-2.01 according to parallel tests of the same sections of concrete structures by ultrasonic method (GOST 17624-87) and by the method of separation with shear fracture (GOST 22690-88).

During test of steel structural elements there are objective difficulties, the main of which are the following. To carry out the standard test it needs to cut out segments of the exploited construction material, and to make specimen standard sizes. The greatest interest is the most loaded place where the weakening of the cross section cut is associated with maximum risk of destruction of the structure itself. If you consider that obtaining reliable information is possible only on the basis of the processing is quite a representative sample (10-15 samples of elements of the same grade of steel), these difficulties become even more obvious. Furthermore, during prolonged operation the metal characteristics change from the viewpoint of strength and plastic, and reduce their carrying capacity, causing even greater doubts about the appropriateness of carrying out destructive tests.

In this regard, we have developed a unique system of non-destructive control of strength of metal, which to date has tens of approbation under the control of the mechanical characteristics of responsible metal structures and allows to simultaneously determine the main mechanical properties: hardness, yield strength, tensile strength, elongation at any place design. The instrument [9] for determining the mechanical properties of steels (Fig. 1), which consist of handle 1, spring 2, striker 3, body 4, indenter holder 5, sensor 6 and 9, body part 7, damper 8.

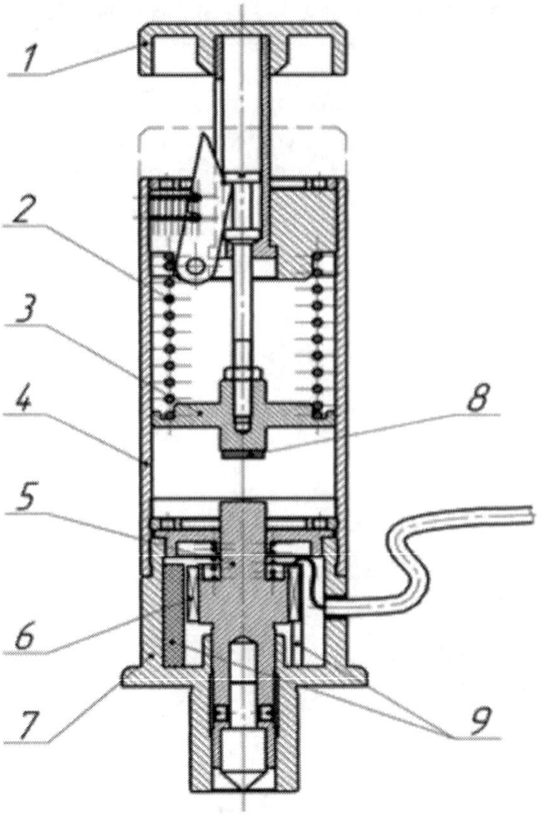

Fig. 1. The devise for mechanical properties measurement.

During measurement striker 3 hits over the indenter holder 5 and sensor 6 registers signal of velocity *V(t)*. The dependences *S(t)* and *W(t)* are obtained by integrating and differentiating [10].

3 Investigation of dynamic effects on the structure of the pedestrian crossing when driving a motor vehicle

Vibration measuring 12-channel device VK-12 (Fig. 2) is developed and implemented on the basis of the E14-440 module manufactured by the company "L-Card" and is designed for receiving, amplifying and converting mechanical oscillations of the elements of the structure under dynamic influences, Including at the noise level (microseismic vibrations) and signal transmission in digital form for further processing in IBM PC.

During the experimental studies accelerometer sensors were fixed in the centre of the ceiling panel with the help of anchor bolts. This type of fastening provides the best matching of the sensor and over covering.

In the course of experimental studies, the level of vibration velocities from the following sources was investigated:

• the impact of a car with a mass of 20.5 tons at a speed of 30 km/h and 50 km/h;

• hit on the supporting beam.

For experimental determination of the frequencies of the Flexural vibrations of a beam was used the results of the registration of the accelerations of vibrations under the impact. The impact corresponds to a Gaussian pulse

$$p(t) = Pe^{-a^2t^2}, \quad (1)$$

where P - the peak value of impact load , $2/a$– time pulse.

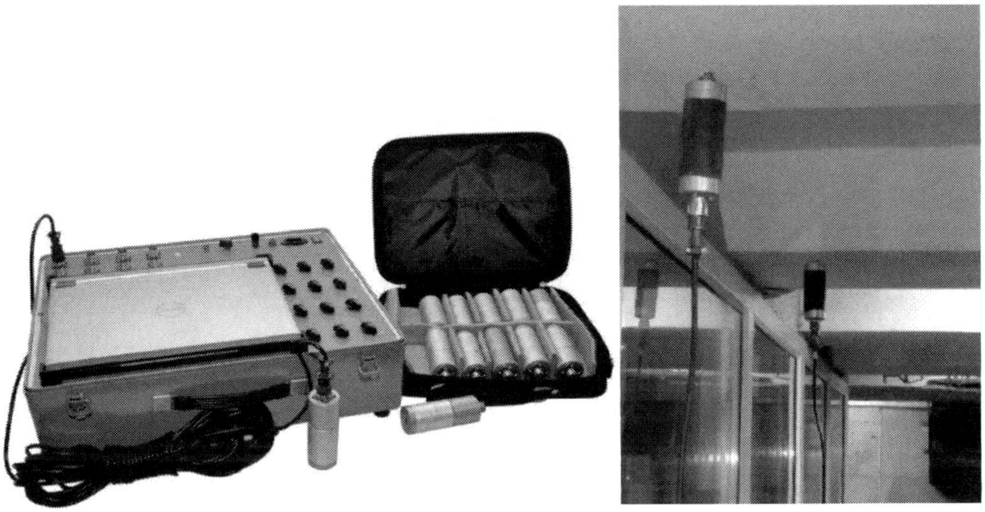

Fig. 2. Mobile vibro-measuring complex.

The spectrum of the Gaussian pulse is defined by the formula

$$S(\omega) = P\int_{-\infty}^{\infty} e^{-a^2t^2}e^{-i\omega t}dt = \frac{P\sqrt{\pi}}{a}e^{-\left(\frac{\omega}{2a}\right)^2}, \quad (2)$$

where $\omega=2\pi f$– circular frequency of oscillation, f is the linear frequency.

With an external effect on the structure, the spectrum of the detected response depends on the spectrum of the external action and the spectral characteristics of the structure. This dependence is expressed by the formula

$$S_o = S_b \cdot S_к, \quad (3)$$

Where S_o is the response spectrum, S_b is the external impact spectrum, $S_к$ is spectrum of natural frequencies of the structure. Then the spectrum of the construction is determined by the formula

$$S_к = \frac{S_o}{S_в} \quad (4)$$

Thus, it follows from (4) that under impact conditions the response spectrum will differ very little from the intrinsic spectrum of the structure.To determine the dynamic response to an arbitrary dynamic effect, it is necessary:

1. Perform a test record of the response to the dynamic impact in question at an arbitrary location (for example, at a low-intensity auto-horn).
2. Perform the same registration of the response to the reference impact.
3. Get the response of the structure by the formula

$$V(t) = \frac{1}{2\pi}\int_{-\infty}^{\infty}\frac{\overline{V_{nc}}(\omega)}{\overline{V_{sc}}(\omega)}\cdot\overline{V_s}(\omega)\cdot e^{i\omega t}dt, \tag{5}$$

Where $\overline{V_{sc}}(\omega)$is the response velocity spectrum for the standard effect,$\overline{V_{nc}}(\omega)$is the response velocity spectrum for the nonstationary effect under study, $\overline{V_s}(\omega)$is the velocity spectrum of the response to the reference effect recorded on the elements of the pedestrian crossing structure.

4 Results and discussion

Visual inspection was conducted with the establishment of damages and drawing up of a defective sheet. As a result, a set of cracks with a width of 0.3 to 1 mm was revealed, several traces of soaking, several exposures of working reinforcement of reinforced concrete structures, recommendations for restoration.

When geodetic work on each of the tunnels, the leveling of the ceiling was leveled. For branched underground pedestrian crossings, the rooms were further divided into sectors. The work was carried out by the method of geometric leveling, using the leveling GEOBOX N7-26. Leveling was carried out on the transverse beams, each of which was taken at three points

The non-destructive testing of steel strength was subjected to armature elements of the reinforcement of the slabs of the coating. A grinding machine stripped the armature metal and the mechanical properties were determined using the "Strength" system: hardness (HB), yield stress (σ_y), tensile strength (σ_b) and elongation (δ). The survey showed that for the wall panel reinforcement the average values of HB = 150, σ_y = 344 MPa, σ_b = 437 MPa, δ = 23% correspond to the requirements for the strength class of the A400 armature (AII).The strength characteristics obtained correspond to the requirements for the strength of reinforcing steels specified in the standard structure.

To determine the spectral characteristics of the response of structural elements to the dynamic effects from the transport passage 3 sensors accelerometers were installed in the zones with the greatest load of the beam.

Analysis of the response to impact showed that a resonance peak occursat a frequency of 45 Hz (Fig. 3). This result is in good agreement with the results obtained from the experimental full-scale studies of the oscillations of ceiling panel under the action of a shock load. It should be noted that the resonance obtained as a result of the calculation is somewhat lower than that obtained experimentally. Hence it can be concluded that in panel on which the registration was carried out, there were no processes that could worsen the dynamic properties of the panel.

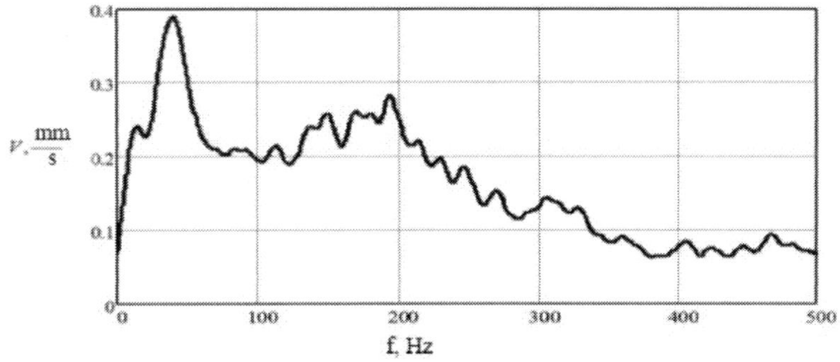

Fig.3. The amplitude-frequency characteristics of vibration velocities of the response of the slab.

5 Conclusions

As a result of the studies, the following conclusions can be drawn:
• maximum dynamic stresses from the impact of individual units of passing vehicles do not exceed 15 kPa;
• Total stresses when passing vehicles do not exceed 16 MPa
• the maximum speed of ceiling panel from the impact of individual units of passing vehicles did not exceed 0.4 mm / s;
• under the existing speed limits, the greatest dynamic stresses under the influence of the machine flow will not exceed 30 kPa;
• frequency spectrum of impact from passing vehicles is within the limits of 1-20 Hz;
• the intrinsic frequency of bending oscillations of the panel is about 40 Hz.
• the natural frequency of the panel lies outside the range of frequencies that are excited in the floors when driving. Due to this, there will be no resonance of panel during the passage of vehicles

Studies have shown that the transition is in a workable state. Required activities:
A) in accordance with the defective sheet, to strengthen the elements of the entrance design according to a specially executed project;
B) in places of strong soaking, dismantle the lining, inspect the technical condition of the panels and perform the necessary repairs;
C) in the zone where the reinforcement is exposed, clean the reinforcement with metal brushes and repair the protective layer of concrete with a repair composition;
D) carry out work to eliminate the flow of rainwater;
E) in the places where the cracks are opened, install beacons.

Thus, a comprehensive method for diagnosing defects in underground structures has been developed, including a number of authoring techniques for vibration diagnostics of the dynamic characteristics of structural elements, determining their resonant frequencies and the complex of mechanical characteristics of steel bearing structures.

References

1. A.N. Beskopylny, A.A. Veremeenko, B.M. Yazyev, MATEC Web of Conferences **106,** 04004 (2017)
2. A.N. Beskopylny, I.G. Kadomtsev, N.I. Beskopylnaya, Theoretical foundations of civil engineering **83-84** (2005)
3. A.S. Chepurnenko, V.I. Andreev, B.M. Yazyev, A.N. Beskopylny, MATEC Web of Conferences **67**, 06059 (2016)

4. A.N. Beskopylny, A.A. Lyapin, V.I. Andreev MATEC Web of Conferences **117,** 00018 (2017)
5. F. Ascione, F. Ceroni, R. F. De Masi, F. De'Rossi, M.R. Pecce, Applied Energy **185** Part 2, 1517-1528 (2017)
6. M. Sedec, A. Serwa, The Egyptian Journal of Remote Sensing and Space Science **19**, issue 2, 273-283 (2016)
7. K. Alexander, Procedia Engineering **20**,1-11 (2011)
8. M. Kraus, T. Vondráčková, V. Nývlt , MATEC Web of Conferences **93,** 03004 (2017)
9. D.M. Belenkii, A.N. Beskopylny, N.N. Beskopylnyi, E.K. Polibin, B.A. Pesenko Method for determining mechanical characteristics and the device for its implementation. Patent No. 2079831 (1997)
10. D.M. Belenkii, A.N. Beskopylnyi, N.L., L.G. Shamraev, Industrial laboratory **64** (5), 340-343 (1998)

Insufficient dimensional stability of bitumen sheets as a source of flat roof defects

Jan Plachý[1,*], *Jana* Vysoká[1] , and *Radek* Vejmelka[1]

[1]Institute of Technology and Business in České Budějovice, 370 01 České Budějovice, Czech Republic

Abstract. One of the frequent sources of flat roof faults is failure of the top layer of a multilayer bitumen sheet system in cross-connections. One of the possible causes of this disorder is failure to observe technological discipline. The paper deals with how the full melting of the bitumen sheet influences its possible shrinkage during the roof exposure. Bitumen sheets with a polyester carrier which is not reinforced with fiberglass fibers are burdened with shrinkage effect. The preload is introduced into these belts during production. The aim is to show how is the effect on loosening the preload in the carrier layer with full-scale smelting during the realization of the waterproofing layer. The preload loosening experiment, expressed by the dimensional stability test according to ČSN EN 1107, was performed in a laboratory on two bitumen sheets.

1 Introduction

One of the frequent failures that arise on roof claddings is the release of bitumen sheets (BS) in transverse joints due to volume changes from the thermal load. These changes are manifested by the longitudinal shrinkage of the BS (Figure 1). One of the possible causes of this disorder is a failure to observe the technological discipline. In the case of multilayer systems, the top layer is not full-scale smelted on the substrate BS. The shrinkage effect is primarily due to the BS with a polyester carrier which is not reinforced with fiberglass fibers. The BS is preloaded during production, which gradually releases each additional thermal stress. This thermal stress is achieved by full-scale melting and exposure to sunlight. The aim of the paper is to show how the effect on the loosening of the preload in the carrier layer with full-scale smelting during the realization of the waterproofing layer.

The issue of volume changes BS has been dealt with in great detail [1] which, among other things, states that the most suitable for flat roofs are BS with composite carrier liner made of polyester fibers which are reinforced with fiberglass fibers. It also highlights the effect of the BS for volume changes. Another author [2] notes the force effects that are introduced into the carrier during the BS production. It is also possible to mention the group of authors [3], which notes the relation between the dimensional stability of the carrier and

* Corresponding author: plachy@mail.vstecb.cz

the BS. In conclusion, the authors state that the influence of the covering material on the volume stability is small.

Although all authors [1, 2] and [3] agree that composite inserts are the most suitable for roofing BS, it is now possible to meet BS that contain a carrier only from polyester fibers. The reason for this composition is primarily economic aspects. None of the above mentioned authors dealt with dimensional stability of the BS after releasing the stress during full-scale smelting.

BS are tested for dimensional stability - volume changes according to ČSN EN 1107-1 [4]. UEAtc technical guide for the evaluation of roofing waterproofing systems comprising bitumen sheets modified by polymers APP or SBS [5] indicates that the values for dimensional stability have to be less than 0.5% in the case of multiple layer systems, and less than 0.3% for single layer systems. The abovementioned dimensional stability values for BS also take into account its requirements as ČSN 730605-1 [6]. The authors of this paper are of the opinion that full-surface smelting BS, which is intended for full-surface smelting application, has a decisive influence on the dimensional stability of the BS. Thus, a hypothesis was made that when the BS is fully smelted, most of the preload is loosened in the BS. The shrinkage value represents 50-75% of the total preload.

Fig. 1. Bitumen sheet failure.

2 Material and methods of testing

2.1 Materials

For the tests, two BS samples were selected with a polyester carrier with a different basis weight. Both BS samples are designated as the topsheet of a multilayer system. The mode of application of BS is selected for full-scale fusion, see Table 1.

Table 1. Bitumen sheets – Dimensions and characteristics. Signification: E – the mass modified by elastomers, PY – polyester fleece.

Characteristics of the test sample / number of test sample	Modification type of bitumen coating matter	Nominal thickness [mm]	Reinforcement	
			material	Area weight [g/m²]
T 180 80	E	4.2	PY	180
T 230 80	E	5.2	PY	230

2.2 Methods of testing

The test specimens were tested according to EN 1107-1 [4]. This standard allows two methodologies to be used for testing - optical, which is more accurate [1,3] and mechanical. For economic reasons, the mechanical method was chosen using the sliding scale as shown in Figure 2. The exposure time of the test samples is at 80 °C for 24 hours, see [24]. Based on experience, see [7], further measurement intervals of 6, 12, 48, 72 and 96 hours were determined. The steady state in terms of dimensional stability was 96 hours. For each sample, nine test specimens were always prepared in the longitudinal direction of BS production. Since the major failures in this type of BS are achieved primarily in the longitudinal direction, they have not been tested in the transverse direction of the BS.

An increase in the number of test specimens from five to nine, which is prescribed [4], was carried out for better statistical evaluation. Samples simulating full-scale smelting were labeled with T 180 n80 and T 230 n80. The simulation of full-scale smelting was achieved by heating the BS with a propan-butane burner (PB burner) after the measuring points were prepared from a distance of 100 mm (about 5 seconds during the melting of the separation film) for a short time. The PB burner with a diameter of 60 mm and a power of 58 kW was used. The measuring points were protected against flame by an aluminum foil.

Fig. 2. The testing specimen for the dimensional stability of bitumen sheets according to ČSN EN 1107-1 [4].

2.3 Evaluation

Four types of samples were processed for evaluation. Nine test specimens were tested for each sample. The resulting value of the dimensional stability is given as the mean of the valid measurements, including the standard deviation, which tells how the values of the statistical set differ from the mean value. Was evaluated:

- percentage of initial preload applied to bitumen sheet for samples T180 n80, T 230 n80 which are released by heating the bitumen sheet by PB burner
- changes in the dimensional stability of the bitumen sheet in time-dependence for samples T180-80, T230 80, where 24 h/80 °C exposure was determined as the reference temperature according to ČSN EN 1107-1 [4], the dimensional stability of the bitumen sheet at this temperature represented 100%
- determining the dimensional stability of the bitumen sheet in relation to the requirements of ČSN 730605-1 [6].

3 Results

The measurement results are shown in Figure 3. The standard deviations are collected in Table 2.

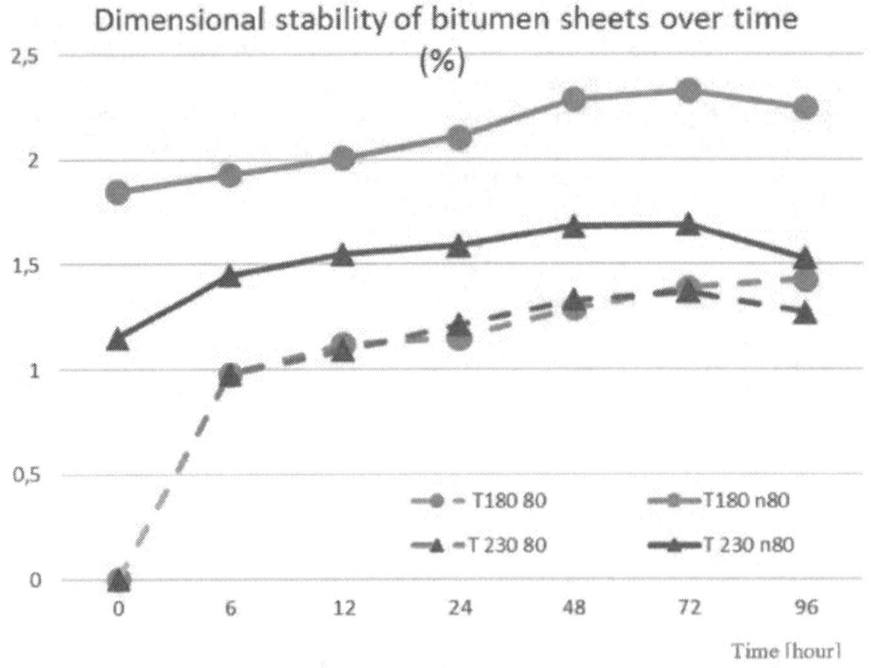

Fig. 3. The dimensional stability of bitumen sheets over time.

Table 2. Standard deviations.

Sample/exposition [h]	0	6	12	24	48	72	96
T180 80		0.206	0.232	0.236	0.198	0.194	0.2
T180 n80	0.356	0.368	0.372	0.256	0.373	0.363	0.378
T 230 80		0.115	0.109	0.0092	0.161	0.145	0.102
T 230 n80	0.124	0.143	0.131	0.127	0.114	0.119	0.119

4 Discussion

The hypothesis asserting that 50 - 75% of the total preload will be released when the test specimen is heated by the PB burner, has been confirmed. For sample T180 n80, 79.4% of overall shrinkage was achieved. For sample T230 n80, 68.0% of overall shrinkage was achieved. For both samples, maximum shrinkage was achieved after 72 hours. It is possible to assume that another thermal exposure already exerted an extension of the carrier. The shrinkage value in the case of a larger basis carrier liner is smaller, which can be explained by introducing less deformation in the production of BS, which is stronger thanks to the carrier liner.

The results transferred to Figure 4 demonstrate that the test procedure according to ČSN EN 1107-1 [4] shows that, depending on the type of carrier, there is still 15-24% shrinkage of the bitumen sheet. Conversely, more than 80% shrinkage after 24 hours according to

ČSN EN 1107-1 [4] is achieved after 6 hours of exposure. Depending on the overall shrinkage, the value after 6 hours represents about 70% of the overall shrinkage.

From the point of view of the comparison of the results with the requirements of the standard ČSN 730605-1 [6], all samples had a significantly higher shrinkage than the required 0.5%. Samples that have not been heated by the T180 80, T 230 80 PB burner have shrunk values of 1-1.5%. If the samples T180 n80, T 230 n80 after heating the PB burner are evaluated, the shrinkage is at the tolerance limit of 0.5%. In the case of exposure taking 24 hours the samples meet the standard requirement according to [6] see Table 3.

Table 3. The dimensional stability of bitumen sheets samples after heating with propan-butane burner.

Sample/exposition [h]	Requirement according to [6]	6	12	24	48	72	96
T180 n80	0.5	0.08	0.16	0.26	0.44	0.48	0.4
T 230 n80	0.5	0.3	0.4	0.44	0.53	0.54	0.38

By comparing the shrinkage values of all samples, it is interesting to note that the T180 80, T 230 80 samples at 80 °C do not remove most of the preload that is introduced into the BS during production. This result is determined by the temperature at which BS production takes place. This temperature is up to 180°C, see [4]. On the other hand, the PB flame temperature at smelting, although acting in the short term, reaches 1100 ± 100 °C.

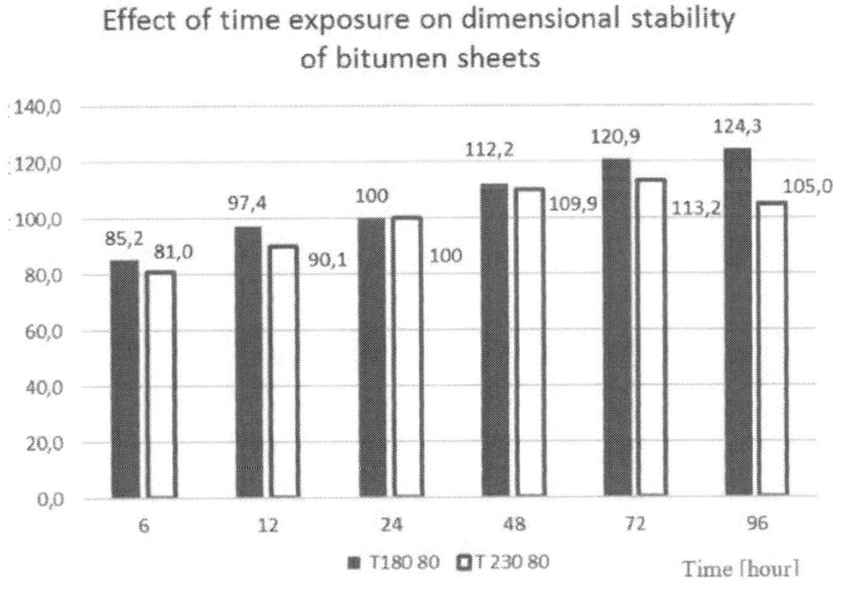

Fig. 4. The effect of time exposure on the dimensional stability of BS.

5 Conclusion

In conclusion, it is possible to state that in the case of bitumen sheets intended for the topsheets in a multilayer system with a non-glass fiber reinforced polyester liner and intended for full-scale smelting, the full-scale smelting is decisive for loosening the preload produced during the production of BS.

Test results according to [4] showed that the exposure of 24 hours could theoretically be reduced to 6 hours because it is only a value for comparison between BS products. The test procedure in [4] applies to samples loosely laid on the substrate. However, this process is used in [8] for topsheets of bitumen sheets without any difference in application to the substrate.

In the next phase of the research, the attention will be paid to the measuring time interval and to bitumen sheets with composite carrier liners made of glass fiber reinforced polyester.

References

1. P. Malych, *Volume changes of bitumen coatings. Volume changes of bitumen membranes* (University of Technology in Bratislava, Faculty of civil engineering, Bratislava, 2001)
2. J. Plachý. *Volume changes of bitumen sheets* (Roof Symposium, Bratislava, 2010)
3. J. G. Lopes, J. R. Correia, M. X. B. Construction and Building Materials **25 (1)**, 3229-3235 (2011)
4. *ČSN EN 1107-1 Flexible sheets for waterproofing - Part 1: Bitumen sheets for roof waterproofing - Bitumen sheets for roof waterproofing - Determination of dimensional stability* (Prague, 2000)
5. The European Union for Agrément (UEAtc). UEAtc technical guide for the assessment of roof waterproofing systems made of reinforced APP or SBS polymer modified bitumen sheets (UEAtc, Paris, 2001)
6. *ČSN 730605-1 Waterproofing of construction works – Flexible sheets for waterproofing – Requirements for use of bitumen sheets*
7. J. Plachý, *Bitumen sheet joints under the influence of temperature and time in the composition of single-membrane roofs* (Brno University of Technology, Faculty of civil engineering, Brno, 2005)
8. ČSN EN 13707. *Flexible sheets for waterproofing - Reinforced bitumen sheets for roof waterproofing - Definitions and characteristics* (Prague, 2014)

Performance of selected polyurethane joint sealants in concrete structures

Barbora Nečasová[1,*], *Pavel* Liška[1], and *Jiří* Šlanhof[1]

[1]Brno University of Technology, Faculty of Civil Engineering, Veveří 331/95, Brno 602 00, Czech Republic

Abstract. Often underestimated and neglected, joint sealants in concrete structures play an important role in protecting the structure against its premature deterioration. The focus of the presented paper lays upon the issue of sealing of joints in concrete structures. Concrete is one of the most difficult materials to seal since there are always small particles of dust and the material itself. In this paper, the joint will be viewed as traditional expansion joint. Two representatives of flexible one component polyurethane sealants were selected. The main reason for this selection was that conventional repair materials are usually very rigid and easily incline towards the formation of cracks and gaps. The aim of this paper is not only to verify the suitability of selected materials but also to demonstrate that polyurethane sealants might be a better and long-lasting solution for filling of joints and patching of cracks within concrete structures.

1 Introduction

From time immemorial, building structures have been designed such that they are able to withstand the contrary effects of the environment for a certain time. The terms "for a certain time" and "the surrounding environment" are used for a purpose here, and that is because the phenomenon of ageing buildings is currently an increasingly discussed problem that is faced by every larger city [1-5]. The materials used in the buildings (such as sealants) age just as the buildings themselves do. Their ageing can affect their mechanical properties and functionality. The surrounding environment affects the materials in the same way, therefore, they are no longer able to fulfil the task for which they were designed, which may also result in the degradation of the building itself.

Currently, there are countless sealants, filling adhesives and materials on the market, so it is not surprising that designers often scratch their heads over the question of what sealant to use in the design, whether the selected sealant is suitable for use in the proposed combination of materials etc. Unfortunately, this lack of information and knowledge can greatly accelerate the ageing of the selected sealant and irreversibly damage the whole structure. While some products have a service life as low as three or four years, see Fig. 1 (C, D), today we already know that some silicones can last up to 25 years in the case of a correct installation [6].

* Corresponding author: necasova.b@fce.vutbr.cz

In this paper, the authors focused on the sealants of concrete structures, as this is the most used building material at present. Concrete is a composite material, i.e. an "agglomerate" of more raw materials, and it is generally known that after plastics, it is the material that has the most complicated material properties for sealing [7, 8]. It is very common that the joints between individual concrete elements must be protected from environmental influences, the permeation of moisture and so forth, see examples in Fig. 1 (part A and B).

Fig. 1. Joint between concrete elements (A, B) and common types of joint sealant failure (C, D).

An inappropriately selected filling material may not adhere sufficiently to the concrete structure and thus not fulfil the function for which it was designed, see Fig. 1 (part C and D). The aim of the authors of this article was to test two polyurethane sealants from the same manufacturer which are, on the one hand, designed for sealing concrete structures, but are also commonly available in the market.

2 Methodology

The presented work comprised two parts. Firstly, the maximum elongation of the tested sealants at break was evaluated by extension test. The elongation at break as well as maximum force applied were recorded and calculated. Secondly, the failure modes of all samples that appeared during testing were observed to express the compatibility of tested materials. However, the main aim of this case study was to compare results obtained from tests performed on samples that were manufactured in perfectly clean environment of laboratory, and on samples that were prepared, cured and stored in an external environment.

2.1 Materials

The selection of materials for presented case study was based on an already completed project. The results of this project [7, 8] showed that only polyurethane sealants are able to, a certain level, fulfil the high requirements for sealing of concrete structures. Hence, two representatives of high strength one-component polyurethane sealants from the same manufacturer were selected.

Table 1. Mechanical properties of selected sealants.

Property/Sealant	**Type I**	**Type II**
Tear Strength at Break [N/mm^2]	8	8
Shore A	37	37
Secant Modulus /100 % elongation [N/mm^2]	0,6	0,6
Elongation at Break [%]	> 600	> 700
Elastic Recovery [%]	> 90	> 80
Service Temperature [°C]	- 40 up to + 70	

Both selected representatives are adhesive even without the use of penetration coatings and/or activation agents. However, according to the information that are described in technical data sheets, it is advisable to use an overcoat for optimum adhesion in critical and high-performance applications. Moreover, in the case of concrete structures, the application of the improving liquid is actually recommended. Comparison of mechanical properties can be seen in the Table 1.

The information presented in Table 1. show that both selected sealants have very similar material properties, however, the first selected sealant, hereinafter Type I, can be classified as sealant for non-structural use in joints in buildings and pedestrian walkways [9]. It should provide perfect performance when joining concrete structures or elements. Therefore, this sealant has been designed predominantly for the purpose which was tested in the presented case study. On the contrary, Type II sealant can be classified as an universal filling material. According to the technical data sheet [9] this product should provide an excellent adhesion on all cement-based materials, brick, ceramics, glass, metals, wood etc. Therefore, worse adhesion to concrete elements can be expected.

The concrete elements were prepared from dry mix concrete B20 for strength class C16/20. The geometry of all elements was 40 × 40 mm in cross section and 160 mm long.

2.2 Methods

Although joint sealing may seem like a simple task, the method and discipline of this work is very demanding. When applying selected sealants, it is necessary to adhere to technological discipline and proceed according to the manufacturer's instructions. The discipline of preparation and application relates mainly to the preparation of the substrate, i.e. the sealed structure. The substrate must be clean, dry, homogeneous, and free of oils, grease, dust and of free or non-cohesive particles. Therefore, the preparation of the samples and the subsequent application of the sealant was carried out according to the principles and procedures specified by the producer of the individual system. Generally, the edges of all samples were cleaned of dust using the cleaning product from the system before the actual sealing was performed. To achieve maximum adhesion, primer was used in all cases.

The same preparation and application procedure was used for the test samples prepared in the laboratory as well as for those prepared in the exterior. Three test samples were produced for each exposure period for the testing of normal stress.

The used method for determination of joint sealants performance was a 'real joint' method [10]. The aim of the method is to test samples that were prepared the same way as if the joint was installed in real structure [10]. Hence, the sealant joint of 10 mm thickness and 10 mm width was prepared. The preparation and geometry of test samples can be seen in Fig. 2.

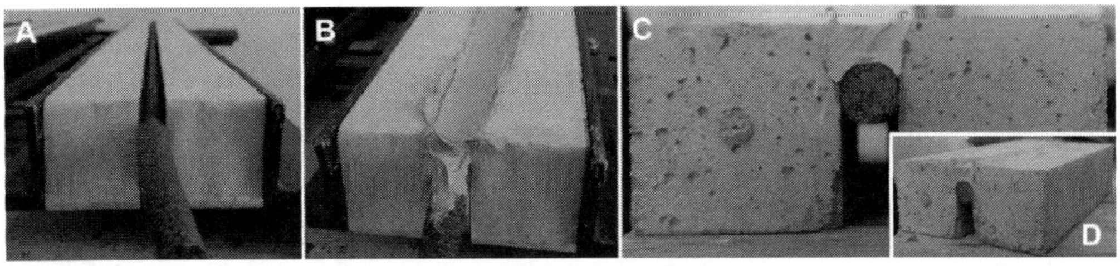

Fig. 2. Specimen manufacturing (A, B) and geometry (C, D).

After preparation of test samples always follows the curing period. However, in this case study, the length of the curing period was the key element, since some samples were manufactured, cured and stored in laboratory, where they were left in a perfectly dry and clean environment (at 23 ± 2 °C/ 50 ± 10% humidity), for a specific period, so that they aged sufficiently. On the contrary, some samples were manufactured, cured and stored in the environment with direct exposure to weather changes. Test samples were tested after 1, 3, 6 and 12 months of curing. Results of samples tested after 1 month, i. e. 28 days, were also used as a reference standard for the evaluation of test data.

Before the testing itself, the measuring of the width of the joint was performed after c. 48 hours, see chapter Results. Subsequently, the test samples were placed into test moulds, see Fig. 4 (A) and extended at a speed of approximately (5.5 ± 0.7) mm/minute. The testing was performed in the following steps:
- Step 1.: Extension of the joint to 100 % of the original width.
- Step 2.: Extension of the sample for the period of 24 hours.
- Step 3.: After 24 hours, extension of the sample until the break.
- Step 4.: Recording of failure modes.

3 Results

The width of the joint was measured before the commencement of the testing so that it was possible to determine whether there occurred any deformation of the sealants during the exposure period. The width of the real joint was measured in three points with the use of Vernier callipers. The measured values were averaged and a representative width of the sealed joint was obtained. Based on this average value, a so-called normalized value of the joint width was determined according to equation (1). The normalized value less than 1 indicated that the joint sealant was in a compressive state and a normalized value close to 0 indicates more compression of the joint sealant with respect to the original width. Values higher than 1 prove the elongation of the real joint.

$$NV = (l_1 / l_0) \qquad (1)$$

where NV is normalized value [-], l_0 is the original joint width measured after 48 hours after manufacturing in mm and l_1 is the joint width after specific exposure period and before testing in mm.

Another objective was to establish the size of the joint deformation which in this case is represented by its relative elongation that can be expressed in % as joint tensibility. The tensibility was determined based on an equation (2).

$$\delta = (\Delta l / l_1) \times 100\,\% \qquad (2)$$

where δ is tensibility in %, Δl is a change of joint width in mm and l_1 is the joint width measured before testing in mm.

The measured values were subsequently used to determine the approximate stress-strain curve, to find the relation between joint deformation and the level of extension resistance. The formula for calculation of an average force needed for the joint deformation was derived from a standard formulation of Hook's law, see equation (3).

$$\Delta l = (F \times l_1)/(E \times A) = (\sigma \times l_1) / E \qquad (3)$$

where Δl is a change of joint width in mm, l_1 is the joint width measured before testing in mm, F is the maximum force in N, E is modulus of elasticity in N/mm^2, A is bonded area in mm^2 and σ is strain in N/mm^2.

Table 2. Comparison of test results.

Property/Sealant	**Exposure period**	**Type I**		**Type II**	
Storing Environment [-]		LAB	EXT	LAB	EXT
Average Maximum Elongation at Break [mm]	1 month	50,67	**65,64**	54,27	**61,83**
	3 months	**62,43**	57,27	**57,97**	47,45
	6 months	**58,99**	57,97	**48,90**	44,25
	12 months	**65,55**	58,67	41,97	**54,21**
Average Tensibility δ [%]	1 month	352,6	**464,1**	391,3	**419,7**
	3 months	**432,5**	423,2	**448,7**	358,3
	6 months	**479,3**	469,7	**367,7**	331,7
	12 months	**544,4**	476,6	314,7	**427,3**
Average Maximum Force [N]	1 month	3384,8	**4455,4**	3757,0	**4028,8**
	3 months	**4152,1**	4062,7	**4307,2**	3439,5
	6 months	**4600,9**	4509,6	**3530,3**	3184,7
	12 months	**5226,2**	4575,7	3021,2	**4101,7**
Average Stress Strain [N/mm^2]	1 month	2,12	**2,78**	2,35	**2,52**
	3 months	**2,60**	2,54	**2,69**	2,15
	6 months	**2,88**	2,82	**2,21**	1,99
	12 months	**3,27**	2,86	1,89	**2,56**

4 Analysis

As can be seen from the presented results, the aim of the authors was not only to record the maximum force needed to break the test specimens, but also to assess the way in which all the cracked joints break, since in the case of selected sealants, the maximum forces are reached only rarely when the sample breaks.

From the results of failure modes presented in Fig. 4., can be seen that for all test specimens left in a clean laboratory environment, only one manner of failure mode was recorded, namely the cohesive failure. As can be seen from Fig. 3 (C), the sealed joint almost always cracked right at its centre at an elongation of up to 500 %. As was expected, the test specimens that were left in the laboratory environment were completely protected against weathering and the effects of weather changes and their ageing was thus not accelerated. Nevertheless, shrinkage was monitored for all test samples. This was confirmed by the calculation of the normalized value. Eventhough, the differences of compression between sealants tested in laboratory and in exterior have not been considerably high, the normalized value showed that samples stored in exterior compressed a little bit more, approximately about 10 %, than those stored in interior.

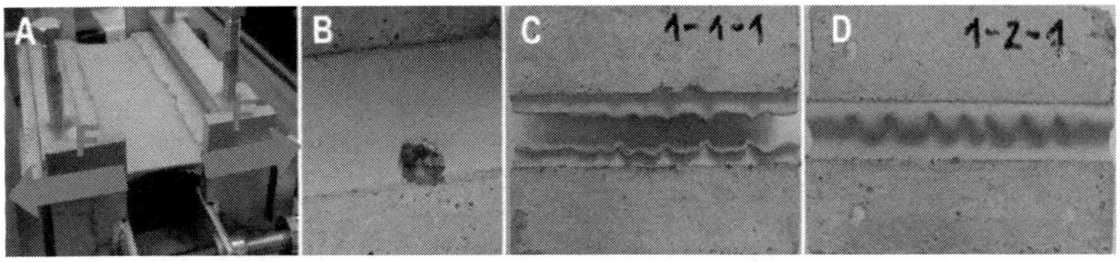

Fig. 3. Test samples inserted in moulds (A) and examples of failures (B-D).

The results obtained from the laboratory measurements were compared with the results that were measured for the samples produced and left in an outdoor environment, see the results presented in Table 2. and Fig. 4. Here, however, the same failure modes as well as similar elongations were recorded, only in two cases no failures were recorded and permanent deformation of c. 40 % had been measured, see Fig. 3. (D) with an example of permanent deformation of joint sealant. In addition to the dominant number of cohesive failures only two samples embodied a combination of adhesive and cohesive failure, see Fig. 3. (B). Eventhough, the effects of ageing were monitored for all the samples, which should have an effect on the reduction of the elasticity of the applied sealant, no effect was observed in any case, and thus also any effect of ageing and of external environment have not been monitored. Moreover, the external environment has not had any influence on the resulting maximum elongation which can be seen in Table 2. where the maximum values had been highlighted.

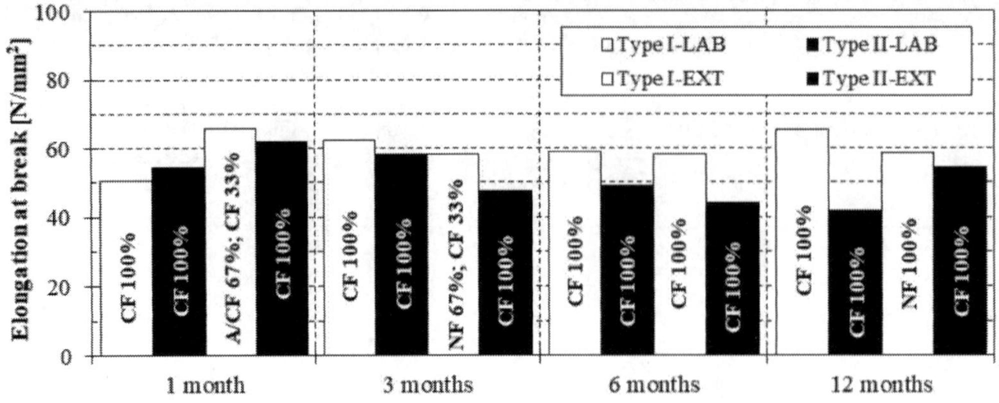

Fig. 4. Comparison of ultimate elongations at break of test samples.

5 Conclusion

The measured results demonstrate the suitability of the selected sealants for use in sealing concrete structures. Contrary to the initial expectations, no significant differences were found between the sealants used. While the only good properties for sealing concrete structures were confirmed for the Type I sealant, for the Type II sealant, the results of the case study demonstrated that even a universal sealant can be used to seal more complicated materials and structures without any problems.

However, the truth is such that this sealant is only an exception and, as has been mentioned already in the introduction to this paper, every combination of the proposed materials, a sealant and concrete structure in this case, should be tested at least in the way presented here before its application.

References

1. D.C.W. Ho, Property Management **11** (3), 6 (1993)
2. S.H. Ng, P.K. Kam, R.W.M. Pong, J. Environ. Psychol. **25** (3), 14 (2005)
3. D.J. Stobbelaar, B. Pedroli, Landscape Research **36** (3), 19 (2011)
4. D.C.W. Ho, Y. Yau, Ch. Law, E. Luisman, Urbani Izziv **23** (1), 15 (2012)
5. H.O. Nigro, S.E. González Císaro, Cities **56**, 6 (2016)
6. J.C. Beech, A.T. Wolf, *Durability of Building Sealants* (Routledge, 2013)
7. J. Šlanhof, M. Šimáčková, P. Liška, B. Nečasová, *Proceedings of Construction Technology and Management (CTM 2014)*, 9 (2014)
8. J. Šlanhof, B. Nečasová, P. Liška, M. Šimáčková, AMM **824**, 8 (2016)
9. Sika CZ, s.r.o., Product data sheet. Available on http://sika.cz (2017)
10. B. Nečasová, P. Liška, J. Šlanhof, Proc. Eng. **109** (2017)

MATEC Web of Conferences **146**, 03001 (2018) https://doi.org/10.1051/matecconf/201814603001
Building Defects 2017

Interior surface materials as sources of indoor hygiene defects

Ingrid Juhasova Senitkova[1,*]

[1]Institute of Technology and Business in České Budějovice, Department of Civil Engineering, Okružní 517/10, 37001 České Budějovice Czech Republic

Abstract. The building design and material selection has to respect the environmental requirements of indoor air quality and indoor hygiene parameters. As the first stage of building design is the most important for final indoor air quality more and more by clear constructional architecture the benefit of environmental safety, good indoor air quality for health and productivity are dominant. Nowadays the interior surface materials have received greatest attention as sources of indoor air pollution. The major pollutants in indoor air are volatile organic compounds emitted from materials and building products. Volatile organic compounds, or VOCs, are among the most complex and troubling indoor air pollutants. Manufactured and synthesized products often release large quantities of VOCs, some of the compounds they release, such as benzene, styrene, formaldehyde, and toluene, may be irritating, toxic, or even carcinogenic. VOCs emitted can become attached to other surfaces in the space, especially fabrics, and then be re-emitted over time. The impact of interior surface materials on air quality and indoor hygiene is discussed within the paper. The building design concerning to heath of occupants and hygienic performance regime can be realized only by interdisciplinary team of professionals respecting the environmental building and indoor design.

1 Introduction

Buildings are designed to provide shelter from the outdoor elements and to provide a safe and healthy indoor environment at work and home. If a building fails, the results can be structural damage, poor indoor air quality, health complaints, high absenteeism, and poor worker productivity, damage to belongings and manufactured goods, and costly remediation. The indoor environment may influence the health, comfort and productivity of occupants in buildings. Comfort can be defined as that condition of mind which expresses satisfaction with the environment. Health and comfort in the built environment is a multidisciplinary issue involving various specialists who bring together expertise on health, comfort, air quality and environmental monitoring and control. Environmental quality of indoors is the conceptual system principle of architectural, constructional and environmental design development.

* Corresponding author: 14667@mail.vstecb.cz

Environmental safety building is determined with mass and environs aspect. Mass aspect is determined by the system Building-Material-Subject and environs aspect is influenced by Building-Fluid-Subject system. Environmental factors are determined by operative perceived comfort and safety criteria of indoor physical, chemical and biological compounds. Sustainable building design is doubtless task of mass and energy flow reduction. The mass flow reductions mostly consist in material recovery. The problems with indoor air quality occur when energy flow reduction works together with changed material base composition and used services technology became dominant. There is no contradiction between indoor air quality and energy performance, but the more mass is changed the more energy flow is needed [1].

Growing attention is being paid to indoor environment as one of the main health and well-being factors. Indoor air quality problems may arise from one or more of the following causes: indoor environment - inadequate temperature, humidity, lighting; excessive noise, indoor air contaminants - chemicals, dusts molds, fungi, gases, vapors, odors and insufficient outdoor air.

Many of the materials used in buildings, either as interior materials or as furnishings, are the main sources of indoor air pollution in addition to those caused by humans and their activities and HVAC systems. The important source is the indoor surface materials of building itself, which in many cases can result in great amount of indoor volatile organic compounds (VOCs). It is especially building material finishes (floors, walls and ceilings) which are considered to have an important role also of perceived indoor air quality [2]. The research project is focused to indoor air sciences especially to perceived air quality observation focused on selected chemicals occurrence in public buildings have shown that mainly office buildings have high level of pollution related to interior materials [3].

Also hygiene problems may mainly develop as a result of emissions from sources inside the building such as emissions from building materials, furniture and equipment. Additionally lack of outdoor air for HVAC system and defective HVAC systems as well as the pollutants present in the outdoor air entering the building may also cause the indoor air quality problems.

2 Nature and source of indoor chemicals

Chemical contaminants of the indoor air can occur as gases and vapours (inorganic and organic) and particulates. Their presence in the indoor environment is the result of entry into the building from the outdoor environment or their generation within the building. The relative importance of these indoor and outdoor origins differs for different pollutants and may vary over time. The major chemical pollutants commonly found in the indoor air are the following: organic compounds, carbon dioxide, carbon monoxide, nitrogen oxides, sulphur dioxide and ozone.

Carbon dioxide (CO_2) is a metabolic product and often used as an indicator of the general level of air pollution related to the presence of humans indoors Carbon monoxide (CO), nitrogen oxides (NO_x) and sulphur dioxide (SO_2) are inorganic combustion gases formed predominantly during the combustion of fuels and ozone (O_3), is a product of photochemical reactions in polluted atmospheres but may also be released by some indoor sources. Elevated levels of pollutants generated by combustion, particularly nitrogen dioxide and carbon monoxide in indoor spaces, usually result from unvented, improperly vented or poorly maintained combustion appliances and the smoking of tobacco products. Unvented kerosene and gas space heaters emit significant quantities of CO, CO_2, NO_x, SO_2, particulates and formaldehyde. Gas cooking stoves and ovens also release these products directly into the indoor air. Under normal operating conditions, vented gas-fired forced air heaters and water heaters should not release combustion products into the indoor air. However flue gas spillage and back drafting can occur with faulty appliances when the room is depressurized by competing exhaust systems and under certain meteorological conditions.

The indoor research studies in various types of buildings were carried out within the research projects. The investigation of carbon dioxides, nitrogen oxides, particles and volatile organic compounds concerning to building materials, furnishing, occupancy and other indoor operation sources were the main subjects of the interest. Present results (Fig.1) confirmed the highest priority of chemicals especially volatile organic compounds which are closely associated with odors and characterize the occurrence of odorants in indoor air. Odors are perhaps the most pressing of all air quality problems. The presence of odors in air may or may not represent a health risk, but no other air pollution problems will provoke such an immediate response. The most risk group of buildings are school buildings (the air quality was under basic hygiene level) and office buildings (the highest level of pollution related to interior materials). The perceived air quality assessment only by indicator of occupant's presence (CO_2) has shown as insufficient. Objective assessment should be at least extended of TVOC concentrations [4].

Fig. 1. Indoor air chemicals ranking [4].

Organic compounds originate from a variety of indoor sources and outdoors. Hundreds of organic chemicals occur in indoor air although most are present at very low concentrations. It is generally known that these can be grouped according to their boiling points and one widely used classification, shown in the Table 1, identifies four groups of organic compounds: very volatile organic compounds (VVOC), volatile (VOC), semi-volatile (SVOC), and organic compounds associated with particulate matter (POM). Particle-phase organics are dissolved in or adsorbed on particulate matter. They may occur in both the vapour and particle phase depending on their volatility. For example, polyaromatic hydrocarbons (PAHs) consisting of two fused benzene rings (e.g., naphthalene) are found principally in the vapour phase and those consisting of five rings (e.g., benz(a)pyrene) are found predominantly in the particle phase.

A major pollutant in indoor air are volatile organic compounds (VOCs) emitted from materials and building products [5]. Definition by U.S. EPA is: VOC means a hydrocarbon or derivative of hydrocarbon that has a vapor pressure greater than 0.01 kPa (0.002 psi) at a temperature of 20°C and pressure of 102.9 kPa (14.7 psi). Excluded compounds are methane, ethane, methylene chloride, 1,1,1-trichloroethane (methyl chloroform), trichlorofluoromethane, dichlorodifluoromethane (CFC-12), chlorodifluoromethane (CFC-22), trifluoromethane (FC-23), trichlorotrifluoroethane (CFC-113), dichlorotetrafluoroethane (CFC-114), and chloropentafluoroethane (CFC-115) [6]. Indoor air pollution by VOCs is a particularly variable phenomenon: type and concentration of VOCs may change from one indoor space to another and concentrations may vary between different locations within the same space.

Table 1. Organic compounds classification.

Description	Boiling range (°C)
Very volatile organic compounds (VVOC)	<0 to 50-100
Volatile organic compounds (VOC)	50-100 to 240-260
Semi volatile organic compounds (SVOC)	240-260 to 380-400
Organic compounds associated with particulate matter or particulate organic mater (POM)	380

Building materials can have a major impact on air quality and can affect occupants, especially sensitive ones including children, the elderly, and patients who are immune-compromised or have respiratory problems. Careful selection of materials along with proper ventilation, operation, and maintenance can improve air quality.

Classification of health effects of indoor air pollution is possible to introduce from comfort point of view. The cause of sick building syndrome remains unknown, although it is likely that there are connections between psychological factors, inadequate ventilation and the level of hygiene in the building. Factors identified as possible causative agents are briefly summarized in Table 2.

Table 2. Selected indoor pollutants and health effects [7].

Pollutants	Source	Health effects
CO_2	Humans	SBS
VOC_s	Interior materials	SBS, acute risk, chronic risk
CO	Combustion	acute risk - respiratory, neurology
NO_x	Combustion	Acute risk - respiratory
O_3	Laser printers	Acute risk - rhinitis
Particles	Materials	Chronic risk - cancer

3 Volatile organic compounds measurements

Concentration of total volatile organic compounds (TVOC) expressed as concentration of toluene was measured with ppbRAE 3000, which is a photoionization detector (PID) with UV lamp. This device has a measuring range of 1 ppb to 10000 ppm and measurement accuracy ±3%. Three-second response time allows real-time monitoring with this instrument. Ultra-fast electronic nose called zNose®, which is based on combination of gas chromatography and surface acoustic wave detector was used for sampling and analysis of individual VOCs. Manufacturer of this device specifies the standard deviation < 2%.

Building materials and interior surfaces release a wide range of other VOCs which have been the subject of increasing. The emission can be a complex mixture of individual compounds, though a few may be dominant. A study of selected building materials identified different chemical species. These VOCs were primarily aliphatic and aromatic hydrocarbons, their oxygen derivatives and terpenes. The compounds with the highest steady-state emission concentrations, in decreasing order, were toluene, m-xylene, terpene, n-butylacetate, n-butanol, n-hexane, p-xylene, ethoxyethylacetate, n-heptane and o-xylene. The complexity of emission has resulted in indoor concentrations often being reported as the total volatile organic compound (TVOC) concentration or release. Table 3 gives examples of rates of TVOC emission for a range of interior building materials. These show that significant differences in emissions exist between products, which mean that if adequate data were available materials could be selected at the planning stage to minimize the VOC release in newly constructed buildings.

Table 3. TVOC concentrations associated with various floor and wall coverings.

Typ of materials	**Concentrations (mg/m^3)**
wallpaper – vinyl, paper	1.24
wallpaper – printed paper	0.87
Hessian	0.36
Wallpaper –textile	2.11
Linoleum	4.95
PVC flooring	19.12
soft plastic flooring	3.75
Laminate flooring	5.27
Wooden flooring	0.74
Carpeting	2.89
Cork flooring	0.56

Many VOCs are present in the indoor environment at concentrations ranging from approximately 2 to 20 mg/m^3. Building materials make a significant contribution to indoor concentrations and new homes are likely to have a greater number of compounds exceeding 100 mg/m^3. Renovation and painting contribute to significantly higher VOC levels. Concentrations of compounds such as ethyl acetate, 1,1,1-trichloroethane and limonene can exceed 20 mg/m^3 during occupant activities, and during residents' absence the concentration of a range of VOCs may decrease by about 50%. Specific cases of elevated concentrations of contaminants due to materials and furnishings being associated with occupant complaints have been described. These include white spirit from injected damp-proof courses, naphthalene from products containing coal tar, ethylhexanol from vinyl flooring and formaldehyde from wood-based products.

4 Perceived indoor discomfort

In the case of perceived discomfort in public buildings, the questionnaire as subjective assessment tool includes characteristics covering age, gender, type of work and type of work. Questions also ask about users' perceived comfort related with physical factors in the indoor environment covering temperature, indoor air quality, lightning and noise; including users' assessment of productivity in relation to perceived factors. Important part of the questionnaire includes questions related with tested buildings negative factors affected primarily psychical well-being of users. Occupants would then ask about other unknown factors with impact on their well-being, sick building syndrome symptoms and work performance.The SBS symptoms reported after one hour long testing procedure had significantly increased occurrence with decreased ventilation rate. The participants noticed mainly symptoms as dry nose, headache, fatigue, decreasing in reaction speed, inability to concentrate. Several seldom reported symptoms were also noticed. Percentage of individual SBS symptoms occurrence is presented in the Figure 2. Classification of health effects of indoor air quality is possible to introduce from acceptability point of view. The cause of sick building syndrome remains unknown, although it is likely that there are connection between psychological factors, inadequate ventilation and the level of hygiene in the building. Factors identified as possible causative agents are briefly summarized as temperature, humidity, light, noise and odors.

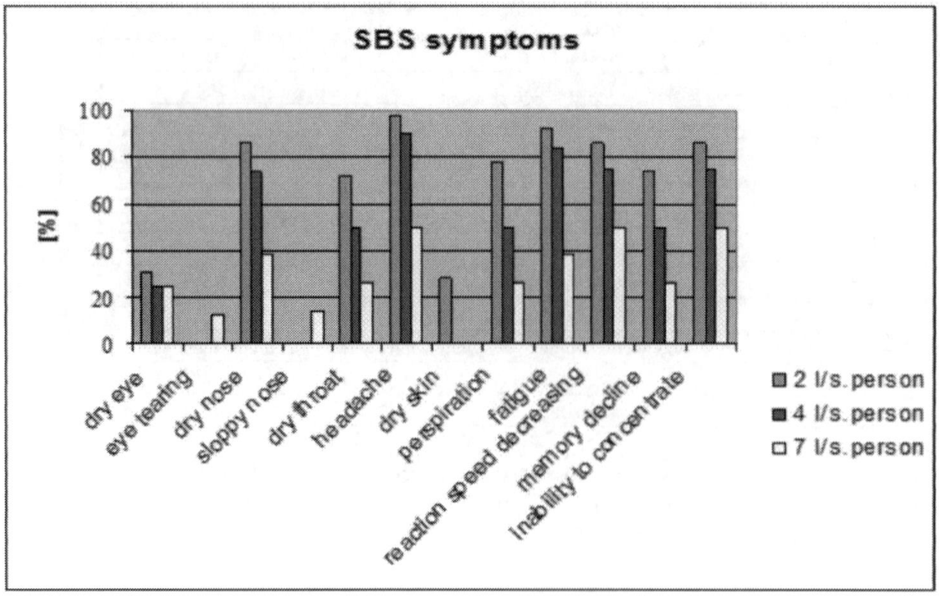

Fig. 2. Percentage of individual SBS symptoms occurrence [8].

5 Conclusions

Even all indoor environmental standards are met the users are usually not satisfied and perceived discomfort is occurred in the smart buildings. The most frequently cause of discomfort in smart buildings is overrun of intelligence. There are physical and psychological factors that influenced building users' comfort. Poor indoor air quality is believed to be responsible for a substantial part of the sick building syndrome that has been reported, but the indoor environment is extremely complex and other factors also undoubtedly contribute to the problem. Provided an environmental and economic will within our society will became aware of the real needs of solving the healthy buildings problem – we are able to offer the models and strategies the tools, and criteria, and finally the construction technology and an architectural idiom to design, to build, and to maintain a sustainable buildings on short and long term as well as affordable. A health impact assessment and an environment impact assessment are needed to judge the real quality of a building and particularly indoor design.

The author is grateful to national grant agency for supporting the indoor air sciences research projects.

References

1. I. Juhasova Senitkova, MATEC Web of Conferences **93**, 03001 (2017)
2. I. Senitkova, *Material Science Engineering C* **36,** 6 (2014)
3. I. Senitkova, T.Tomcik, *Advanced Science Letters* **19**, 9 (2013)
4. I. Senitkova, Selected Scientific Papers **2**, 35-48, (2007)
5. O.Wilke, O. Jann, D. Brödner, Indoor Air **14**, 98-107, (2004)
6. United States Environmental Protection Agency. SIP – NWAPA580
7. I. Senitkova, M. Bucakova, *Selected Scientific Papers* **3**, 45-56 (2008)
8. I. Senitkova, M. Badida, ISEC-7, 1-6 (2013)

Design without thermal bridges

Irina Maltseva[1,*], *Aleksandr* Elokhov[2] , *Konstantin* Tkachuk[1] , and *Kseniia* Maltceva[1]

[1]URFU, Department of Civil Engineering, 620002 Yekaterinburg, Russia
[2]Institute of Passive house, Production Department, 117405 Moscow, Russia

Abstract. The main focus of the article is on the quality design and implementation of the thermal insulation of buildings. Shells of buildings consist not only of conventional structures (wall, roof, ceiling), but also of edges, corners, joints and places of disruption of the outer thermal insulation shell integrity (due to through passage of pipelines, ventilation ducts and etc.). In all these places, heat losses are generally increased in comparison with conventional surfaces. Some of them are easily taken into account by calculation, using a special method proposed by the authors. Other thermal bridges due to unfavorable structural details can be avoided by observing the rules and recommendations that are classified and discussed in detail in this article.

1 Introduction

At the present time, the efficient use of energy is a worldwide concept. Many countries, even those with large reserves of natural resources such as natural gas, oil, etc., have realized the importance of preserving their resources so that they can be used for a longer period. Energy consumption is a priority problem in the domestic and world construction. Energy is consumed for various purposes, for example, heating hot water, using household appliances, but the dominant are the costs of heating in winter and air conditioning in the summer, which is usually about 70% of the total energy consumption. Most of this energy is obtained from fossil fuels, and its consumption leads to the corresponding emissions of pollutants. In most cases, this can be avoided through the implementation of energy efficient technologies and energy conservation measures. To this end, the EU has developed a program to increase energy efficiency by 20% until 2020. Various studies [1] confirm that this goal is achievable. Currently, the main direction of Russia's government policy are: conducting energy audits, developing energy efficiency standards for certain types of products, revising old standards and adopting new standards for existing and newly constructed buildings for various purposes, promoting energy efficiency principles [2]. In this regard, the issue of building energy-efficient buildings in Russia is becoming one of the key, and the problem of rational use of energy resources is becoming increasingly important. An innovative and very strict policy in terms of energy consumption in buildings for various purposes (for example, the Directive on the construction of buildings with zero energy consumption) has contributed to the development of high-level competence in the

* Corresponding author: i.n.maltceva@urfu.ru

development of energy-efficient buildings [3]. Today, there is an opportunity, both in Russia and in European countries, to design and construct buildings with a sufficiently low energy consumption, which is achieved by a correct solution of the thermal envelope of buildings and their ventilation systems. In the article we will consider the rules and recommendations for the design of thermal protection of buildings.

2 Thermal protection of building

The main principles of good thermal insulation are based on the need to determine closed thermal (heat-insulating) shell covering of the comfort zone. This insulation should be high quality and fit tightly without gaps and openings. It is also necessary to avoid complex forms of the outer thermal insulation of the building [4,5]. All rooms, which temperature in winter should be + 21 ° C, are inside this shell, which is interrupted only in the places where the windows are installed and has high thermal insulation characteristics throughout the circuit. The minimum thickness of the insulation, with a coefficient of thermal conductivity $\lambda = 0.04$ W/(m·K), in the Urals region should be of about 200 mm in every point of the thermal insulation shell. In most cases it is desirable even to improve the heat insulation of opaque external building elements up to $R_0 = 10$ m^2·K/W, which corresponds to the equivalent thickness of the effective insulation about 400 mm. In the future, more and more vacuum panels will be used, with the use of which a high thermal insulation effect is possible with relatively small wall thicknesses. Complex designs must be avoided: they are not practical and mainly expensive. Thermal insulation shell can be designed to be practical and very simple[6]. Observance of the four rules given below helps to reduce the heat loss caused by thermal bridges [7,8].

The first rule is avoiding thermal bridges is to exclude opening holes in the outer shell of buildings unless extreme necessity.

The second rule is the laying of thermal insulation. If it is not possible to avoid holes in the heat-insulating layer, then it is necessary to maximize the heat transfer resistance of the enclosure as a whole, for example, to use porous concrete in combination with mineral wool insulation.

The third rule is tight junction (for joints). The location of the insulation in the joints of building elements and structures must be without empty spaces, i.e. joint must be completely isolated.

The fourth rule relates to the geometry of the building: it is necessary, if possible, to choose edges with obtuse angles (more than 90 °).

Due to careful design and conscientious implementation, it is possible to significantly reduce the heat loss caused by thermal bridges.

3 Thermal bridges

This title hides a very important concept for any civil engineer, as well as a person who wants his house to be comfortable and cozy. Any erected building, especially in our Ural region, should be warm. Thermal bridges are such areas in the building structure, through which there is a large heat transfer. Shells of buildings consist not only of conventional structures (wall, roof, ceiling), but also of edges, corners, joints and places of disruption of the outer thermal insulation shell integrity (due to through passage of pipelines, ventilation ducts and etc.). In all these places, heat losses are generally increased in comparison with conventional surfaces. When designing the thermal insulation shell of energy-efficient buildings, it is necessary to carefully develop and calculate all the nodes at the design stage and during the construction.

When choosing materials for the thermal shell coefficients ψ (psi) and χ (chi) are controlled. Their values (in external dimensions) should be ψ <0.01 W/(m·K) and χ <0.01 W/K, or the following condition must be fulfilled: the sum of the specific transmission heat losses through the outer shell of the building without thermal bridges should be less than the amount of losses with taking into account the thermal bridges. That means, the sum of all heat losses from thermal bridges is less than zero. This is quite possible, because all areas of the filler structures for calculation in PHPP are taken from the outside dimensions.

Specific transmission loss

$$H_T = \sum f_j A_j U_j + \sum f_j \Psi_j l_j + \sum f_j X_j \quad (1)$$

Where:
$\sum f_j A_j U_j$ = specific transmission loss through the outer shell
$\sum f_j \Psi_j l_j$ = specific transmission loss through linear thermal bridges (TB)
$\sum f_j X_j$ = specific transmission loss through point TB

Design without thermal bridges

$$H_{norm} = \sum f_j A_j U_j \geq H_T \quad (2)$$

The appearance of particular thermal bridges can hardly be avoided: for example, increased heat loss along the outer edge of the building. These and other so-called "geometric" thermal bridges can be easily taken into account by using the "Passive House Planning Package" (PHPP) program, which allows to fully account for heat flows through all constructions and structural layers of the thermal insulation shell [9]. According to the PHPP method, heat losses are calculated for the external surfaces of the corresponding structures. Losses should be taken into account for linear heat transfer coefficients ψ > 0.01 W/(m·K), and it is possible to conduct a full account of all heat losses.

To do this, there are coefficients that take into account the heat loss through linear thermal bridges and relate to the external dimensions, ψ. They can be entered in the "Areas" worksheet, which is in the PHPP program (Fig. 1).

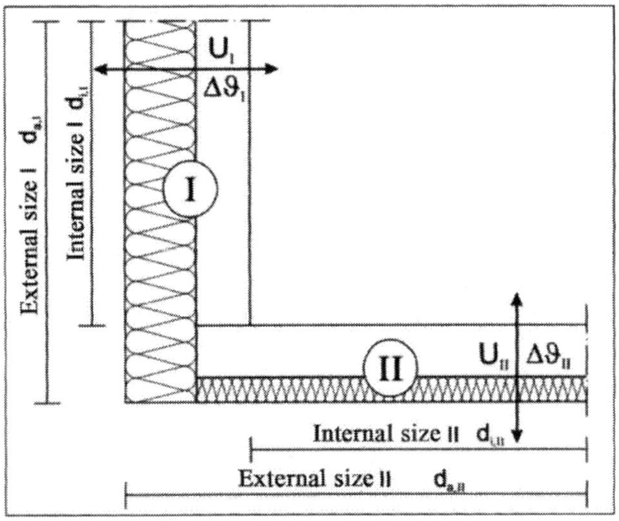

Fig. 1. Dimensions of building elements for calculation with taking into account thermal bridges.

Other thermal bridges are caused by unfavorable structural details: for example, if the balcony plate is not thermally separated from the concrete ceiling (it is better to use a separate foundation for the balconies), the additional heat loss of this "cold rib" becomes significant. The most unreliable joints are in connections between the wall with ceilings and roofing, esconson of windows, places under the windowsills, exits to the balcony or veranda. Such constructive thermal bridges in the thermal insulation shell should be avoided, especially in passive buildings. In addition to poorly connected material, the cause of "cold bridges" may be insufficient thickness of the thermal insulation layer.

Thermal bridges in the existing building lead to:
- the decrease in indoor temperature during the cold seasons and overheating of premises in hot seasons;
- the formation of condensate on the inner surface of the filler structure;
- the significant increase in the expenditure of resources for heating the house;
- the increased probability of dampness and mold on the inner walls of the house (Fig.2);
- the development of harmful bacteria, and this is dangerous for human health.

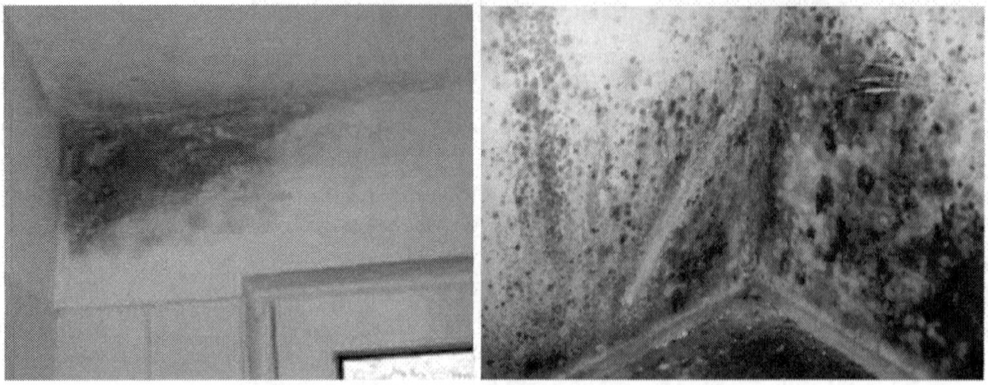

Fig. 2. Dampness, fungus and black mold on walls and ceilings.

It is very hard to remove mold. In most cases, changing the wallpaper or repainting the wall does not lead to anything, mold gets out and again takes its place. Treatment with special addings is needed to completely remove the mold from this place. And the cold bridges themselves are usually removed by implementation of additional thermal insulation. To control the mold and dampness it is important to look for problem areas, check the internal surfaces to understand where there is a heat outflow. Thermal imager helps a lot to check this. On the thermogram in the corners there is an intense heat outflow that cools the wall and air inside the room (Fig. 3).

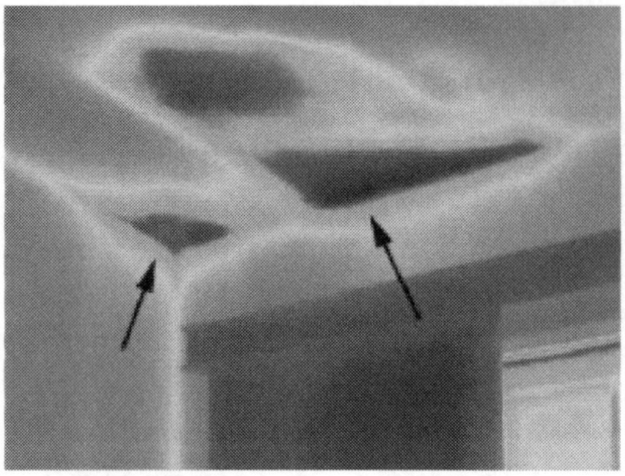

Fig. 3. Thermogram of the joint of the wall and the ceiling, obtained with the thermal imager.

Such buildings need reconstruction and, most importantly, the elimination of thermal bridges. The main thing is to construct continuous thermal insulation shell. This requirement seems obvious, but in fact it is not always fulfilled when building a house. And for energy-efficient buildings, it is of particular importance, since the formation of a large number of linear thermal bridges calls into question the whole idea. To determine the level of criticality of each thermal bridge, special calculations are made (Fig. 4).

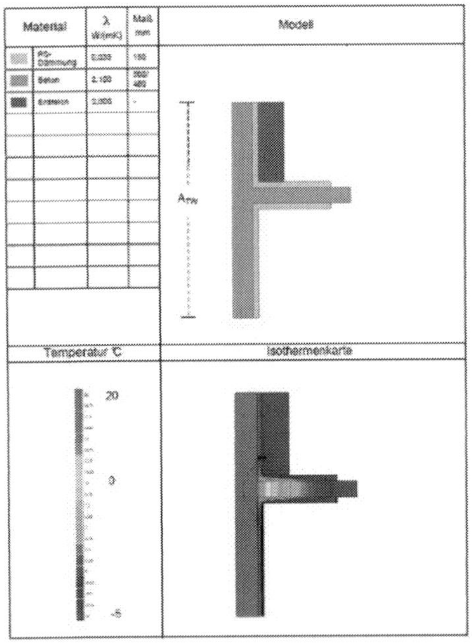

Ergebnisse			
Bezugstemperaturdifferenz des Wärmedurchgangskoeffizienten	Δθ	30	K
linearer Wärmedurchgangskoeffizient	Ψ	0,305	W/(mK)
minimale Innentemperatur bei 5°C/-5°C	θ_{min}	19,4 / 18,4	°C

Stairs wall to the basement ceiling;
insulation on the external side
(10x200 cm)
10_KD-TW_insulation outside

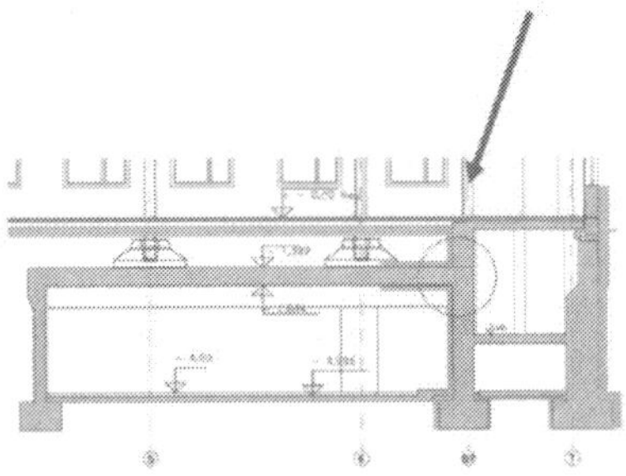

Fig. 4. An example of calculation of a geometric thermal bridge during the reconstruction of an old building according to the passive house standard.

4 Conclusion

To sum up, it is necessary to emphasize that the construction of new energy efficient buildings or the reconstruction of existing without thermal bridges begins at the design stage. The basis is a complex approach. All parameters are determined by the calculation method «Passive House Planning Package» (PHPP). Closed thermal insulation shell is obligatory element throughout the entire contour of the building with an increased thickness of insulation, especially in the Ural-Siberian region of Russia. Certified materials designed specifically for the passive house allow to ease the work of designers and protect them from unnecessary errors [10]. Most manufacturers of building materials and products, as well as producers of engineering equipment, have developed together with the staff of the Passive House Institute a whole base of certified goods [11]. Following these recommendations, we will improve the quality of people's lives, we will get multi-comfortable human habitat without harmful bacteria, fungus and mold.

References

1. V. Bianco, F. Scarpa, L.A. Tagliafico, Energy Conversion and Management **87**, 754–764 (2014)
2. Roadmap. EU-Russia Energy Cooperation until 2050 (March 2013)

3. T. Maltby, Energy Policy **55**,435-444 (2013)
4. V. Gagarin, ABOK, 1-3 (2009)
5. S.G. Sheina, D.K. Upennikov, Internet-journal "NAUKOVODENIE" **7**, 3 (2015)
6. M.M. Brodach, N.V. Shilkin, VIII International Scientific Conference «Walls and facades. Actual problems of thermophysics»: collection of materials, 191-196 (2003)
7. V. Feist, *Basic provisions for the design of passive houses (Translation from German with additions edited by A.E. Elokhov),* KONTI PRINT, **144** (2015)
8. R. Perminov, I. Maltceva, N. Kaganovich, A. Yelohov, AMM **725-726**, 1389-1394 (2015)
9. Available on http://www.passiv-rus.ru
10. Schöck Design Guide: Solutions to Prevent Thermal Bridging (June 2014)
11. Passive House Database. Available on http://www.passivhausprojekte.de/index.php?lang=en

Analysis of interaction of the soil temperature field and structures at the level of the basement of the building

Petr Kacálek[1,*], and *Tomáš* Petříček[1]

[1]Brno University of Technology, Faculty of Civil Engineering, Institute of Building Structures, Veveří 95, 602 00 Brno, Czech Republic

Abstract. The article describes an analysis of the effect of soil type on temperature field distribution at the level of family house basement. Regarding engineering calculations (heat loss calculation, designing and evaluating constructions), a simple and generally accepted method to determine the course of temperature field adjacent to basement walls is missing. The methods specified in European standards are, in terms of calculation of heat flows in soil, problematic. There is currently an insufficient number of measured values, which could be integrated in a simple calculation method. A certain option concerns continual measuring of temperatures in soil for buildings in situ. The article uses the measurements of soil temperatures in winter in pre-determined depths at the building parts in contact with soil. The measured data from a two-year period were integrated in the model on the basis of statistical evaluation. The main aim of the article is to generate an impulse to create unified outcomes of temperatures of certain soils for pre-selected depths, building structures, and parameters of indoor and outdoor environment.

1 Introduction

The distribution of temperatures in soil under buildings is currently relatively little examined field of civil engineering physics, even though it is closely related to accurate calculations of heat technology.

Despite being different from air temperature, soil temperature reflects long-term trends in air temperature, and the temperature differences on the surface are then transferred deeper. Soil has relatively low heat conductivity, reacts slower to climatic conditions, and its underground reaction to changes in external effects is also slow. The temperature of soil is different from air temperature; in most cases the soil temperature is higher, and their difference then depends on precipitation conditions, snow cover, vegetation type and changes over time, soil quality, heat accumulation, etc.

Other factors that affect soil temperatures include well designed surrounding building structures. They particularly concern correctly selected heat insulation, waterproofing [1]

* Corresponding author: kacalek.p@fce.vutbr.cz

(locally much used bituminous sheets), and in many cases well designed details (at the level of building entrances [2], well made waterproofing joints [3], etc.).

The faults in waterproofing have negative effect on heat conductivity coefficient of selected material; subsequently, the heat conductivity coefficient fails to meet the selected standard requirements of the designed construction material. Building constructions must be repaired (using direct or indirect methods) and the whole process results in changes in the temperature field of the evaluated structure [4].

The distribution of temperatures in soil in civil engineering is particularly used for a calculation of specific heat flux of buildings with and without cellar.

2 Options for modelling of soil temperatures at buildings

The real measuring of soil temperature of buildings is a not an easy task. Such measurement can generally be performed only with devices that allow recording of temperatures in certain depths. However, this is not always the case and, therefore, in the majority of cases we need to make an estimate of how high the temperatures in soil at a building can be.

The designed soil temperatures, or estimates of soil temperatures in °C, adjacent to vertical or horizontal building structures in winter are shown in Table 1.

Table 1. Designed soil temperature in winter, adjacent to building structures according to ČSN 73 0540-3 [5].

Location of adjacent soil layer		Temperature of adjacent soil in °C at designed ambient air temperature	
		$\theta_e > -15$ °C	$\theta_e \leq -15$ °C
at vertical wall	up to depth of 1 m	- 3	- 6
	in depth of 1 to 2 m	0	- 3
	in depth of 2 to 3 m	+ 3	0

The estimates of temperatures for structures adjacent to soils are allowed according to the existing valid standard ČSN 73 0540, particularly regarding the calculation of specific heat flux H_T (estimate of temperature is a part of temperature reduction coefficient b for structures in contact with soil).

Variant estimate of temperatures of soil under buildings is also provided by standard ČSN EN ISO 10211-1 Thermal bridges in building construction – General calculation methods [6].

This article describes the issue of temperature field of soils around basement part of a building. The main aim of the article focuses on extending the knowledge in this issue.

The practice part (Fig. 1) consists of measurements of a real building in 2007 – 2009. Based on the statistically evaluated data and existing legislation, a theoretical simulation in Ansys software was made which simulates the real measurement at the building. As a part of the simulation, the trend in temperatures at the interface of soil and building structures is shown together with the application of different boundary conditions.

3 Experimental measurement

The experimental measurement was performed for a real family house construction (hollow core bricks CD INA-A, pre-wall CPP). The construction is on the northeast side, along the whole length, partially under the terrain; there is a downslope towards the west. The surroundings include a rolling terrain, the lowest point is 328 m above sea level, and the highest point of the locality is 524 m above sea level.

The following values were monitored during the experimental measurement on the real structure (internal air temperature, relative humidity of internal air, external air temperature, relative humidity of external air, height of snow cover, soil temperature in pre-selected depths, soil heat conductivity coefficient, pre-wall heat conductivity coefficient, external walls heat conductivity coefficient). [7]

Fig. 1. View of probes placed in soil at family house structure.

4 Measurement and evaluation of measured values

Values were recorded continuously without interruption for two years. The device recorded temperatures every 15 minutes. The setting of this time interval gives us sufficient overview of individual temperatures. In order to provide clearer overview of statistical files, individual boundary conditions are represented by temperature median values. The external and internal environment will be characterized by the median values of external air temperature + 2.1 °C and median values of internal air temperature + 11 °C shown in Table 2.

Table 2. Soil temperatures (sandy clay) [7].

	Median values of external air temperature + 2.1 °C		
	Depth of measurement (m)	Arithmetic average (°C)	Temperature median (°C)
Median value of internal air temperature + 11.0 °C	0.40	7.53	7.90
	0.80	9.29	9.30
	1.20	9.45	9.70
	1.50	10.71	10.60

5 Model of temperature field of foundation soil based on measurements at real structure

If we are interested in temperatures on building structures in contact with soil, it is necessary to take into account effects of more-dimensional heat conductivity in soil under the structure. In that case we need to create a model and determine the temperature distribution by numeric solution of more-dimensional temperature field by the network method or finite elements method. [7]

The simulation model shown in Fig. 2 is based on the values measured during an in-situ survey of the building. The modelling considered all values as isotropic material.

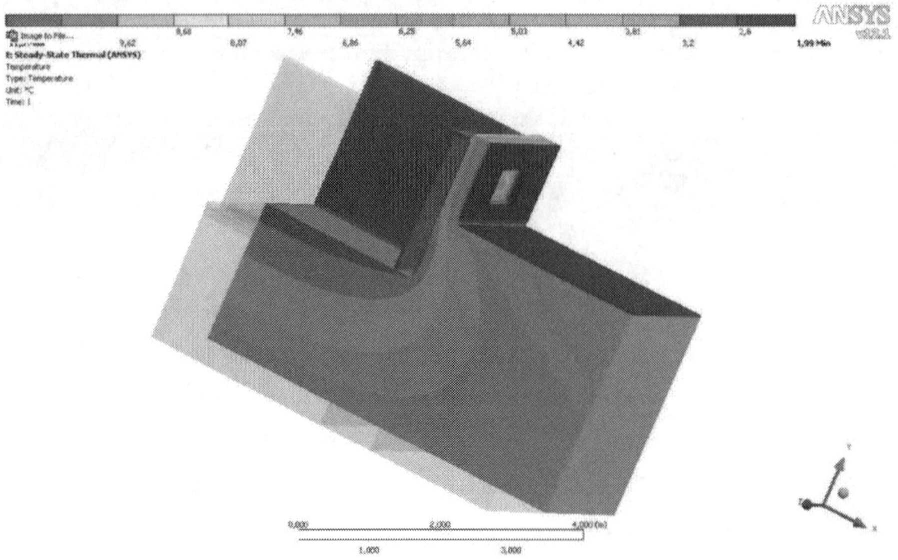

Fig. 2: Model of a part of structure.

6 Influence of indoor air temperature changes on temperature field in the basement structures

The simulation takes into account the design standard temperatures of internal as well as external air while using the model from the measuring and modelling of the real structure.

External environment will be characterized by a temperature field with design temperature of – 15 °C, design internal temperature for winter environment of + 21 °C and + 10 °C.

Tables 3 and 4 show temperatures in selected depths for individual soil types.

Table 3. Design temperature of external air - 15 °C, design temperature of internal air + 10.0 °C; (* (W/m.K)) – value in brackets for coefficient of heat penetration through given soil.

	Design temperature of external air – 15 °C				
	Measurement depth	**Sandy clay (2.76; 1.68) real model**	Soils and clays (1.5)	Dry sand (2.0)	Wet sand (2.7)
	(m)	(°C)	(°C)	(°C)	(°C)
Design internal air temperature + 10.0 °C	0.40	**4.22**	4.32	4.80	5.39
	0.80	**6.07**	6.12	6.42	6.08
	1.20	**7.40**	7.42	7.65	7.92
	1.50	**8.10**	7.08	8.33	8.61

Table 4. Design temperature of external air - 15 °C, design temperature of internal air + 21.0 °C; (* (W/m.K)) – value in brackets for coefficient of heat penetration through given soil.

	Design temperature of external air – 15 °C				
	Measurement depth	**Sandy clay (2.76; 1.68) real model**	Soils and clays (1.5)	Dry sand (2.0)	Wet sand (2.7)
	(m)	(°C)	(°C)	(°C)	(°C)
Design internal air temperature + 21.0 °C	0.40	**8.89**	9.43	9.39	9.41
	0.80	**10.85**	11.19	10.91	10.70
	1.20	**11.76**	12.02	11.68	11.39
	1.50	**11.86**	12.04	11.84	11.66

7 Conclusions

The existing results of temperatures in soil at the structure show that the main effect on soil temperature is internal air temperature, where the underground building parts transfer substantial amount of heat to the soil. As mentioned above, the model and the simulation were based on a real building structure. It is necessary to bear in mind that the structures were not insulated and could not have been used as sufficient insulators. The influence of sufficient insulation in the interface structure – soil on soil temperature can be subjected to another experiment.

It is not currently apparent from the measurement what role is played by snow cover and vegetation in the resulting temperature. Good quality of the waterproofing layer of the substructure may have a substantial effect on the resulting temperatures [1, 3].

It is obvious from closer comparison of Tables 3 and 4 with Table 1 that the temperatures in the depth up to 1 m and temperatures from 1 m to 2 m differ significantly.

Experimental and theoretical simulations are used to offer us a better idea of soil temperatures around structures. After verifying the given measurement at other structures within several years, it is partially possible to make the existing values of adjacent soil temperatures more accurate.

This paper has been worked out under the project No. LO1408 "AdMaS UP - Advanced Materials, Structures and Technologies", supported by Ministry of Education, Youth and Sports under the „National Sustainability Programme I".

References

1. J. Plachý, V. Petránek, Z. Caha, AMR **897**, 103-106 (2014),
2. R. Smolka, J. Sobotka, J. Šulistová, MATEC Web of Conferences **93**, 01009 (2017)
3. T. Petříček, P. Kacálek, T. Hlavačka, MATEC Web of Conferences **93**, 03007 (2017)
4. J. Sobotka, R. Smolka, MATEC Web of Conferences **93**, 01008 (2017)
5. *CSN 73 0540-3 Thermal protection of buildings - Part 3: Design value quantities* (Czech office for standards, metrology and testing, Prague, 2005)
6. *CSN EN ISO 10211 (73 0551) Thermal bridges in building construction – Heat flows and surface temperatures – Detailed calculations* (Czech office for standards, metrology and testing, Prague, 2009)
7. P. Kacálek, T. Petříček, R. Smolka, AMR **889**, 58-61 (2014)

Risk assessments of contemporary accidents in construction industry

Michal Kraus[1,*]

[1] Institute of Technology and Business in České Budějovice, Department of Civil Engineering, 370 01 České Budějovice, Czech Republic

Abstract. The contribution deals with the risk assessments of contemporary accidents in construction in the Czech Republic. The risk analysis is defined as a systematic and structured data processing to identify potential hazards. The risk is a quantitative or qualitative expression of threat. A number of accidents and human errors are characteristic of the construction industry. A zero risk does not exist. An analysis of 615 incidents is presented which identified the location within the building sector in the Czech Republic. The risk analysis including the Risk Assessment Matrix (RAM) is based on secondary data from the Construction and Technical Prevention System (CTPS) in the Czech Republic from 2010 to 2016. The risk level is determined by the likelihood of occurrence and consequence of defects, faults, and accidents in the construction industry.

1 Introduction

Construction is the process of constructing a building or infrastructure carried out by workers. This process is always associated with certain risks. Due to the rapid development of infrastructure and building sector, the level of risk from the building sector is rising in the Czech Republic. All human actions carry some level of risk or error. There is no zero risk [1, 2]. The building can be defined as all construction works that are built by building or assembly technology, regardless of their construction, building materials, and structures, for purpose of use and duration. Terms such as defects, fault, snag, failure, accidents, and crashes are commonly used. There are a number of definitions of these terms, see [3-6]. Building defect can be characterized as a lack of construction due to incorrect design, inadequate or imperfect execution decreasing the performance of structures [1]. The fault can be defined as a change from the original state.

2 Risk management

Risk management is a discipline used in all social and technical disciplines, such as natural, engineering, political, and economic or social sciences [7]. Risk management is an integral

[*] Corresponding author: info@krausmichal.cz

part of today's sustainable construction. Risk management is systematic process included risk identification, risk assessment, risk control and prevention. Risk assessment includes quality methods and quantitative methods [2].

2.1 Risk assessment matrix

Risk assessment of incidents can be done using a so-called risk matrix. Risk assessment matrix (RAM) is a key tool for risk management. The risk matrix is two-dimensional. Two factors are evaluated - the likelihood and the consequences. Assessment is based on experts with the necessary knowledge and experience. The general risk assessment matrix is shown in Figure 1. According to the ISO 31 000:2009, the likelihood or probability is a chance to something might happen. The consequence is the result of an activity or action that affects the outcome [2, 8, 9].

Obviously, the higher the likelihood of its occurrence and the higher severity of the consequences is, the risk is more significant. The most significant risks are listed in the upper right quadrant of the matrix. Vice versa, the least significant risks are in the lower left quadrant of the matrix. Generally, there are 4 groups of risk - low, moderate, high and extreme.

Table 1. General Risk Assessment Matrix (RAM) [2].

Likelihood	Consequences				
	Insignificant (1)	**Minor (2)**	**Moderate (3)**	**Major (4)**	**Catastrophic (5)**
Almost certain (5)	Low (5)	Moderate (10)	High (15)	Extreme (20)	Extreme (25)
Likely (4)	Low (4)	Moderate (8)	High (12)	Extreme (16)	Extreme (20)
Moderate (3)	Low (3)	Moderate (6)	Moderate (9)	High (12)	High (15)
Unlikely (2)	Low (2)	Low (4)	Moderate (6)	Moderate (8)	Moderate (10)
Rare (1)	Low (1)	Low (2)	Low (3)	Low (4)	Low (5)

3 Incidents in the construction industry

Figure 1 illustrates the serious accidents in the Czech Republic from 2010 to 2016. The basic statical data are taken from the Construction and Technical Prevention System (CTPS) in the Czech Republic [10]. The Construction and Technical Prevention System collects, monitors and analyses serious or repeated defects in construction. A repeated defect is defined by the third occurrence of an identical or similar defect or defect of the same origin.

The Building Law [11] imposes an obligation to immediately notify the occurrence of a defect, failure or accident of a building if there is a loss of life of persons or animals, a threat to life or the health of persons or animals, a threat to the safety of a building or significant property damage (more than 18 000 Euros). Other incidents are not monitored.

Six hundred and fifteen incidents are analyzed in this study. The type of incident, region, and year of origin, building class, functional unit, and severity are monitored. Incidents are divided into 6 basic groups:

- Carbon monoxide poisoning,
- Explosion,

- Fall from height, caving or injured animal,
- Fall from height, caving or personal injury,
- Fire,
- Structural collapse.

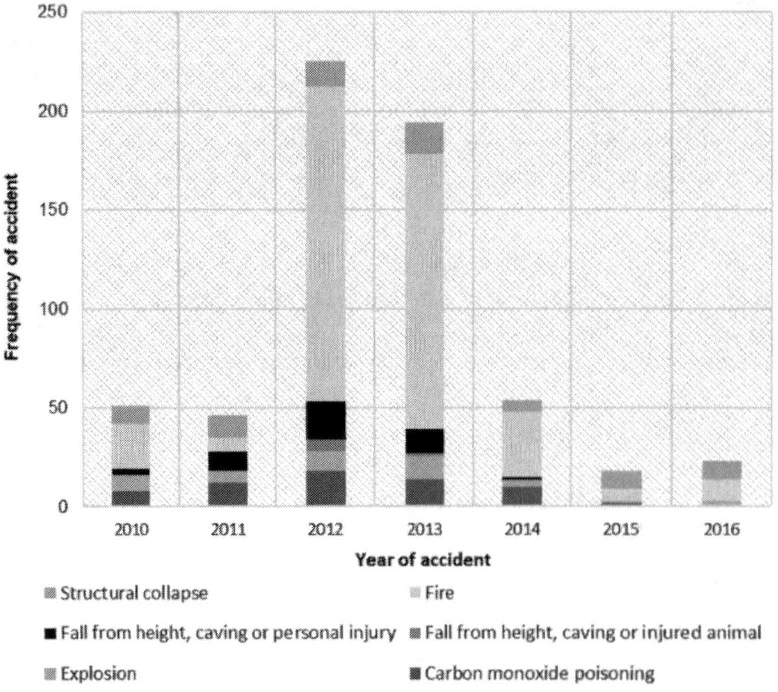

Fig. 1. Overview of accidents by year.

The most serious incidents are monitored in 2012 and 2013. Obviously, the most common incident is a fire. The main causes of fires are the negligent conduct of building users as well as technical faults of functional units of constructions and building equipment [1].

Table 2. Fires by type of building [12].

Buildings	**2010**	**2011**	**2012**	**2013**	**2014**	**2015**	**2016**
Buildings of civil construction	588	794	783	665	478	551	547
Apartment houses	1848	1752	1737	1627	1588	1619	1553
Family houses and other residential buildings	2087	1658	1697	1629	1605	1524	1847
Buildings and halls for production and services	324	339	334	324	324	373	354
Energy production building	79	99	85	98	86	83	97
Garage facilities	141	122	121	121	118	116	22
Buildings for storage	139	145	128	91	85	108	114
Buildings for agriculture	85	84	78	75	80	50	42
Objects under construction	41	61	47	44	33	40	44

Table 2 illustrates fires by the type of buildings. Data are extracted from the Statistical Yearbooks of the Fire Rescue Service of the Czech Republic from 2010 to 2016 [12]. These data include all fires, not only serious ones.

3.1 Incident severity level

The severity of incidents is divided into five groups from least severe (1) to very severe (5):
- Considerable damage (1) without injuries and deaths both animals and people. These incidents are characterized by material damage exceeding € 18,000 (the equivalent of 500,000 Czech crowns).
- Threat to lives and health of animals (2)
- Loss of life of animal (3)
- Threat to lives and health of people (4)
- Loss of life of people (fatality) (5).

According to the figure 2, more than 50% result in personal injury (37%) or death (14%). Almost 50% of the incidents can be classified in the first degree of severity. Injury to animals or their death is minimal.

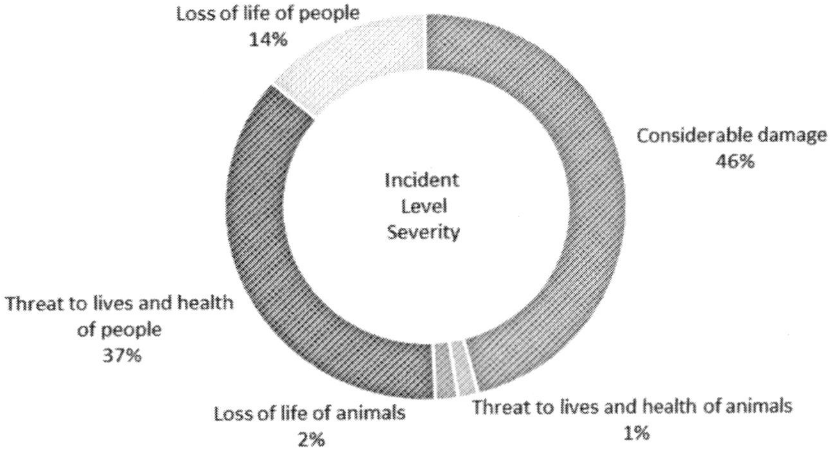

Fig. 2. Incident level severity.

3.2 Risk matrix

The risk level is a value of a risk derived from its likelihood and consequence. The scoring (multiple of the likelihood and the severe) reflects the priority and urgency of adopting measures to reduce risk (Table 3).

The likelihood of an incident (p) is defined as the proportion of the frequency of the type of incident and the number of all incidents monitored. The severity (s) is expressed by:

$$s=\Sigma(s_i \cdot n_i)/n \quad (1)$$

where s is the severity of the type of incident [-],
s_i is the degree of severity of the incident (1-5) [-],
n_i is the frequency of incidents for the degree severity [-],

n is the frequency of the type of incidents [-].
Table 4 shows the results of the risk analysis. In term technical viewpoint, the risk R3 is insignificant and acceptable. This risk can be ignored. The risks R4, R5, and R6 are regarded as moderate. They are not high-priority. It is necessary to pay attention to risks R1 and R2. A particularly R2 represents an extreme risk. This risk has extremely negative consequences and it is the highest priority.

Table 3. Risk assessment table of incidents of contemporary constructions.

Risks		**Likelihood** (1-5)	**Consequence** (1-5)	**Overall Risk** (1-25)	**Risk level**
R1	Carbon monoxide poisoning	3 (p = 0.106)	4 (s =4.185)	12	High
R2	Explosion	2 (p = 0.067)	4 (s =3.610)	8	Extreme
R3	Fall from height, caving or injured animal	1 (p = 0.016)	2 (s =2.000)	2	Low
R4	Fall from height, caving or personal injury	2 (p = 0.072)	4 (s =4.409)	8	Moderate
R5	Fire	5 (p = 0.619)	2 (s =2.124)	10	Moderate
R6	Structural collapse	3 (p = 0.119)	3 (s =3.026)	9	Moderate

Table 4. Risk Assessment Matrix of incidents of contemporary constructions.

Likelihood	**Consequences**				
	Insignificant (1)	**Minor (2)**	**Moderate (3)**	**Major (4)**	**Catastrophic (5)**
Almost certain (5)		R5 Moderate			
Likely (4)				R2 Extreme	
Moderate (3)			R6 Moderate	R1 High	
Unlikely (2)				R4 Moderate	
Rare (1)		R3 Low			

4 Conclusion

The risk is perceived as a measure of the risk of injury or the occurrence (in extreme case to death) or financial damage. According to the semi-quantitative risk assessment, the results clearly indicate the significance of the incidents – defects, faults, and accidents. The fire, structural collapse and personal injury have a moderate degree of risk. The most frequent incidents are fires of buildings or building elements. On the contrary, the smallest probability of occurrence is the fall from a height, caving or injured an animal. The risk of injury or death of animals is low. Extreme risk is the explosion. Carbon dioxide poisoning has a high risk. These risks have a high priority and it is necessary to pay attention to these risks and to minimize them.

References

1. M. Kraus, T. Vondráčková, V. Nývlt, Matec Web of Conferences **93**, 03004 (2017)
2. M. Kraus, P. Bednářová, K. Kubečka, AMM **824**, 657-664 (2016).
3. Act No. 183/2006 Coll., on town planning and the building code (the Building Act)
4. J. Alencastro, A. Fuertes, P. de Wilde, Renewable and Sustainable Energy Reviews **81**, part 1, 883-891 (2018).
5. N. Forcada, M. Macarulla, M. Gangolells, M. Casals. Build Res Inf **42**, 629-640 (2014).
6. M. Macarulla, N. Forcada, M. Casals, M. Gangolells, A. Fuertes, X. Roca. J Constr Eng Manag **139**, 968-976 (2013).
7. R.E. Hester, R.M. Harrison. *Risk Assessment and Risk Management* (Royal Society of Chemistry, 1998)
8. A.M. Doro-on. *Risk Assessment and Security for Pipelines, Tunnels, and Underground Rail and Transit Operations* (CRC Press, 2014)
9. Institute for Spatial Development. Construction and Technical Prevention System. Available on https://www.uur.cz/?id=948
10. Smartsheet. *All the Risk Assessment Matrix Templates You Need.* Available on https://www.smartsheet.com/all-risk-assessment-matrix-templates-you-need
11. *Law no. 183/2006 Sb on Planning and Building Code (Prague, 2006)*
12. Fire Rescue Service of the Czech Republic. *Statistical Yearbooks 2010 – 2016.* available on http://www.hzscr.cz/hasicien/article/statistical+-yearbooks.aspx

Environmentally safe system for treatment of bio corrosion of ETICS

Katarína Minarovičová[1,*], and *Dušan* Dlhý[1]

[1]Department of Building Constructions, Slovak University of Technology in Bratislava, Faculty of Civil Engineering, Radlinského 11, 810 05 Bratislava, Slovakia

Abstract. Effects of microorganisms on building facades are responsible for aesthetic, bio-geophysical and biogeochemical deterioration. The process of cleaning of contaminated facades involves the removal and eradication of micro flora on the surface of insulation using chemical products, killing cells and eliminating all living mass, including organic ingredients. The removal of bio corrosion coating from ETICS structure by means of chemical and preservative substances (biocides) is currently the only effective and most used technology. Uncontrolled leaching of the used biocides is unacceptable. Meantime, new technology for environmentally safe maintenance of ETICS is needed. Scientists have been working on replacement the biocides currently used in facades treatment with eco-friendly biocides that have no negative effects on the environment or human beings. While the eco- treatment will be available, safe dewatering of chemicals being leached from the surface of the facade could be provided by e.g. special drain systems adjusted to the building type, use and age. The paper gives an overview of the problem in context of Slovakia and examples of leaching systems designed for new and renovated buildings.

1 Introduction

World Commission on Environment and Development (WCED) understands the concept of sustainable development as development that meets the needs of the present without compromising the ability of future generations to meet their own needs. This requirement is contrary to uncontrollable and rising demand for chemical products in construction practice and maintenance of buildings.

Maintenance of bio corrosion of existing External Thermal Insulation Composite Systems (ETICS) is mostly based on chemical disposal of microorganisms. It means removal and eradication of micro flora on the surface of insulation using chemical products, killing cells and eliminating all living mass, including organic ingredients. Fact is, that removal of algae coating from ETICS structure by means of chemical and preservative substances (biocides) is currently the only effective technology [1].

* Corresponding author: katarina.minarovicova@stuba.sk

Maintenance and protection cycle of the ETICS related to bio-corrosion resistance is then proposed in the interval of 3-6 years, depending on other factors. The expected life of ETICS is 25-30 years (ETAG 2004) so this treatment should be repeated [2].

Fig. 1. Typical façade with bio corrosion. Deteriorated areas reflect climate conditions of the environment (direction of prevailing winds, defects in plumbing, sealing etc.). Bratislava, Lamač. Author of a picture: K. Minarovičová.

Studies in the field of ETICS colonization by microorganisms have been oriented to assess the significance of colonization [3-4]. The chemicals used for microorganisms´ elimination on the facades, either as bounded, encapsulated in pigments and plaster or loose, un-encapsulated agents, reducing their development and growth, containing mainly toxic heavy metal compounds, are based on the basis of active silver. Their durability in applications is controversial; some research results show the opposite, bio accumulative effect and the negative impact of toxicity to the environment. This situation is unsustainable [5]. Biocidal products can pose risks to humans, animals and the environment. As a result, the EU has set up strict rules and procedures to ensure a high level of protection for human health, animal health and the environment.

European legislation on chemicals has changed in recent years and four regulations have been introduced to ensure the free transfer of chemicals in the European Union and a high level of protection for human health and the environment [6]:

• REACH - Regulation on the Registration, Evaluation, Authorization and Restriction of Chemical Substances;
• CLP - Regulation on Classification, Labelling and Packaging of Substances and mixtures;
• BPR - Biocidal Products Regulation;
• PIC - Prior Informed Consent Regulation (concerning the export and import of hazardous chemicals)

The Biocidal Product Regulation (BPR, Regulation (EU) 528/2012) concerns the placing on the market and use of biocidal products, which are used to protect humans, animals, materials or articles against harmful organisms, like pests or bacteria, by the action of the active substances contained in the biocidal product. This regulation aims to improve the functioning of the biocidal products market in the EU, while ensuring a high level of protection for humans and the environment. It repealed the Biocidal Products Directive (Directive 98/8/EC).

In Annex V to the BPR the biocidal products are classified into 22 biocidal product-types, grouped in four main areas. The most used products in the building industry are preservative agents PT7, PT8, PT10 and PT12 - in group- preservatives. There is recorded more than 1000 types of biocides.

2 Sustainable ETICS maintenance

There are 2 basic types of technologies for removal of microorganisms [7]:

1. Elimination of microorganisms by biocides, which is used in combination with subsequent washing of the cleaned surfaces with power water,
2. Application of biocides, contaminated surface should be dry cleaned if needed or washed with pressurized water before application of biocides.

When using both technologies it leads to rinse off chemically active substances to the environment at an early stage during the washing of dead residues of biofilm from the surface of the facade, followed by washing by rainwater or rinsing as a long-lasting washing of the façade surface.

Maintenance of ETICS may be operative or planned; in practice they are often combined. Operational disposal is in fact high-efficiency decontamination according to required technological procedures. The planned maintenance is related to a professional planning approach. It should be based on the ETICS manual of use, which contains information on the maintenance intervals and the renewal of the surface biocide protection [8].

3 Sustainable leaching systems

The concept of a safe drain of chemicals leaching from the facade must take into account building structural system of the building, perimeter wall construction, the purpose of the building, arrangement of house surroundings, financial and technological context of bio corrosion elimination. Drain system must be designed with regard to the amount of water required to wash the facade (for regular cleaning of the facade biocide solution) or on the amount of precipitation water in a given area (spontaneous leaching of biocides) [9-10].

The collected water must be piped into collection containers, respectively tanks; they will be transported for processing or disposal in accordance with applicable legislation.

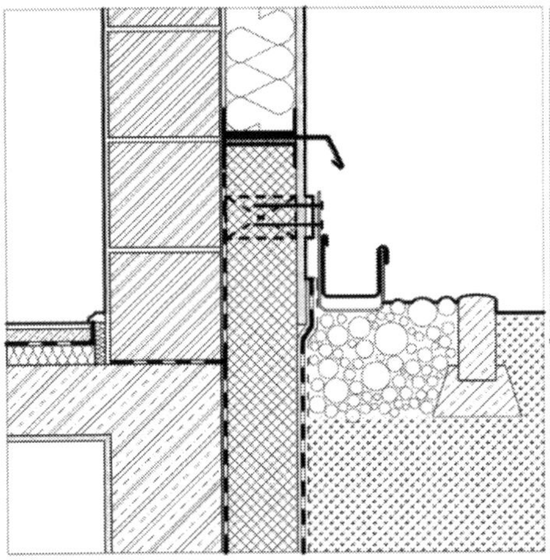

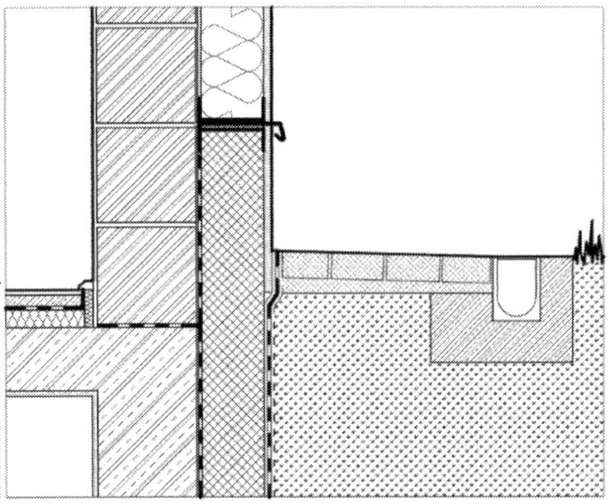

Fig. 2. Gutter leaching system (temporary structure). **Fig. 3.** Line drainage channels (permanent).

Leaching system may be applicable as:

a) special biocide-resistant foils at the level of the perimeter sidewalk, the problem can be the collection of contaminated water from these areas,

b) flat collection mechanism - like in the previous case - under the surface of the thermal insulation (e.g. underpass)

c) gutter at the plinth profile of thermal insulation structure (installation of hooks carrying gutters sloping into the collecting containers. During construction it would be necessary to prepare profiles for the anchoring of the hooks, that will be settled during regular cleaning and the contaminated water should be collected in collecting receptacles and prepare for further processing - Fig. 2),

d) linear drains around the house - at the level of the surrounding terrain (the line drainage channels around the house that remain part of the house surroundings with the possibility of collecting the water-contaminated water should be collected in collecting vessels and prepared for further processing or through filtering and decontamination shafts the system will be connected to the sewer system, shafts or drain into the surrounding green - Fig. 3). This system ensures the permanent collection of chemically active substances washed from the surface of the insulation and is part of the house pavement. The installation of line drainage or filtration drains is part of planned maintenance.

4 Conclusions

Drain construction system will have to adapt to the design of the facade and to its close surroundings. Uncontrolled leaching of the used biocides is unacceptable. We have to develop technology, which controls the release of biocides into the environment, especially in soil and water. Research on reducing the use of biocides by supplementation with enzymes and bio dispersants is in progress. It is needed to replace the biocides currently used in facades treatment with eco-friendly biocides and have no negative effects on the environment or human beings.

The "polluter pays" principle is becoming a common policy (Directive 2004/35/CE of the European Parliament and of the Council of 21 April 2004 on environmental liability with regard to the prevention and remedying of environmental damage) [11-12]. On the other hand, we believe that if the current trend is back to the nature, maybe in this area it would be easier to replace the effort to regulate biological organisms on ETICS by the desire to exploit their natural characteristics in the built environment.

There is a wide range of materials used for ETICS finishes. It is important to take into account reliability and maintainability of the construction in the whole life cycle of a building. Operation and maintenance should be a significant element of the life cycle cost of a building.

The paper partly originated within the framework of the Scientific Grant Agency of VEGA 1/0685/16 project.

References

1. V. Helmuth. *Fassadensanierung, Praxisbeispiele, Produkteigenschaften, Schutzfunktionen* (Practical guide, Berlin 2011)
2. N. Antošová. *Selected technologies for protection and maintenance of ETICS.* (Brno, Tribun EU, 2011)
3. W. Hofbauer, K. Breuer, K. Sedlbauer. Bauphysik **6**, 25 (2003)
4. K. Breuer, et al. Bauphysik **34**, 170-182 (2012)
5. K. Wrede, et al. *European Coatings Conferences Novel biocide technology V*, (2011)
6. ECHA. Available on https://echa.europa.eu
7. N. Antošová. *Analysis of the knowledge of causes and technologies of ETICS biocorosion treatment and a model of their resistance* (STU Bratislava, 2014)
8. Palková, M. Palko. *Progressive construction materials 2015/2* (STU Bratislava, 2015)
9. K. Minarovičová, N. Antošová. AMM **820**, 194-199 (2016)
10. K. Breuer et al. Bauphysik **34**, 19-23 (2012)
11. CCHLP. Available on http://www.cchlp.sk
12. OECD. Available on http://www.oecd.org/env/ehs/pesticides-biocides/biocidesprogramme.html

Building objectives and incorrect use of regulatory standards

Miroslav Šůra[1,*]

[1]CTU Prague, Faculty of Civil Engineering, Thakurova 7, 166 29 Prague, Czech Republic

Abstract. European and other regulatory standards affect, besides everyday life, also planning, designing, and other building processes. Regulations alter, sometimes threaten, or even completely prevent, the achievement of the intended objectives of investors and other participants in building processes. Implementation of regulative standards may result in problems that cause various defects to planning-design-construction objectives. Such regulatory issues are known to have already caused ethical, social and financial burdens in other areas of human activity. In planning and construction processes, various participants may be the source of misuse problems – unintentionally or otherwise. A model of categories of problems is presented (and has been tested) – of misuse problems that result from potential conflicts between (EU) regulatory standards and objectives of participants in building processes. The categories have been compiled and compared also according to sources and actors of the problems. Protective strategies against incorrect application and/or misuse of regulations are in the process of formulation and verification, taking into account sources and actors in the problem categories. Finally, it helps and protects participants – especially as regards matters of technology, design phase, public attitudes, and money.

1 Introduction

We try to describe and categorize possible threats that may appear in association with application of regulations imposed on building processes. The concern was mainly focused on categorizing threats and problems related to regulations that are voluntarily adopted by (some of) the various building process participants. Compulsory regulations, however, are apt to raise threats that can be classified within the same categories, as stated below. We focus on external ("non-technological") threats and problems, originating in surroundings of the building processes—related to participants of any type: law makers, developers, investors, owners of products, adopters or applicators, etc.

To identify categories of the problems and threats, analogies found in various other areas were explored to this end, too. Finally, this builds the base for suggested ways how to avoid or resolve such problems and threats.

* Corresponding author: sura@fsv.cvut.cz

1.1 Questions, Quest & Examples

Building activities require a lot of intellectual work and other work (such as administrative or clerical work) in order to get the desired results. Further, such work is being performed within an environment of huge mass of juristic and other regulations. Can everyone understand and conduct the work adequately, with equal insight and comprehension of the objectives?

Where do the problems originate, what are the threats? Threat of unintentionally acting improperly, without full understanding of the objectives, using misleading interpretation, or in extreme cases acting with intention to misuse regulations and/or additional restrictions for side-effects—these are just some examples.

2 Implementation phase threats

Each party in a building process [1] may expect different effects. Participants [2] directly engaged in the project introduction obviously share mutually closer and more accurate views on (adopted) regulations than those joining the building process only in its later stages. Possible causes of this threat: insufficient experience, unrealistic expectations. If only partial information is available to some of the participants, they might disregard such information—this danger may lessen as more experience is acquired by the various parties.

Some of the adverse financial effects of regulations—whether mandatory or adopted regulations—may be detected in a relatively short period of time, and these may be subjected to continual correction by removing the cause: by modifying impacts, by readjusting expectations, by completing information (database) etc. And of course, the threat of adverse financial effects may be reduced by training participants, in-depth analysis, preparatory work, etc.

Long-lasting adverse (not only financial) effects may substantially influence participants' negative attitude towards future projects. This might be avoided by not allowing accumulation of adverse effects, but dealing with regulation drawbacks quickly and peremptorily.

3 Intentions and actual constraints conflict

The threats arise from the possibility that some of the participants exploit building activities or implement building industry processes without regard to the actual regulations. Although other participants do their best and follow actual constraints imposed by these regulations, a single participant can degrade the quality of final results. Typically such a participant's activities are not closely linked with other participants in a fairly short time period; as a consequence such inadequate irregularities may stay undetected for quite a long time and violate the due constraints.

The common case is that part of the project implementation differs from the original project, and these differences are not properly coordinated with other processes—only some of the processes are modified both with respect to appropriate regulations and project objectives.

4 Threat of institutionalization

It is human nature to act in such a way as to achieve—subjectively—the highest possible (not necessarily financial) benefits. No wonder that participants in building process will tend to apply regulations so as to maximize their individual subjective benefits.

Any restrictions of state space available to a participant for his/her operations can only decrease his/her attainable benefits, never increase it. Such restrictions usually occur if some interest group has the ability to restrict the participant's set of available actions. One of or several cooperating participants may form such a group if he/she/it succeeds in taking over the control of the building process or of its part. Restrictions imposed on the weaker parties by such an interest group would not probably be very lasting if the weaker parties grow stronger as they gain more experience.

Long lasting effects of decreasing participants' benefits arise in case of interest groups that are usually external to building activities. Typical instance of such an external, institutional, interest group is legislation that can impose such regulations and constraints on building processes that reduce the manoeuvring space and disturb the balanced relations among participants. As a result, some participants are discriminated against others, which reduces their maximum achievable benefits. While the preferred participants collect their extra, unfair advantages.

5 Statutory interventions

This threat arises as one of the ramifications of 'good-doers' activities.

This threat is enabled by easiness of menial, unqualified, derogative and/or financially destabilizing interferences as practiced by authorities or subjects possessing delegated competences. These interferences may arise from actual or shamming statutory competences to impose regulations and constraints to building processes. Alternatively, the 'perpetrators of good' may push authorities to such interferences, and induce the above mentioned unfavourable ramifications in otherwise rewarding [3] building activities.

The goals of such interferences may be status build-up of the intervenor, pursuit for conveniences and benefits, or simply getting a backhander (racketeering). However, the most dangerous is probably one kind of good-doers—those ignorant enough to believe others to be incompetent and themselves to know better than most of the rest of humankind.

And again, building industry activities seem to be unprotected against this kind of threat. As a rule, to protect anything from 'perpetrators of good' is a rather difficult task.

6 Inconsistent perception of meaning

The terms (designations) and concepts in the context of unequal perception of meaning can be divided into two groups, with fuzzy dividing borders or with terms belonging to both groups. One of the groups contains terms used *within* building activities (professional terminology), the other group contains terms used in communication *about* building activities (including the lay public).

Inconsistent perception of meaning of building terminology should be rather fully eliminated in the long run—otherwise misunderstanding would be the source of various defects in the building process. The strict meaning of professional terms is consensually settled and the inexactness in communication disappears.

Inconsistent perception of meaning in communication *about* building activities is a different problem. This communication is performed mostly in natural language. Some inaccuracy in communication and unequal interpretation by various participants cannot be fully removed. This may easily cause unintended misunderstanding; such threat can be decreased by coordinated effort of participants, by reducing the inaccuracy in communication. Permanent threat lies in the fact that it enables intentional and managed misunderstanding—one or several participants, in their individual earnings efforts, deliberately seek for divergent interpretations and meanings of terms. Increased inexactness is used to induce unfavourable effects and defects, and this may end up in a court which is asked to explain the "proper" meaning.

7 Threat of redefinitions

This threat arises when regulations and constraints imposed on building project are interpreted by the participants having stronger positions so as to serve their own benefits [4, 5].

7.1 Threat of reinterpretation

One of participants, having a sufficiently strong position, may enforce changes in interpretation (meaning) of competences, goals, processes specifications, and so on, to depart it from customary standard interpretation of regulations and constraints. Such changes may adversely influence effectiveness of building processes as well as moral attitude of co-participants.

7.2 Threat of reformulation

One of participants, in a sufficiently strong position, may enforce changes in wording of competences, goals, processes specification, and so on, wording that is in contradiction to what has been previously agreed upon with respect to actual constraints and regulations. Likewise, this may also adversely influence effectiveness of building processes as well as moral attitude of co-participants.

8 Customs and traditions threat

Participants in building activities (planners, owners, investors, contractors, suppliers, facility management...) usually have gained experience and habits in their previous activities [6]. However, the environment, conditions and tenets of construction industry change rapidly, and to prevent participants from using insufficient and inadequate procedures, from using their old customs, beliefs or doctrines is ever more important. Neglecting this may degrade processes in the construction industry.

9 Legal requirements misuse – rent-seeking threat

This threat lies in setting up lobbies ('rent-seeking' groups) that push legal processes ending up in passing laws and regulations, constraints of which only the rent-seekers can comply with—thus creating effective monopolies or cartels. Or, these rent-seeking lobbies may transform existing legal requirements to suit their monopoly, enforce using their monopolistic claims, gadgets or tools, at the expense of others [4, 5]. This limits (temporary only, in the better case) or destroys the building process objectives.

This type of threats is a well-known phenomenon and is 'exploited' in various areas, currently by needless fees, charges or just 'protection money' in such 'popular' domains as fighting against 'global warming', or 'climate change', unfounded and superfluous 'environmental protection', social and community projects (rather regularly failing to achieve the declared goals, like social housing construction), unavailing and wasted away support of various 'minority' and 'disadvantaged' groups, and many others. Currently, building industry activities seem to be unprotected from this kind of misuse, too.

And yes, rent-seeking threat is often bound closely to statutory interventions threat.

10 Encumbering participants

This threat arises where authorized or accredited but corruptible body is in a position to (easily) generate superfluous requirements on participants. The goal of encumbering may be e.g. to prevent implementation of some objectives of a building project, in favour of the possibility to implement a competing project.

This threat seems to be similar to the one mentioned in 5 Statutory interventions, however it is not identical.

11 Introducing conflicts externally

One of the reasons why participants in a building industry project (sometimes) do not 'speak the same language' [7] is that they act at different areas or levels of detail.

The threat lies in the possibility that one of the participants is forced to acquire dominant position towards the others—forced by external factors or powers.

This threat also arises if the parties are forced, by external factors, to 'battle for dollars'—to compete for a share of a limited resource pool.

12 Case study

Real names are omitted to avoid lawsuits for slander, libel or defamation. Therefore, the following should be considered a "possibility" even though it is listed as "facts".

It is a case of a settlement/town, with rural character and population slightly—or substantially—exceeding 5.000. Local authorities were assigned a task to build a wastewater treatment plant (WTP). With regard to relevant regulations, three locations were selected as rather equally suitable—an old unused local industrial facility, and two of undeveloped greenfield sites.

The winning WTP project was chosen to be placed on the old unused local industrial facility site. In addition to statutory regulations, the choice was based on subsidy (by EU funds) and other financial affordability. This imposed additional restrictions—a claim for a gift from public money (additional subsidy money) for revitalizing parts of the old industrial facility, while for greenfield sites no such subsidy was available.

A few years later, due to growing population both on working and non-working members, another task was set—to build health and care centre (HCC) and sports and recreational facilities (SRF). These had to be built on one of the greenfields although the old industrial facility would be much more suitable if still available and not already used for the WTP. It should be noted there were only highly unclear visions of HCC&SRF at the time of WTP construction. But in retrospect, other uses for the old industrial facility could and should have been anticipated, and so it would have been better to swap the locations in case of WTP.

Some of the incorrect and disadvantageous use of regulations and constraints were identified as follows:

- Intentions and actual constraints conflict: although there are only unclear visions of HCC&SRF at the time of WTP construction, a better use for the brownfield could and should be foreseen.
- Threat of institutionalization: subsidies (claims for other people's money) perceived and handled in a restrictive manner (old industrial facility preference) is an inappropriate approach from a long-term perspective.

- Threat of reinterpretation: subsidies redefined to "recoverable advantage/money" discriminate the greenfield sites, adversely influence and cause defects of future projects, and effectivity and success thereof.
- Legal requirements misuse – rent-seeking threat: expanding WTP project to include and exploit subsidies highly discriminates the undeveloped greenfield sites as an option, but generates extra profit for the brownfield owners supporters (potential rent-seekers).
- Statutory interventions: working hand-in-hand with rent-seeking, this actually causes a synergistic effect eliminating undeveloped greenfield sites as options. But "gaining subsidies" is a tool for status build-up of bureaucracy and local politicians.
- Customs and traditions threat: The custom of yielding to restrictions/regulations without rational reconsidering, and the tradition of preferring brownfield to greenfield – both turn out to be counterproductive.

13 Conclusions

Incorrect use of regulations harms and causes defects in plans, building objectives and building process outcomes, and public attitude and view thereof. Being aware of these threats is the first step to avoiding them. Focusing attention of professionals and public on various sources of such threats has the potential to put (some of) these sources out of action. For this purpose, however, the threats need to be recognized in a real-world environment. We identify and define features of various threats and classify the threats according to such distinguishable features. Then the features and the corresponding threat can be discovered more easily in an actual case, and eventually the threat and its sources are successfully eliminated.

Support from various European projects and foundations is gratefully acknowledged.

References

1. B. Blackwell, *Industrial strategy: government and industry in partnership.* Available on https://www.gov.uk/government/uploads/system/uploads/attachment_data/file/34710/12-1327-building-information-modelling.pdf
2. Team-of-authors, *SMART (summer 2014).* Available on http://www.sps.cz/RDS/_PDFDoc_2014/Smart-Leto%20-2014.pdf
3. Legrand gr., *Green building–Ecological construction.* Available on http://www.legrand.com/EN/green-building-description_12850.html
4. Ch. W. Mills, *White collar: the American middle classes* (Oxford University Press, New York, 2002)
5. P. Hampl, *White collar revisited*, PhD thesis (Charles University, Prague, 2015)
6. S. Plous, *The psychology of judgement and decision making* (McGraw-Hill, Inc., New York, 1993)
7. T. Casey, *Residential Energy Efficiency: Taking It To The Next Level.* Available on https://greenbuildingelements.com/2017/05/31/residential-energy-efficiency-taking-next-level

Solution of problems of ETICS with biocorrosion

Barbora Beláňiová[1,*], *Naďa* Antošová[1], and *Lucia* Šupejová[1]

[1]Faculty of Civil Engineering in Bratislava, STU, Department of Building Technology, Radlinského 11, 810 05 Bratislava, Slovakia

Abstract. To find solutions for maintenance and repair of existing structures ETICS becomes the current another issue of improvement thermal insulation proprieties of external cladding, which need to be answered necessarily till December 2020. Double thermal insulation technology can be achieved by jointly increasing of thermal insulation properties and by repairing surface porosity ETICS. The aim of the research is the analysis of alternative insulation materials in plaster repair by technology "double thermal insulation" ETICS in terms of technology, humidity and temperature conditions in the layers. Technology of double thermal insulation is analyzed in the context with solution biocorrosion surface thermal insulation. In the end of the research there is obtained the data from the comparisson of technology double thermal insulation with different insulation materials expressed by graph. Obtained data can be the basis for decision to choosing a repair, to solution for the structural details of double thermal insulation and also for experimentally assess the lifespan of the double ETICS.

1 Introduction

In Slovakia the construction of contact insulation is used for approximately 25 years. During the time, various faults appear in older thermal insulation systems, but they do not have to threaten the general function of the structure. The presence of microorganisms in building environments, which according to the analysis threatens the construction from the inside, is also possible [13]. The most common visible thermal contact faults are cracks in each direction, springing-off and mechanical damage, separation of the surface layer from other layers, separation of the reinforcement layer together with the surface plaster from insulation, microorganisms contamination and biological attack of ETICS layers.

The cause of the biological attack of thermal insulation surfaces is not completely known. The research shows that acrylic and silicone plasters are more attacked. This material causes that the drying time of the surface of the plaster on the north and west sides is longer than the silicate plaster. The long-term wet surface of the plaster creates suitable conditions for biofilm formation. According to [13], certain types of microorganisms on the

[1] Corresponding author: barborabelaniova@gmail.com

surfaces of building materials produce mycotoxins and play their role within the so - SBS Building Syndrome, who is global problem. Micromycotes act by their physicochemical activities on building materials, including plasters, and degrade them.

Elimination of micro-organisms is the most often used for chemical surface cleaning, but it has a short lifespan and it needs to be repeated regularly. The construction of double thermal insulation could be a lasting solution to this problem. When performing a double thermal insulation, a modern plaster with modified properties that are resistant to microorganisms will be used on the surface. At the same time, meeting the increasing requirements on the Energy Efficiency of Buildings (EEB) by 2020. [2] However, there are several open and unresolved issues in addressing biocorrosion by double thermal insulation. One of them is the risk of the new colonization of "double ETICS" surfaces by microorganisms for using plasters of the original generation and also the threat of microorganisms development in the thermal insulation layer. Such a risk is real in ensuring long-term moisture at the construction site where microorganisms have been recorded on the original surface. The role of long-term research is to investigate the lifespan of the double structure in terms of bio-corrosion, with a new colonization of microorganisms on the surface or renewed colonization in the set of original and new thermal insulation.

2 Methodology of work

The research is divided into several parts. The first part describes the principle and rules of implementation technology of double thermal insulation. In the second part, the thermal-technical and moisture parameters of the double structure are examined for using an insulating layer made of EPS - polystyrene, MW - mineral wool and facades panel PUR - polyurethane foam.

The properties of the different insulating materials have been analyzed so that the co-operation condition with the duplicate system is met, which means that the new contact insulation must have the same properties as the original insulation. For comparison, one type of external cladding of a panel apartment building with the most prevalent type of thermal insulation and thickness of thermal insulation was selected. At the same time, the thickness of the new insulation layer was designed to meet the 2020 requirements. [2]

Table 1. Development of standardized requirement of heat transfer coefficient [2, 7].

Building construction	Boundary conductance U [W/(m^2.K)]			
	U_{max} from 1.2.1997	U_n from 1.1.2013	U_N=U_{r1} from 1.1.2016	U_N=U_{r2} from 1.1.2021
External Cladding	0,46	0,32	0,22	0,15
Structural floor under the outside	0,30	0,20	0,15	0,10

The third part evaluates the indicators of the individual double thermal insulation systems with regard to the possibility of moisture in the structure as a potential for development of the microorganisms in the layer. The results are the basis for the proposed methodology for examining the lifespan of the double thermal insulation structure from the perspective of biocorrosion.

3 Principle and rules of double construction of ETICS

The essence of double thermal insulation is the addition of another thermal insulation layer, reinforcement layer and new plaster, without deforming the previous layers. The double thermal insulation is currently classified as the usage of a conventional thermal insulation system under specific conditions where the original supporting structure and thermal insulation remained [1].

Before realization the construction of the additional ETICS, it is therefore necessary to consider the original structure in accordance with the applicable standards. The requirements for the underlying structure, the properties of the new external thermal insulation composite, the static load bearing capacity of the load-bearing structure and the static design of double insulation are assessed. In addition to other tests, the chemical and physical compatibility of inbuilt and newly applied materials have to be verified. In particular, the mutual compatibility of cement adhesives designed for bonding and dispersion plasters used for old insulation and, in case of inappropriate conclusions, to propose appropriate measures.

3.1 Original construction of ETICS

In the research and theoretical assessment, the panel construction system BA-BC, which was realized according to the type of "Prefabricated residential houses of Bauring-Camus T-BA-BC" panel system, was used. It is a typical representative of mass housing construction in Slovakia. The original construction of the solved problem is characterized as a box high-frequency carrier system. Constructional depth of floor is 2700 mm and the vertical clearance is 2500 mm. The external cladding is made of layered panels which are in the front side and gavel 260 mm thick. They consist of a 130 mm indoor reinforced concrete layer reinforced with a welded mesh of a thermal insulation layer, a 50 mm foam polystyrene and a 80 mm reinforced concrete layer. The concrete layers are interconnected by a space anchor with diagonals of anticorrosive steel Ø 8 mm and spacing needles Ø 2 mm [12].

4 Specification of ETICS doubled construction

It follows from the principles of double thermal insulation that the new insulating material must have the same or better diffusion properties as the original material. The fixation of the new thermal insulation layer of the ETICS is done by means of anchors which ensure the transfer of load in their axes by anchoring them into the supporting layer of the original structure. It is necessary to use anchors with a full-metal screw element, with a plastic spray, when doubling the thermal insulation layer [6]. The maximum thickness of the ETICS thermal insulating layer is limited. The minimum thickness of the new ETICS thermal insulation material is 50 mm. Thickness must be considered by thermo-technical assessment. For greater insulation thicknesses is double thermal insulation need to be taken into account by increased expansion movements, especially at the edges of the ETICS, on the corners [6]. Strength in tension perpendicular to the plane of the new ETICS thermal insulating material for EPS boards min. 100 kPa and in the case of MW boards min. 10 kPa. During realization of double structure it must by observed, the maximum weight of the thermal insulating material, including plasters and adhesives, have to be respected.

4.1 Alternative solutions of double thermal insulation

The realization of double thermal insulation was assessed in terms of thermal and technical properties of the structure and humidity regime in three alternative solutions. Alternative I involves the addition of thermal insulating material from EPS in hr. 80 mm. The bulk density of the polystyrene is 40 kg/m^3 and the diffusion resistance factor is 53.10^9 m/s. [8, 11]. In Alternative II, the thermal insulation properties of the mineral wool at a thickness of

50 mm were used in the double insulation. The bulk density of the mineral wool is 45 kg/m^3 and the diffusion resistance factor is 3.3 10^9 m/s. [10, 11]. In Alternative III, the thermal insulation properties of 40 mm polyurethane foam boards were used in double insulation. The bulk density of the PUR boards is 45 kg/m^3 and the diffusion resistance factor is 4.25 10^9 m/s. [9, 11]. The detailed composition of each of the double thermal insulation technologies compared is shown in the following table. 2:

Table 2. Composition of double thermal insulation in three alternatives [8-10].

no.	Alternative I	Thickness	Alternative II	Thickness	Alternative III	Thickness
	EPS - polystyrene foam	t	ISOVER TF PROFI - MIN. Wool	t	TPD-polyurethane boards	t
	Name of material	m	Name of material	m	Name of material	m
1	Lime-cement plaster	0.010	Lime-cement plaster	0.010	Lime-cement plaster	0.010
2	Ex. reinforced concrete panel	0.130	Ex. reinforced concrete panel	0.130	Ex. reinforced concrete panel	0.130
3	Polystyrene foam	0.050	Polystyrene foam	0.050	Polystyrene foam	0.050
4	Ex. reinforced concrete panel	0.080	Ex. reinforced concrete panel	0.080	Ex. reinforced concrete panel	0.080
5	Lime-cement plaster	0.010	Lime-cement plaster	0.010	Lime-cement plaster	0.010
6	WEBER.THERM EXCLUSIVE	0.004	WEBER.THERM EXCLUSIVE	0.004	WEBER.THERM EXCLUSIVE	0.004
7	Polystyrene foam	0.080	Polystyrene foam	0.080	Polystyrene foam	0.080
8	WEBER.THERM EXCLUSIVE	0.004	Reinforcing grid	-	Reinforcing grid	-
9	Reinforcing grid	-	Undercoating - WEBER 700	-	Undercoating - WEBER 700	-
10	Undercoating - WEBER 700	-	WEBER.PAS EXCLUSIVE	0.002	WEBER.PAS EXCLUSIVE	0.002
11	WEBER.PAS EXCLUSIVE	0.002	WEBER.THERM EXCLUSIVE	0.004	NEW-THERM ST04	0.004
12	WEBER.THERM EXCLUSIVE	0.004	ISOVER TF PROFI - MIN. WOOL	0.050	TPD-polyurethane boards	0.040
13	EPS - polystyrene foam	0.060	WEBER.THERM EXCLUSIVE	0.004	NEW-THERM ST04	0.004
14	WEBER.THERM EXCLUSIVE	0.004	Reinforcing grid	-	Leno fiberglass fabric R 117 A 1017	0.470
15	Reinforcing grid	-	Undercoating - WEBER 700	-	NEW-THERM® PGM	0.001
16	Undercoating - WEBER 700	-	Weber.MIN	0.002	NOVALITH - cover plaster (silicate)	0.002
17	Weber.MIN	0.002				

From the point of view of the distribution of the temperature field and the humidity regime in the thermal insulation, the values of the thermal field in the cross-sectional area and in the selected critical detail at the level of the window sill were considered and evaluated. The purpose of the assessment was to locate the moisture in the structure of the new structure as a possible source of water supply to residual microorganisms that may be located on the original thermal insulation construction.

Table 3. Thermal-technical and moisture parameters "Double ETICS" with EPS insulation.

The results of the calculation of the thermal-technical properties of the double ETICS (Polystyrene foam EPS - 60 mm)							
Thermal resistance of the structure		m².K/W		R =	5,36		
Diffuse resistance of the structure		10^9 m/s		R_d =	83,441		
Heat transfer coefficient		W/(m².K)		U =	0,19		
Internal surface temperature		°C		θ_{si} =	18,8		
Humidity mode		Water vapor does not condense inside the structure					
Construction design in terms of thermal-technical properties							**Evaluation**
Thermal resistance	R =	5,36	>	R_N =	4,4	Normalized value	PASS
Heat transfer coefficient	U =	0,19	<	U_N =	0,22	Normalized value	PASS
Risk of mold	θ_{si} =	18,8	>	$\theta_{si,80} + \Delta\theta_{si}$ =		13,10	PASS
Humidity mode		Amount of diffusing water vapor: G_d = 3.1239.10-8 kg/m².s					
		Water vapor does not condense inside the structure					
		Allowable amount of condensate:					PASS
		M_c =	**0,00 kg/(m².a)**	< M_{cmax} =	**0,50 kg/(m².a)**		

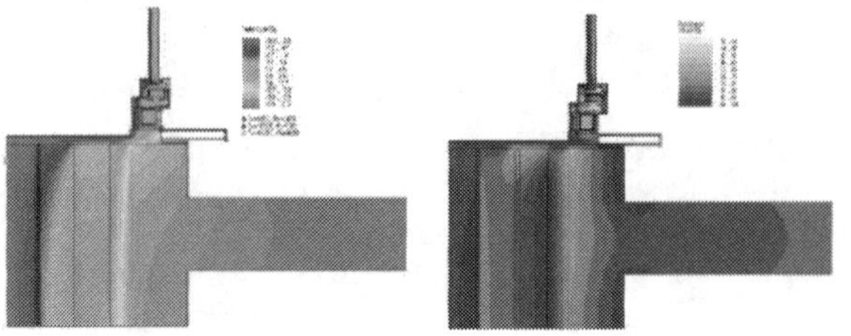

Fig. 1. Layout of temperature and moisture field in double construction (EPS), detail [11].

Table 4. Thermal-technical and moisture parameters "Double ETICS" with MW insulation.

The results of the calculation of the thermal-technical properties of the double ETICS (Mineral wool ISOWER TF PROFI - 50 mm)							
Thermal resistance of the structure	m².K/W			R =	5,16		
Diffuse resistance of the structure	10^9 m/s			R_d =	68,774		
Heat transfer coefficient	W/(m².K)			U =	0,19		
Internal surface temperature	°C			θ_{si} =	18,87		
Humidity mode	Water vapor does not condense inside the structure						
Construction design in terms of thermal-technical properties							**Evaluation**
Thermal resistance	R =	5,16	>	R_N =	4,4	Normalized value	PASS
Heat transfer coefficient	U =	0,19	<	U_N =	0,22	Normalized value	PASS
Risk of mold	θ_{si} =	18,87	>	$\theta_{si,80} + \Delta\theta_{si}$ =		13,10	PASS
Humidity mode	Annual balance sheet:						PASS
	M_c =	**0,068 kg/(m².a)**	<	M_{ev} =	**8,30 kg/(m².a)**		
	Allowable amount of condensate:						PASS
	M_c =	**0,00 kg/(m².a)**	<	$M_{c,max}$ =	**0,50 kg/(m².a)**		

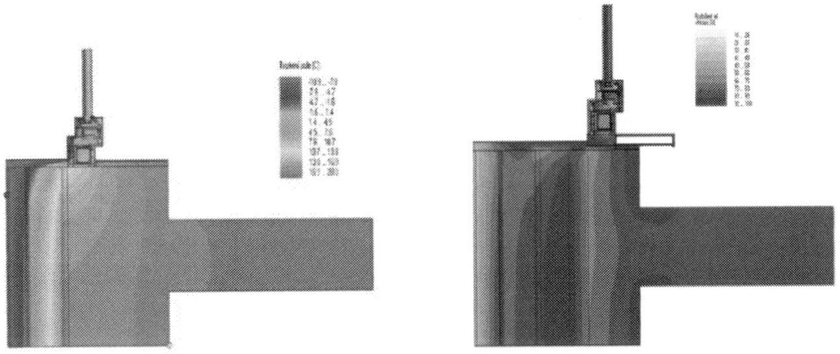

Fig. 2. Layout of temperature and moisture field in double construction (MW), detail [11].

Table 5. Thermal-technical and moisture parameters "Double ETICS" with PUR insulation. (author)

The results of the calculation of the thermal-technical properties of the double ETICS (Boards TPD - PUR - 40 mm)							
Thermal resistance of the structure	m².K/W			R =	5,62		
Diffuse resistance of the structure	10^9 m/s			R_d =	145,89		
Heat transfer coefficient	W/(m².K)			U =	0,18		
Internal surface temperature	°C			θ_{si} =	18,87		
Humidity mode	Water vapor does not condense inside the structure						
Construction design in terms of thermal-technical properties							**Evaluation**
Thermal resistance	R =	5,62	>	R_N =	4,4	Normalized value	PASS
Heat transfer coefficient	U =	0,18	<	U_N =	0,22	Normalized value	PASS
Risk of mold	θ_{si} =	18,87	>	$\theta_{si,80} + \Delta\theta_{si}$ =		13,10	PASS
Humidity mode	Annual balance sheet:						PASS
	M_c =	**0,068 kg/(m².a)**	<	M_{ev} =	**8,30 kg/(m².a)**		
	Allowable amount of condensate:						PASS
	M_c =	**0,00 kg/(m².a)**	<	$M_{c,max}$ =	**0,50 kg/(m².a)**		

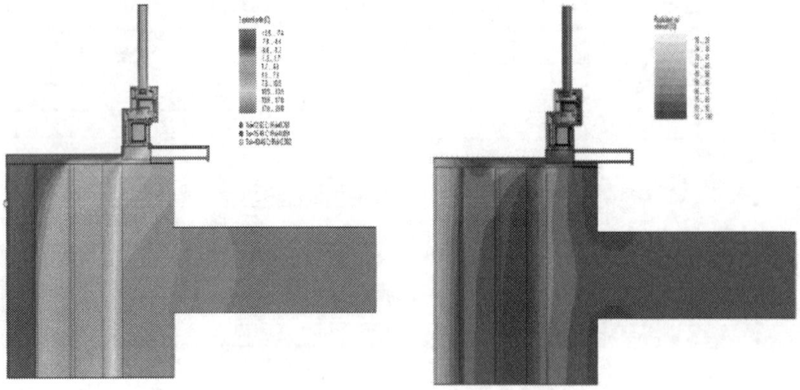

Fig. 3. Layout of temperature and moisture field in double construction (PUR), detail [11].

For the model calculation of the physical properties of the individual alternatives of the double structure, the software was used according to [11].

Discussion and conclusion

Previous results compare the results obtained with the analysis of the thermal-technical and humidity parameters of the double thermal insulation technologies. The economical parameters (cost of realization of individual technologies) and selected parameters from the point of view of impact on the realization performance were also compared.

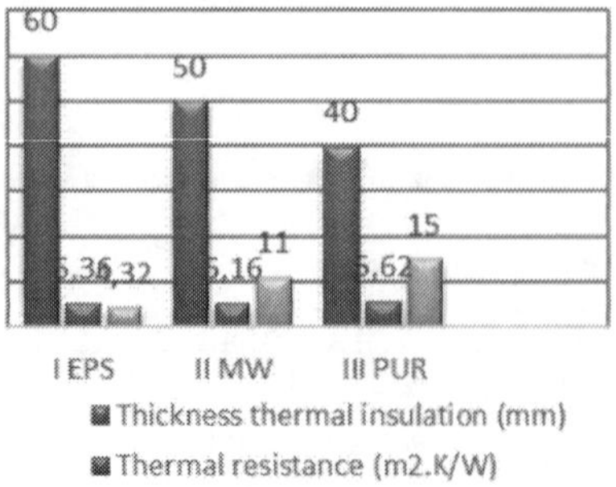

Fig. 4. Comparison of obtained results (thickness, thermal resistance to costs) from analyses.

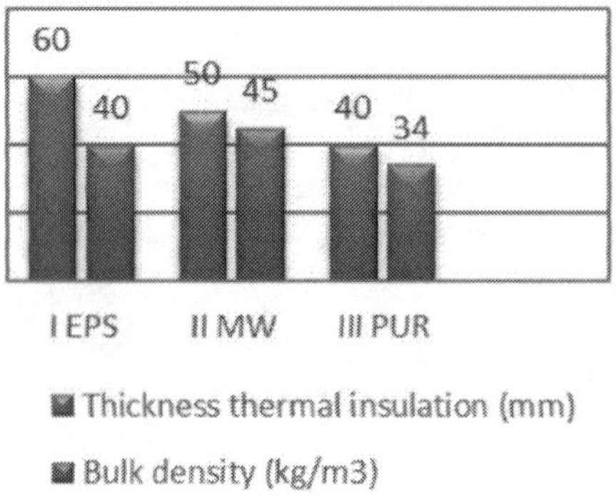

Fig. 5. Comparison of thickness thermal insulation to bulk density from analyzes.

According to the partial conclusions of the previous research, the best double thermal insulation of PUR boards is shown. However, in terms of input direct costs, it is the least advantageous. During deciding which technology is the most appropriate to use, it is necessary to take into account, except the price, the thickness of thermal insulating material, which in the case of PUR foam comes out at almost the same weight insulation per 1m^2 the thinnest. This is an advantage that can be used, for example, to assess the load of the original thermal insulation. However, the decision-making criterion has to be in terms of the development, recovery of microorganisms and the lifespan of individual technologies. We consider the information founded to be the basis for further research in this field and the processing of the basic methodology of the laboratory lifespan survey which consists of:

- From the preparation of "Double ETICS" samples from existing thermal insulation with attacking algae (35x35cm) with multiple types of plaster and ETIC composition to ensure the identity of biological material,
- Simulation of the environment (through SHMU - Slovak Hydrometeorical Institute, detected humidity, precipitation, temperature and wind conditions occurring in a particular area of SR per a multi-year period) and simulation of stress cycles in the climate-chamber Z
- Evaluation of samples in terms of the presence, development, recovery of microorganisms under the activated conditions that they need for growth and evolution.

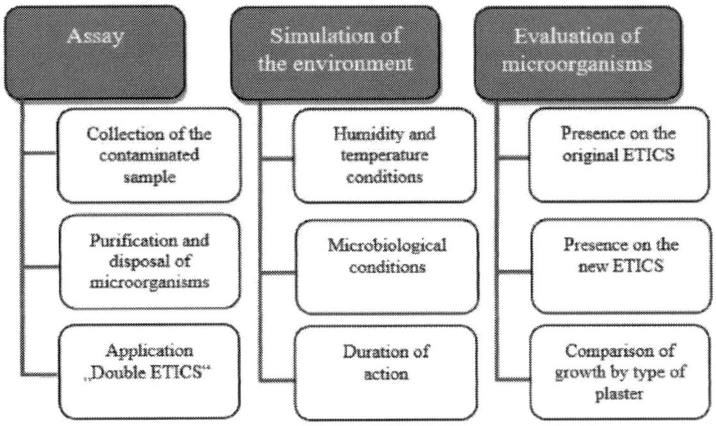

Fig. 6. Method for detecting the life of "bio-corrosion" technology.

We assume that the results of the experimental life test prove the prolonged resistance of thermal insulation against microorganisms and the technology double thermal insulation prevents the development of microorganisms on the original part of the thermal insulation. If the plaster is used on new layers of thermal insulation with biocide protection, we believe that this technology for owners of damaged objects is an interesting solution not only for increasing the thermal protection, but also for the permanent solution of biocorrosion. The double thermal insulation is an effective way of improving the thermal-technical properties of the non-compliant insulated buildings, and simultaneously it helps to solve the problem with algae and microorganisms in the original insulation.

References

1. *STN 73 2901/01 Stage construction external thermal insulation composite system (ETICS)* (2015)

1. *STN 73 0540-2 Thermal protection of buildings. Thermal-technical properties of construction structures and buildings*
2. *Directive of the European Parliament and of the Council no.2012/27/EU on energy efficiency and amending Directives 2009/125/EC and 2010/30/EU and repealing Directives 2004/8!/EC and 2006/32/EC*
3. J. Halásková, European trends to reduce road accidents. Proceedings from the seminar VUT FAST (2007)
4. H. Viner, R. Sinhal, S. Parry, *Review of UK Skid Resistance Policy*. PIARC International Symposium „SURF2004“ (Toronto, 2004)
5. A. Vojtěšek, Comparative measurement of dynamic devices for the measurement of road surface friction coefficients. *Proceedings from the seminar VUT FAST* (2007)
6. *Outside Contact-Imaging Systems ETICS 01 -2014 - Technical Rules of the Guild for Insulation of Buildings of the Czech Republic, O.S.*
7. Technical form EPS FS 70, Producer of ISOVER - Saint-Gobain Construction Products, s.r.o. Available on http://www.isover.sk
8. Technical form TPD PUR30/40, Producer of PCC Moravia. Available on http://www.new-therm.cz
9. Technical form MW-ISOVER TF PROFI, Producer of ISOVER - Saint-Gobain Construction Products, s.r.o. Available on http://www.isover.sk
10. Thermal Reviews Software Svoboda Area and Heat Student Edition.
11. Z. Sternová et. al., *Renovation of apartment buildings. Massive housing construction until 1970.* (Jaga group, Bratislava, 2002)
12. J. Švajlenka, M. Kozlovská, T. Pošiváková, and E. Čonková, Indoor analysis of environment in buildings, Chemical journal, - The official journal of Czech Chemical Society.

The temperature distribution during the microwave drying of wooden elements for structural defects in buildings

Jindřich Sobotka [1,*], and *Radim* Smolka[1]

[1]Brno University of Technology, Faculty of Civil Engineering, Institute of Building Structures, Veveří 95, 602 00 Brno, Czech Republic

Abstract. This paper deals with the theory and application of microwave radiation and experimental optimization of microwave radiation for the distribution of temperature field in the irradiated wooden elements. The article describes the experimental work performed at the Faculty of Civil Engineering, in close cooperation with the company S. P. UNI, s.r.o.. Analysis of the research results was to determine the optimal intensity of microwave emitters and the necessary length of exposure to moist building materials microwave radiation with respect to a particular type of a wooden element. A part of the research was divided into several separate experiments. In each experiment the heating under various conditions was carried out on: I.- samples of the same size, but different weight moisture content, irradiated in the same time periods, II.- samples of the same size, same weight moisture content, irradiated for different time periods.

1 Introduction

Microwave is known as a part of electromagnetic radiation with a frequency from 300 MHz to 300 GHz, corresponding to wavelengths from 1 m to 1 mm. For industrial purposes it is globally determined by the frequency of 2.45 GHz, corresponding to a wavelength of 12.2 cm. Such wavelength is also used in our applications [1].

Microwave Technology (EMW technology) is used in construction mainly for drying of damp structures, or as a way of disposal of biotic pests. In most cases, when heating the material, only the surface temperature of the front side is controlled. Temperatures reached at a depth are only estimated. And it is actually a heating depth and a distribution of temperature field that has a major impact on the speed and quality of the performed remediation. Without sufficient knowledge about how the material is heated, there is a risk of damage of the drying material.

Microwaves belong to the broad spectrum of electromagnetic waves, which also include a visible light region which is governed by Maxwell's equations.

It is applied, that the vectors characterizing the magnetic and electric fields are in every point and in every moment perpendicular to each other and both are perpendicular to the direction of wave motion (Fig. 1).

* Corresponding author: sobotka.j@fce.vutbr.cz

Their amplitudes have a constant ratio and in a vacuum, the electromagnetic waves move with a speed of light and so the following relation is applied:

$$\lambda = \frac{c_\lambda}{f} \quad (1)$$

where: λ length of wave (m)
c_λ velocity of wave propagation ($m.s^{-1}$)
f frequency of microwave radiation (Hz) - (2450 MHz). [1]

In assessing the possibility of using microwave technology in civil engineering practice it is necessary to realize the following aspects:
- the law of conservation of energy always applies;
- the microwave radiation penetrates some materials (glass, porcelain, ceramics, concrete, wood);
- certain material is by EMW radiation reflected (metals);
- by some materials the radiation is "absorbed" (water).

The changeover mechanism of the microwave energy to heat is given by:

$$P = 2\cdot\pi\cdot f\cdot\varepsilon'\cdot\varepsilon''\cdot E^2 \quad (2)$$

where: P energy absorbed per unit volume ($W\cdot m^{-1}$),
f frequency of microwave radiation (Hz) - (2450 MHz),
ε′ permittivity ($F\cdot m^{-1}$),
ε′′ dielectric loss factor of the material,
E intensity of the electric field inside the material ($V\cdot m^{-1}$). [1]

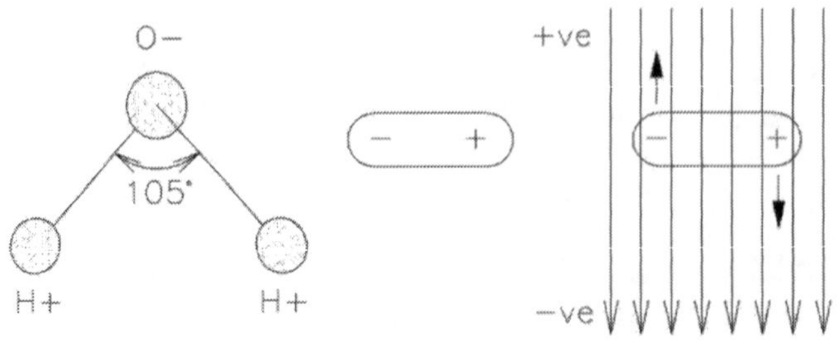

Fig. 1. Water molecule and its orientation in an electric field [5].

In the process, it is necessary to realize that water molecules tend to adjust their polarities changing the polarity of the electromagnetic microwave flux. This leads to the electromagnetic transformation of energy in kinetic energy. Fast movement of molecules and their mutual rubbing is subsequent kinetic energy converted into thermal energy.

Most electromagnetic energy is absorbed and converted to heat energy by the water molecules, which will first come into contact with. The greater the distance of water molecules from the front of the irradiated side is, the less is the flow of electromagnetic energy acting upon them and also its heating.

Following factors have a major impact on the depth of the microwaves penetration:
- water content and its distribution, [3]
- performance of the equipment used. [4] [5]

In the context of long-term research of the usage of microwave technology in the construction industry, which takes place at BUT, Department of Civil Engineering in Brno, we have focused, among other things, on the drying of wooden elements. The entire part

was divided into several independent experiments. In each experiment, the heating under various conditions was performed on:
- samples of the same size, but different moisture content, heated in the same time periods;
- samples of the same size, the same moisture content, heated for various time periods;
- samples of the same size, the same moisture content, heated in the same time periods, but at different distances from the source of EMW radiation;
- samples of the different size, the same moisture content, heated in the same time periods;
- samples composed of several layers.

2 Methods and Material

The SM planed lumber that has been for about three years, placed loosely on the edge was used for the experiment. The individual prisms were numbered and two holes with diameter of 4 mm were drilled on the rear side, so that its bottom had been distanced from the front side 40 mm and 80 mm.

Two beams were used for the experiment. [1] The samples No.1 – 4 were cut from the first beam, the samples No. 5 – 8 were cut from the second beam. The first set of samples No.1 – 4 was stored loosely under a roof for four weeks, so that the samples absorbed only the moisture from the air.

The second set, samples No. 5 - 8, was artificially moistened. The samples were soaked in the water bath for 12 hours and subsequently sealed in PVC containers for 72 hours and taken out about 1 hour before the beginning of irradiation. Hereby was achieved a significant difference in moisture content between the samples No. 1 - 4 and No. 5 - 8.

On average, there was an increase of humidity of 12% in the surface layer of 15 mm. Weight moisture content was measured by digital stylus hygrometer "Greisinger GMR 110". For measurements at the greater depths the thermometer with a surface treatment of the probe "Greisinger GMH 1170" was used.

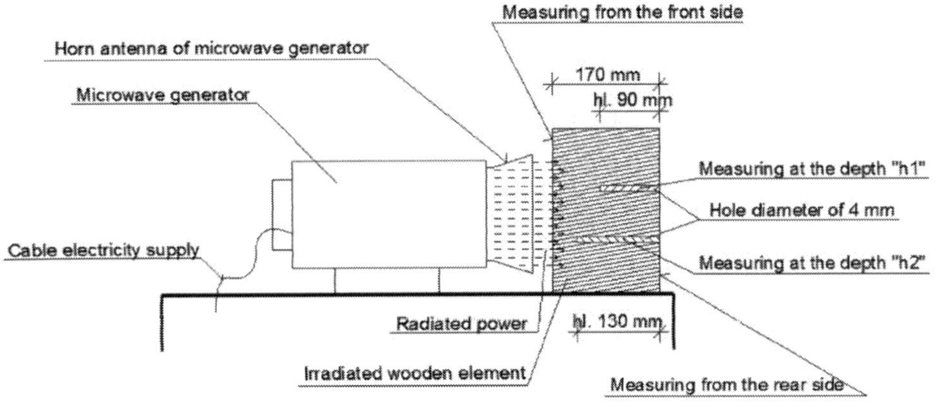

Fig. 2. View of the diagram of the wood element during irradiation.

The irradiation apparatus MB electronic were used, with a power of 1100 W and 750 W radiated power. The heating time was for all samples equal, 10 minutes.

Only one piece of wooden prism was irradiated. The prisms were erected vertically and with dimension of 170 mm in the direction of heat flow = wave (parallel to the antenna of the microwave generator).

Immediately after heating, the surface temperatures on the front and rear side were measured by thermal cameras. Subsequently, the temperature at the bottom of the drilled holes in different depths was determined by the depth thermometer with a special temperature sensor.

3 Results

In the first set of experiments, the samples with air humidity were heated. Measured values are shown in Table 1.

Table 1. Temperature values of wood samples measured in the first set.

Cross section of element [mm]	Moisture content [%]	Time [min]	Front side=reverse [°C]	Temperature at the probe 90 mm [°C]	Temperature at the probe 130 mm [°C]	Rear side=obverse [°C]
140 x 170 n.1	13.,5	10	98.6	79.0	68.2	39.8
140 x 170 n.2	13.45	10	98.2	86.7	66.6	37.3
140 x 170 n.3	13.5	10	91.8	80.6	63.1	36.2
140 x 170 n.4	13.55	10	93.0	71.3	65.3	34,1

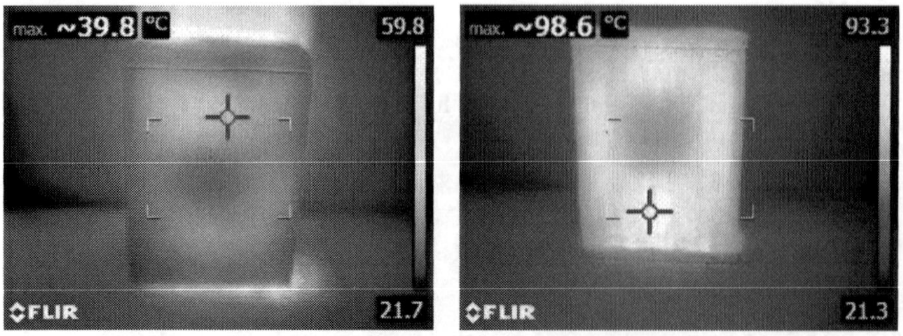

Fig. 3. Example of thermography images of the measured wooden samples of the first set.

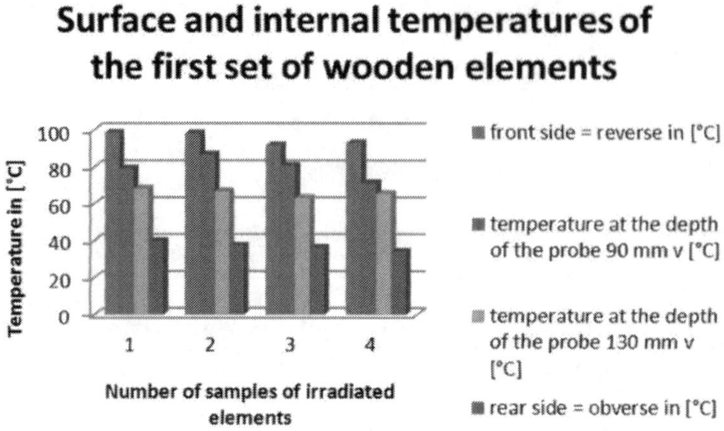

Fig. 4. Measured and evaluated surface and internal temperatures in the first set of samples.

In the second set of experiment the moisten samples were heated. Measured values are given in Table 2.

Table 2. Temperature values of wood samples measured in the second set.

Cross section of element [mm]	Moisture content [%]	Time [min]	Front side=reverse [°C]	Temperature at the probe 90 mm [°C]	Temperature at the probe 130 mm [°C]	Rear side=obverse [°C]
140 x 170 n.5	24.5	10	79.3	67.1	54.1	30.9
140 x 170 n.6	26.1	10	75.1	63.2	49.4	29.6
140 x 170 n.7	24.6	10	85.3	61.7	45.7	31.2
140 x 170 n.8	25.5	10	80.1	73.4	54.2	32.4

Differences in the temperatures measured for the individual samples in the first and second sets of measurements have various causes, such as:
- resin,
- knots,
- size and direction of dried cracks etc.

Therefore, the average values from individual results of samples No. 1 - 4 and No. 5 - 8 were calculated and are shown in Table 3.

Table 3. Average temperature values in both sets.

	Moisture content [%]	Front side=reverse [°C]	Temperature at the probe 90 mm [°C]	Temperature at the probe 130 mm [°C]	Rear side=obverse [°C]
Samples No. 1 - 4	13.5%	95.4	76.9	65.8	36.8
Samples No. 5 - 8	25.175%	79.95	66.35	50.85	31.015

4 Evaluation

All the measurements clearly showed that even in a relatively short time of irradiation, the irradiated areas were heated, but not only at the front side but also in a significant depth of the test prisms.

Primarily the water contained in the wood (surface, bound and free) is heated. [5] Increased surface temperature increases the intensity of evaporation. The temperature increase inside the material, due to the volume expansion generated pressure forces that push water in all directions, including to the surface of the irradiated element. However, there it is necessary to take into account all the factors that influence the transport of moisture in the wood. The effort to accelerate the process of drying moist wood of structural elements could be tempted to try to extend the period of intense heat radiation. In such cases, there is a risk of irreversible damage of the material, such as a shape changes, the emergence of other dried cracks or undesirable chemical changes in composition of wood.

Differences in temperatures measured in the first and second part of the experiment have the following causes:
- for heating of large quantities of water contained in the wet samples, more energy is needed. Lower temperatures are achieved with the same amount of energy. [3]
- greater amount of water in the surface layers of the samples No. 5 - 8 absorbs the greater part of the microwave energy, so that the less energy penetrates to a depth. With the increase of temperature on the front side, the amount of evaporated water is increased, which in turn causes the cooling of the surface. The wetter the surface is, the greater the cooling is. [4]

Differences in the obtained temperatures are shown in Table 3.

5 Conclusion

The results of the experiments clearly showed that EMW is a progressive method of heating of the wood, not only on the surface but also in the depth. [1] [2] Appropriately selected procedure of alternating heating and cooling (stale humidity) allows to aim the optimal heating temperatures and thus to accelerate the drying process.

Under certain conditions, this technology can also be used in the disposal of biotic pests, especially wood-boring insects. Temperatures measured within the test elements highly exceeded the temperatures required to inactivate all of its forms of development. Utilization of EMW technology is subject to extensive knowledge in the area of thermal activity and moisture movement in various building materials.

This paper has been worked out under the project No. LO1408 "AdMaS UP - Advanced Materials, Structures and Technologies", supported by Ministry of Education, Youth and Sports under the „National Sustainability Programme I".

References

1. M. Novotny, K. Suhajda, J. Sobotka, J. Gintar, E. Suhajdova, Wood Res. **59**, 389-400, (2014)
2. J. Sobotka, R. Smolka, MATEC Web Conf. **93**, 01008 (2017)
3. T. Petricek, P. Kacalek, R. Smolka, AMR **1041**, 71-74 (2014)
4. J. Sobotka, R. Kolar, AMM **861**, 295-302 (2017)
5. D. Beckovsky, L. Vackova, T. Beckovska, J. Sobotka, J. Pencik, M. Lavicky, Wood Res. **61**(4), 627-636 (2016)

Modern wood windows now and trends after the year 2020

Milan Palko[1,*], *Adela* Palková[1], and *Peter* Buday[1]

[1]Department of Building Constructions, Slovak University of Technology in Bratislava, Faculty of Civil Engineering, Radlinského 11, 810 05 Bratislava, Slovakia

Abstract. The article deals with the issues of wood windows applicable according to the expected requirements after the year 2020. It analyses the potential of the current top windows in regard to modern design, functionality, and building physical properties. By applying progressive material solutions, it is possible to create windows with proportionally variable frame solutions or even with frameless wing construction. The introduction of vacuum glass systems requires the optimization of the profile system of a wooden window. Modern quality glass systems bring the problem of condensation of water vapour on the exterior surface of the glass. Also, a new phenomenon in the emerging trend of energy-saving buildings and despite the use of quality wooden profile systems is the condensation of water vapour that occurs in the functional joint of the profile system of the wooden window. This fact is increasingly leading to mold growth in functional joints, which subsequently has a negative impact on human health.

1 Windows requirements after 2020

Directive 2010/31/EU of the European Parliament and of the Council on the energy performance of buildings lays down new requirements for the energy performance of buildings, defining as well the intended use of energy. This directive was implemented in the Slovak Republic with an amendment to Act No 300/2012 on the energy performance of buildings and amending Act No 555/2005 on the energy performance of buildings and amending certain acts, which amended Act No 50/1976 on urban and regional planning and the building code. The amended act came into force on 1 January 2013. Implementing Decree No 311/2009 of the Ministry of Construction and Regional Development of the Slovak Republic was superseded in its entirety by Implementing Decree No 364/2012 of the Ministry Transport, Construction and Regional Development of the Slovak Republic of 12 November 2012, which implements Act No 555/2005 on the energy performance of buildings and amending certain acts.

The motto of the recast EU directive is the objective of reducing energy consumption by 20 % and greenhouse gas emissions by 20 % over 1990 levels by the year 2020, and increasing the share of energy from renewable sources to 20 %. The directive defines

* Corresponding author: milan.palko@stuba.sk

various measures through which these objectives will be achieved. In addition to the availability of various forms of energy and its efficient use, the directive also deals with positive environmental influences. The reduction in energy consumption, achieved as a secondary effect of the implementation of the requirements in the operation and management of buildings, should lead to a reduction of adverse influences on the environment.

Some of the requirements on building components arising out of the directive and the act are interpreted in Slovak Technical Standard 73 0540-2:2012 '*Thermal Protection of Buildings. Thermal Performance of Buildings and Components. Part 2: Functional Requirements*'. Requirements for windows are expressed using the heat transfer coefficient 'U_w', with a maximum recommended value after 2020 of $U_w \leq 0.6$ W/(m^2.K) for windows and exterior doors, and $U_w \leq 1.0$ W/(m^2.K) for roof windows, measured or calculated in vertical position. Public buildings must comply with these numbers starting in 2018 [1-5].

2 Profile solutions

Wood-only window frames offer limited possibilities in terms of the requirements on thermal properties. Although a window profile built primarily of can be resolved by increasing the geometric proportions, this does not provide a good solution. A profile solution that is still optimal whilst being wood-dominant is shown in figure 1 (lower right). The insulation properties of the profile are improved with the use of air cavities that have better thermal characteristics than wood. The type of wood selected also plays an important role, as soft woods have better thermal properties than hard woods.

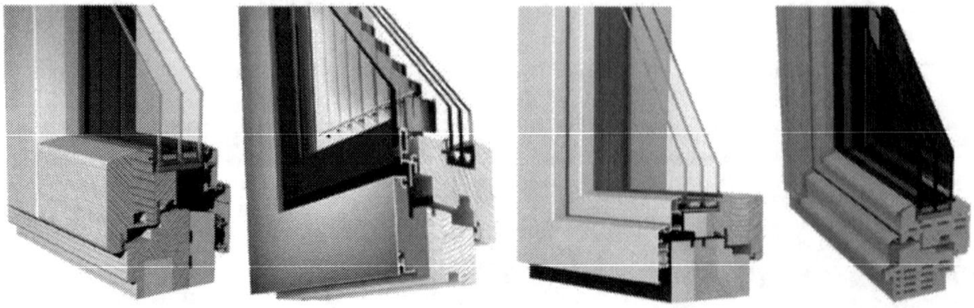

Fig. 1. Current top windows with U_f = 0,7 - 0,8 W/m^2.K.

In some cases the insulation properties of specific types of wood are also affected by the location where the timber grows (varying thicknesses of summer and winter wood). The other examples in figure 1 are a combination of three wood-based materials, effective thermal insulant, and exterior aluminium portion. There are of course profile solutions containing plastic elements [3-5].

3 Glass systems

Advancements in glazing have focussed not only on glass and glass combinations and low-emissivity coatings. Modern technology makes it possible to achieve better glass property values such as thermal insulation, daylight transmittance, sound insulation, and others. A problem arises, however, when attempting to combine all the positive properties into one whole, as some of these properties are mutually exclusive. Research and development has begun to focus on dynamic glazing, where the glazing parameters can be set according to the actual season or the usage of a building (switchable glazing, shading systems, etc.). In today's market, glazing is available with values of U_g = 0.4 to 0.6 W/(m^2.K) for triple glazing with argon or krypton filling.

For thermal properties in products of the highest category, we currently use insulated triple glazing with 4 layers of low-E coating and krypton cavity filling. We are starting to see the first applications of thermal quadruple glazing. The disadvantage of this option is its weight, which makes it difficult to use in the parts of the window that open. Vacuum-insulated glass is the more logical choice, but is not typically available at the moment. Theoretically, it can achieve a U_g heat transfer coefficient of 0.15 W/(m^2 K) in vacuum-insulated double glazing. For vacuum-insulated triple glazing with four low-E coatings (emissivity of 0.05), the U_g value would be less than 0.01 W/(m^2·K). It should be added that the space between the panes would only be a few tenths of a millimetre wider, so the resulting thickness would be within the range of that used today in insulated glass. To eliminate heat conduction in the gas-filled space, there should be a vacuum with a value of 10^{-4} Bar in the space between the panes, and this vacuum must be maintained for the entire life of the vacuum-insulated glass. This means extreme requirements on the air tightness of the special articulation of the edges of the panes. The load of the vacuum on the panes is approximately 10 tonnes per square metre. The pressure must be tamed using small spacers in the cavity between panes, but this increases thermal losses and can affect the transparency of the panes [3 - 6].

4 Design and functionality

There is a demand on modern windows to reduce the visibility of the frame and profile. Contemporary architecture prefers large glass panes without dividers. Likewise, the operational parts and fittings are integrated into profile systems to obtain maximum invisibility. Many of these demands are mutually incompatible in terms of the visual-physical-functional aspects.

5 Potential of segments for heat transfer coefficient

The heat transfer coefficient of window construction can be determined by the following ways:

- Measurement in big climatic chamber (method of enclosed hot chamber),
- Calculation (e.g. according to STN EN ISO 10077-1),
- Detailed computer simulation (analogically as by climatic chamber, but using computer model).

Using the enclosed hot chamber the steady temperature conditions is created, in order to describe transparent product having concrete dimensions based on formula (1):

$$U_w = \frac{\Phi}{A(\theta_i - \theta_e)} \tag{1}$$

Where U_w is heat transfer coefficient of window (W/(m^2.K)), Φ is heat flow rate (W), A is area (m^2), θ_i, θ_e is indoor and outdoor design temperature (°C).

The new approach towards the quantification of thermal characteristics of opening constructions used in European standards included in the Slovak technical standards system (e.g. STN EN ISO 10077-1 Thermal characteristics of windows, doors and shutters. Calculation of heat transfer coefficient) enable designers to show the value of heat transfer coefficient for each window construction based on the values declared by an accredited company using the following formula:

$$U_w = \frac{U_f A_f + U_g A_g + \Psi_g l_g}{A_g + A_f} \quad (2)$$

Where U_w is heat transfer coefficient of window (W/(m^2.K)), U_f is heat transfer coefficient of frame (W/(m^2.K)), U_g is heat transfer coefficient of glass (W/(m^2.K)), Ψ_g is linear thermal transmittance (W/(m.K)), l_g is perimeter of glass (m), A_f is area of frame (m^2), A_g is area of glass (m^2).

The profile solutions and pane systems regularly available on the market cannot achieve the value of $U_w \leq 0.6$ W/(m^2.K) required after 2020 (2018 for public buildings). Even the best products barely attain this threshold. Figure 2, 3, and 4 show the potential values of heat transfer coefficients on the basis of combinations of frames and panes determined by their heat transfer coefficients. The analysis is based on a 1200 x 1500 mm window (the minimum area for which the heat transfer coefficient value is mandatory), with a frame width of 80 mm. Figure 2 is modelled for a linear loss coefficient of glass of Ψ_g=0.08 W/(m.K). Figure 3 is modelled for a linear loss coefficient of glass of Ψ_g=0.06 W/(m.K). Figure 4 is modelled for a linear loss coefficient of glass of Ψ_g=0.04 W/(m.K).

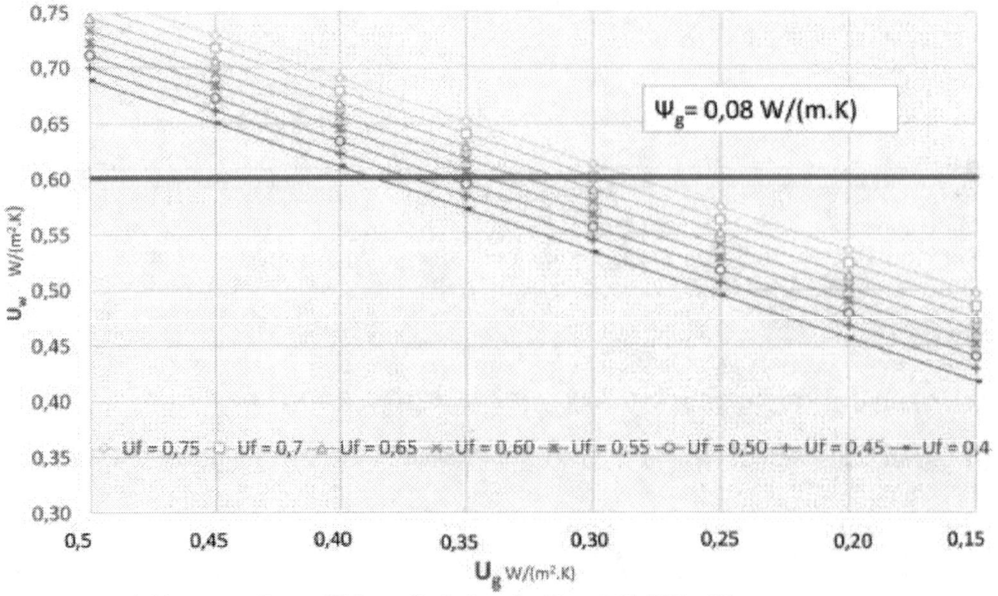

Fig. 2. Potential heat transfer coefficient of window for Ψ_g = 0.08 W/(m.K).

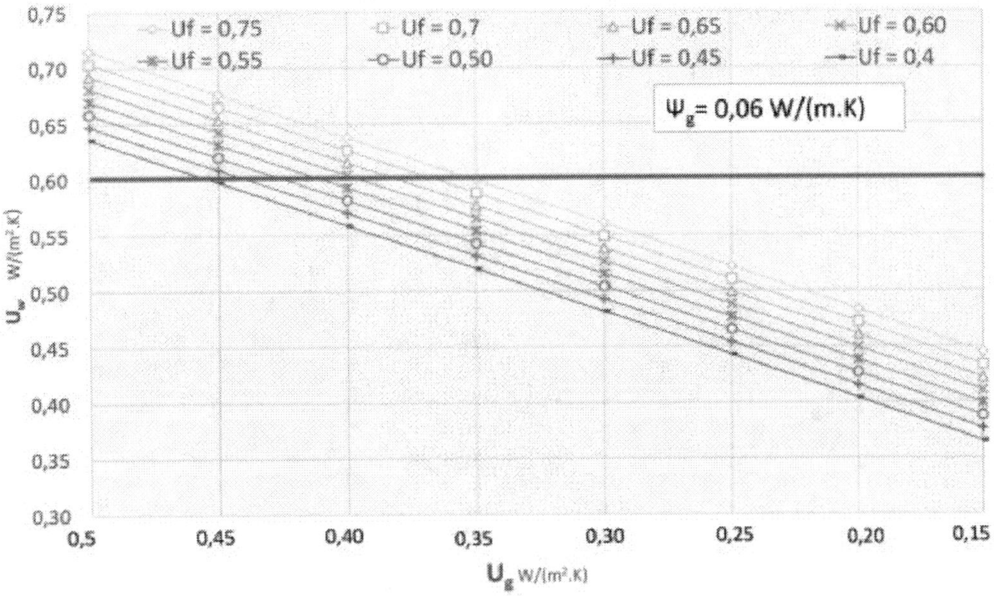

Fig. 3. Potential heat transfer coefficient of window for $\Psi_g = 0.06$ W/(m.K).

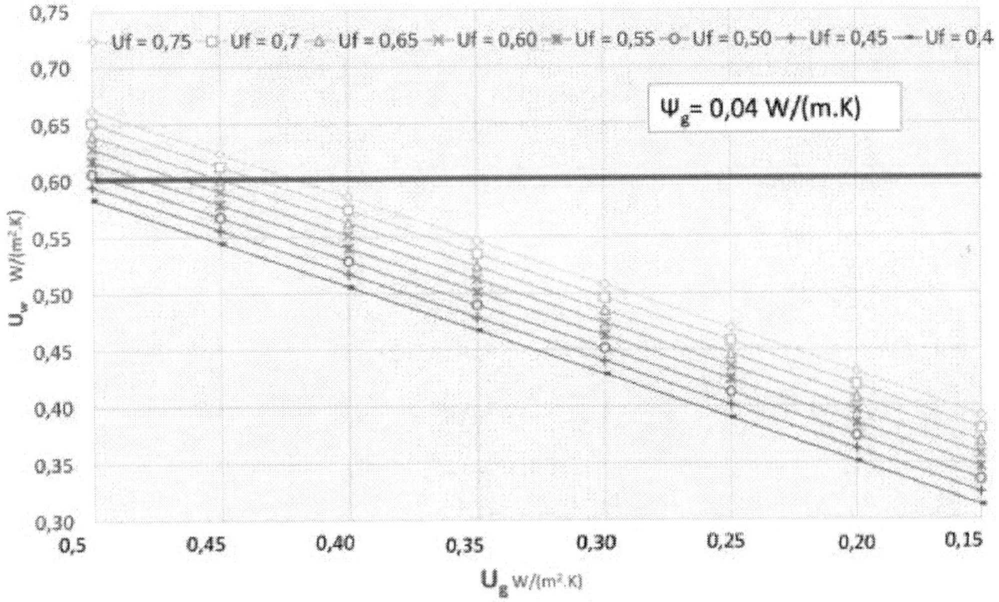

Fig. 4. Potential heat transfer coefficient of window for $\Psi_g = 0.04$ W/(m.K).

6 Problems related to modern windows

With the production of high quality window constructions, there are problems that have not occurred during the previous lower physical quantification of the windows. Ambient environments through indoor and outdoor climate affect the window structure. Trends in lowering the energy demands of a building lead to physical processes that have not been so far in the past. Infiltration through the window structure is eliminated and forced ventilation often causes exfiltration. These changes in physical modes cause condensation or icing in the functional window of the window and thus the risk of mold formation.

In practice, more and more the question arise, "Why are windows misted on the outside?" among the house owners with high quality glass systems. The problem of outer condensation occurs considerably towards the end of summer and early autumn.

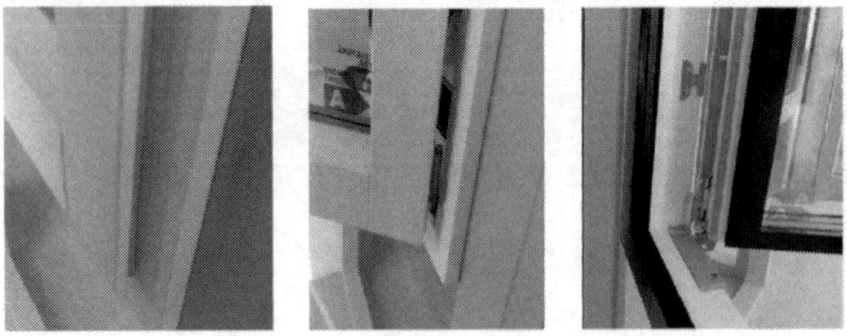

Fig. 5. Hidden all-round fittings.

6.1 Condensation of water vapor on external glazing

Surface condensation of water vapour appears on the external surface of an insulating glazing, if the temperature at the surface of the glazing is substantially lower than the air temperature and if it drops below the dew point. The surface temperature at the outside of the glazing is dependent on:
- heat flow from the inside out,
- transfer of heat by convection with the outside air,
- heat loss by radiation (especially sky component).

Efficient thermal insulation glazing leaks only little heat from the inside out. As a result, the outer pane of the glazing remains relatively cold and therefore leads to a formation of condensation. This is due to the temperature difference between the air and the window area. With the increasing temperature the air can hold more water resulting in a higher relative humidity of the air. If the morning air is warming faster than the window glass, the glass surface is colder compared to the temperature of the air, and the water vapour condensates even in case of quality thermal insulation glazing. Skylights are more affected because they are cooled faster than vertical glazing. Given the above, the increasing of thermo-technical quantification of filling of holes in the outer wall leads to the emergence of new problems related to the hygienic quality of the internal environment or visual discomfort (fogging), and others [7].

Figure 6 shows exterior condensation and frost. Thermal triple glazing in a single-family house is shown on the left. On extremely cold days, frost can develop on the exterior surface. The right side shows the early phases of testing of vacuum-insulated glazing in the laboratory at the Faculty of Civil Engineering at Slovak Technical University. The image shows water vapour condensation on vacuum-insulated glazing from the cold chamber side. The thermal bridges created by the spacers between the panes form an interesting effect on the vacuum-insulated glazing.

Fig. 6. The formation of condensation and icing from the outside of a quality glass system

6.2 Condensation of water vapor in a functional gap

The moisture transport through gap between casement and window frame at condensation temperature is the main cause of condensation of water vapour.
- Air pressure differences (taking water vapour) – infiltration a exfiltration,
- Diffusion of water vapour (different saturation of water vapour in exterior and indoor air).
Condensation of water vapour – phenomena for balancing water vapour partial pressure by interaction of molecules. The diffusing water vapours are moving from places having higher pressure to places with lower pressure [8].

Fig. 7. Condensation and frost of water vapor in a functional gap

Figure 4 shows condensation and frost in the functional gap between window and wood frame. Long-term exposure can damage the surface treatment and as a result, moisture permeates the wood and causes rot. Mould can form in the case of passive windows on buildings where the windows are infrequently opened. Mould is a human carcinogen and creates an unsuitable environment for allergy sufferers and asthmatics. The condensation of water vapor in the joint will improve the acoustic properties slightly [9, 10].

7 Conclusions

The profile solutions and pane systems regularly available on the market cannot achieve the value of $U_w \leq 0.6$ W/(m².K) required after 2020 (2018 for public buildings).In the coming years, manufacturers of profiles, panes, and particularly manufacturers of windows will be faced with the challenge of implementing the new standards. We can expect the introduction of new types of panes and frames. This new added value may also bring with it higher window prices. One possible benefit is that after the new types of windows are installed in buildings constructed after 2020, it is highly likely that we will no longer experience the problems in the interior of buildings that today are frequently attributed to new windows and replacement windows.

This research was supported by Scientific Grant Agency MŠVVŠ SR and SAV under VEGA 1/0685/16. This research was also supported by Apol, s.r.o.

References

1. *STN 73 0540-2 Thermal protection of buildings. Thermal performance of buildings and components. Part 2: Functional requirements* (2012)
2. *STN 73 0540-2/Z1 Thermal protection of buildings. Thermal performance of buildings and components. Part 2: Functional requirements, Change 1* (2016)
3. P. Oravec, Procedia Manufacturing **2**, 348-352 (2015)
4. P. Oravec, AMR **899**, 184-187 (2014)
5. M. Bagoňa, Thermal protection of Building **13**(1) (2010)
6. Izolačný systém na báze vákua. Available on http://www.asb.sk/stavebnictvo/konstrukcie-a-prvky/okna-dvere/systemy-zasklenia-zhladiska-tepelnej-ochrany-budovy
7. A. Palková, M. Palko, AMR **820**, 293-298 (2016)
8. M. Palko, A. Palková, *Proceedings of 24th SVSFEM ANSYS Users´ Group Meeting and Conference 2016,* 77-84 (2016)
9. D. Dlhý, K. Minarovičová, AMR **899**, 487-490 (2014)
10. D. Dlhý, *ATF 2013: 2nd Conference on Acoustics, Light and Thermal Physics in Architecture and Building Structures.Book of proceedings*. 21-24 (2013)

MATEC Web of Conferences **146**, 03010 (2018) https://doi.org/10.1051/matecconf/201814603010
Building Defects 2017

Source data for modeling of thermal engineering calculations

Pavlína Charvátová[1, *], and *Roman* Šubrt[2]

[1]VŠTE, České Budějovice, email: pavlina@e-c.cz
[2] Energy Consulting, z.s. Alešova 21, 370 01 České Budějovice, CZ, e-mail: roman@e-c.cz

Abstract. Increasing demands on thermal insulation. Their more accurate assessment by computers lead to increasingly bigger differences between computational models and reality. The result is an increasingly problematic optimization of building design. One of the key initial parameters is climatological data.

1 Modeling of thermal performance of buildings

Quantification of the heat loss of buildings started with central heating installations, as until then there were local heaters in each heated room which dimensioning was empirically done.

The first dimensioning of heating was necessary only after the start of central heating, although the first known hot-air central heating was already in the 1st century Before Christ in Rome. The dimensioning of heating in the Czech Republic has been modified since the middle of the last century by standards, eg. ČSN 10503 from 1949. Later, this dimensioning was guided by the new standard ČSN 06 0210 - Calculation of thermal losses of buildings with central heating. This standard originally dates back to 1961 and gradually changed the way knowledge of heating increased, and also as did the influence of the Soviet Union in the Czech Republic. It was canceled in November 2008, when there was a transition to European standards, and at the same time calculations were made in order to reduce energy consumption for heating. At the moment, the calculation of energy demand for heating goes according to ČSN EN ISO 13 790 Energy performance of buildings - Calculation of energy consumption for heating and cooling.

Little by little, however, it is found that the calculations made under this standard deviate significantly from real consumption. Paradoxically, there is even a situation where the building is better heat-insulated, the difference between the real energy consumption for heating and the calculative assumption is worse in the sense that real consumption is higher than predicted.

* Corresponding author: charvatova@mail.vstecb.cz

2 Possible causes of differences in computational building behavior and real energy consumption for heating

Possible causes are many. Big role plays the difference between the projected state of a building from the real state, the absence of some thermal bridges or bridges in the calculation, the different ventilation modes, the different solar gains, the difference between the actual climatological conditions from the real, the neglect of the thermal inertia of the building and many others.

Certainly, one of the major inaccuracies is the modeling of the thermal behavior of the subsoil under the construction according to the standard, which does not entirely correspond to the reality.

At the beginning of the last century and in the times before, people did not deal with the temperature in the soil, which was especially important for them to freeze the water under the terrain.

In the Czech Lands, the so-called non-freezing depth was determined empirically, i.e. the depth during which the liquid phase of the water was not frozen during the winter, to 100 cm for the exterior temperature - 20 °C.

For thermal technical calculations, the computational outdoor temperature of the adjacent soil to the building structure was defined in ČSN 06 0210 [1]. This was modified by Table A6 (see Table 1), assuming that the soil temperature is constant for the calculation outside temperature and that the heat from the building passes through the structure into the ground.

Table 1. Table A6 according to ČSN 06 0210.

Location of the adjacent soil layer	The temperature of the adjacent soil t_{ez} (°C) at the calculation temperature t_e		
	-12°C	-15 °C	-18 °C
under the floor	+5	+5	+5
- at the vertical wall: - to a depth of 1 m	-3	-3	-6
at a depth of 1 to 2 m	0	0	-3
at a depth of 2 až 3 m	+3	+3	0
at a depth 3 m	Like under the floor		

ČSN EN ISO 13 370 [2] according to which the assessment is currently being carried out continues in a different way, namely assuming that the heat escapes by the soil from the building to the environment, see Figure 1. For the soil, the general properties are defined in the standard; Either the specific properties of the soil can be used, or the Table 1 of this standard is used (see Table 2). If neither the approximate soil quality is known, it is proceed as if it were sand.

Table 2. soil characteristics according to ČSN EN ISO 13 370.

Description	Thermal conductivity λ W/(m.K)	Volume heat capacity ρc J/(m^3.K)
clay and clays	1,5	3,0× 10^6
sands and gravel	2,0	2,0× 10^6
homogeneous rock	3,5	2,0× 10^6

Fig. 1. Heat flow diagram according to ČSN EN ISO 13 370.

With a simple look at the picture, it is clear that there where there is no building, the temperature would have to be well below the freezing temperature at a depth of 1 meter, which does not correspond to historical experience. It is therefore evident that the heat flow from the building works by earth in a completely different way and it is influenced outside the building by many other aspects such as geothermal heat, heat accumulation of rocks, heat transfer through groundwater flow, petrography, earth anomalies such as karst formations, mining constructions and similar.

3 Climatological data

The climatological data are given by the standard and are based on the long-term temperature average from the measurements between 1900 and 1950. The peculiarity is that the average temperature does not count as an arithmetic mean during the day, but by the decree 194/2007 Coll. [3] is the average daily outdoor air temperature of a quarter of the sum of outdoor temperatures measured in the shade, excluding the radiance of the surrounding areas at 7.00, 14.00 and 21.00, with the temperature measured at 21.00 being counted twice.

For real calculations, real climatological data can also be used, which can be ordered from the Czech Hydrometeorological Authority, or it can be uses e.g. data stored on the MPO website [4].

4 Thermal gains

Heat gains in buildings can be divided into internal heat gains that are generated by people, heat generation by facilities such as lighting, technology, hot water preparation, and solar thermal gains.

Solar thermal gains are again quantified based on standard values based on long-term averages. Precise modeling of solar thermal gains again encounters several problems. The first is the substantial deviations of real sunlight from long-term averages [4], the second is that we only consider transparent solar constructions in the modeling of solar gains, while heat gains also take place through non-transparent constructions.

5 Temperatures in the soil

To accurately model the thermal behavior of the building, it is necessary to know the temperature of the soil in addition to climatic conditions. The Czech Republic has quite a few territories, but there are also places with a distinctly different geological subsoil. From the point of view of the thermal behavior of buildings, there will surely be a difference between locations with a higher geothermal temperature such as the region Karlovarsko or some parts of South Moravia and between lower geothermal temperatures such as limestone karst. However, relatively little space was devoted to the problem of the temperature of the building's subsoil. Several measurements were made in probes under buildings, but this condition was not confronted with the state outside the built-up part and was not compared with the calculation premise. The basis for further calculations of the actual course of temperature under the building is the provision of climatological data for the real temperature in the subsoil without the influence of the building, as well as for the thermal modelling of the above-ground building parts.

For this reason, I started to work with the Geophysical Institute of the Academy of Sciences of the Czech Republic, which has given to my initial data.

On the site of the Geophysical Institute of the Academy of Sciences of the Czech Republic in Prague - Spořilov, it is monitored since 1994 the temperature at selected depths in two experimental boreholes, 150 meters deep (GFU-1) and 40 meters (GFU-2), together with the measurement of air temperature and other meteorological parameters experiment. Since 2011, the drilling site has a depth of 50 meters (GFU-3) at the local Geopark. An experimental field for continuous heat transfer monitoring was put into operation in June 2002.

5.1 Drill holes GFU-1 and GFU-2

These are two separate drill holes close together, 140 meters and 40 meters deep. They are drilled in Ordovician sediments covered with the upper layer of clay. Deeper drill hole is designed for temperature logging with a portable probe hanging on the cable. The second drill hole is equipped with thermistor sensors at depths of 0; 0.05; 0.1; 0.2; 0.5; 0.75; 1; 1,5; 2; 2.5; 3; 4; 5; 7.5; 10; 15; 20; 25; 30; 35; 38.3 meters. The air temperature is monitored at 0.05 m and 2 m high. This station also registers solar radiation and rainfall. The data is transmitted online to the desktop computer. Temperature data from depths below 0.5 are stored 7 times a day, air temperatures and soil temperature to a depth of 0.5 m, as well as sunlight, are stored every half hour. The rain monitor has its own answering machine.

5.2 Drill hole GFU-3

It is a new drill hole from 2011. The depth of this drilling hole is 50 m

5.3 Experimental field

In June 2002 was pun itno operation an experimental field for continuous monitoring heat transfer. The set of models includes grassy and bare clay soil, sand and asphalt. Each model is 150 cm in diameter and 60 cm deep. It contains temperature sensors at 5 levels below the surface (0.02, 0.05, 0.1, 0.2, 0.5 m) and temperature sensors at 5 cm and at 2 meters above the surface. It also measures soil moisture, direct and reflected sunlight, wind speed, rainfall. The height of the snow cover is monitored by a television camera. The data obtained serves to determine the heat transfer between the air and the earth surface, the determination of the diffusivity of the subsurface materials, the influence of humidity, the wind cooling. The results will then make it possible to correct the relationship between the depth and climate history of the past.

6 Conclusions

From the results of temperature measurements in the soil, which cannot be completely mentioned here, because of the large amount of data, it is easy to deduce that the temperature fluctuations occur most at the surface and at least at greater depths. Graph 1 shows the course of soil temperatures in individual depths during 2013 measured at half an hour. Table 3 lists minimum and maximum values for each measurement depth. The temperature fluctuations just below the terrain are 33 °C, but at a depth of 5 meters it is already only 3 °C, with the difference decreasing with increasing depth until finally at a depth of 49.7 m the difference between the highest and the lowest temperature is only two hundred thousand Kelvin. This fact justifies the assumption that for a real modelling of the thermal behavior of buildings it would be appropriate to choose a constant temperature at a certain depth below the terrain and with this assignment to continue the calculations. Optimum depth and optimum temperature will be subject to further investigation.

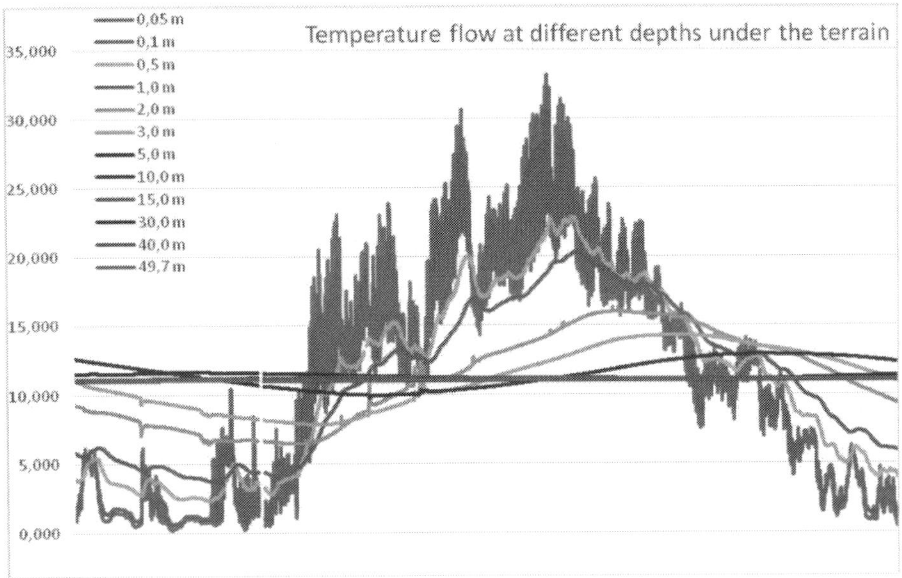

Fig. 2. Temperature flow at different depths under the terrain.

Table 3. Temperature in individual depths [°C].

	0.05 m	0.1 m	0.5 m	1m	2 m	3 m	5 m	10 m	15 m	3 0m	40 m	49.7 m
max	33.157	30.496	22.880	20.372	16,327	14,283	12,871	11,630	11,244	11,031	11,037	11,043
min	0.173	0.577	2.329	3.688	6,371	7,821	9,775	10,982	11,081	11,023	10,991	11,024

References

1. *ČSN 06 0210, Calculation of thermal losses of buildings at central heating* (Czech Office for Normalization, Prague, 1993)
2. *ČSN EN ISO 13 370, Thermal behavior of buildings - Heat transfer by soil - Calculation methods* (Czech Office for Standards, Metrology and Testing, Prague, 2009)
3. *Decree No. 194/2007 Coll. laying down rules for heating and hot water supply, specific heat consumption figures for heating and hot water production, and requirements for the fitting of internal heat installations by buildings regulating and registering the supply of heat energy* (Collection of Law, Prague, 2007)
4. Climatological data, Avalable on http://www.mpo-efekt.cz/uploads/7799f3fd595eeee1fa66875530f33e8a/klimatologicka-data_do-06_2016-verze2.xlsx

MATEC Web of Conferences **146**, 03011 (2018) https://doi.org/10.1051/matecconf/201814603011
Building Defects 2017

Use of technical hemp in the construction industry

Petra Nováková[1,*]

[1]The Institute of Technology and Businesses in Ceske Budejovice, Czech Republic

Abstract. Current requirements for building insulation are constantly increasing, not only in terms of the technical properties of the material used but also in terms of the environmental burden. This article focuses on the use of technical hemp in the building industry, describes individual products, its properties and uses. Part of the article is also the experience of realization and use of the building along with photographs.

1 Introduction

Cannabis is a plant with a rich history, a variety of uses in a number of industries, and a great potential for the future. We can learn about the possibilities of cannabis use in the field of construction from history. Today, the general greening trend determines this plant as an alternative building material.

In the past, cannabis was most often used for the production of ropes, netting, textiles, later in China for paper production. Hemp oil has been used in cosmetics and also in colour or light production. In construction, cannabis was used in conjunction with raw materials that could be harvested or produced directly at the site. It was often soil and matter-derived plants.

Cannabis was also part of the original half-timbered buildings in Alsace and other regions of Germany and France. It is also possible to see the detailed instructions on the production of hemp blocks of the year 1916 on the Internet [1].

2 The three basic cannabis genera

Rum Cannabis (Cannabis ruderalis Janisch.) is annual weed, growing freely on rum and dumps, without significant narcotic effects.

Indian Cannabis (Cannabis indica Lam.) is forbidden to grow for its high content of psychoactive substances called tetrahydrocannabinols (THC). The species has high narcotic effects because the THC content is greater than 0.3 % in dry matter.

Cannabis, so called technical hemp (Cannabis sativa L.), is a thermophilic crop that is demanding for water, soil, nutrients, and agro-technology. It is a dioecious, or even a simple old fiber and cultural plant, growing in height from 2 to 6 meters. It suppresses the growth of weeds, has reclamation and anti-erosion influence, drains soil from poisonous substances and

* Corresponding author: petra.novakova@mail.vstecb.cz

heavy metals. There is no need for pesticides or herbicides during growing, that are harmful to the environment.

It contains about 23 % of the fibers and about 75 % of the woody mass - the so-called shive. The growing period lasts 100 to 120 days, and at least 1/2 hectares of woody material are grown on 1 hectare of growing area over 1 ha of forest growing for several decades [2].

3 Technical hemp

Technical cannabis is a versatile organic crop. From hemp can produce up to 25 thousand 100 percent recyclable products. The basic usable raw material is biomass (fiber, shive) and seed.

Stalk contains plant pulp - a wood part - suitable for biofuel production - pellets, briquettes, petrol, charcoal, methanol and electricity. In the world, cannabis raw materials are used in the food, chemical and textile industries, in medicine, cosmetics etc. In the Czech Republic, hemp seeds are mainly grown for the production of short fibers for technical use, nonwovens, paper and automotive, and for construction purposes.

It is also very important to use woody masses of shive - both for construction and especially for energy use in the form of bio-pellets and biobriquettes. Their heating value is 9 % at a humidity of about 16.5-18 MJ/kg. They are especially suitable for gasifying boilers for wood, fireplace stoves and fireplaces. These briquettes are pressed under high pressure without any binders or other harmful substances and are well usable in other types of stoves and solid fuel boilers [3].

Another product, cannabis seeds, contains 30-35 % fatty acids with a high fat content. Cannabis oil obtained from cannabis pressing can be used as a motor fuel, for the production of paints and varnishes for natural wood preservation, or can be used in the food and canning industry. Cannabis seeds are used in the food industry, fishermen and bird breeders. Crops that remain after pressing the seeds are used as dietetic feed for livestock or as fuel [3].

Fig. 1. Cannabis sativa [1].

Fig. 2. Hemp bricks [4].

4 The use of cannabis in the construction industry

Like other natural materials, the use of cannabis is on the rise as well. Cannabis is a renewable strategic crop with a tradition of growing in our climatic conditions and high yields per hectare. Cannabis processing for cannabis and shive is done mechanically, without any polluting processes. Hemp fiber excels in its strength, durability and natural protection against pests. Threads and cannabis shavers have been embossed to reinforce the structure since the beginnings of construction, as well as grain husks, bran and animal fibers. Nowadays, hemp is used mainly in the building industry for the production of insulation, so-called cannabis concrete and varnishes [3].

4.1 Hemp insulation

This natural material provides a good microclimate with natural protection against mold, rot and pests and creates a healthy environment in the interior of the home. Thermal insulating and acoustic properties of cannabis are comparable to conventional insulations, but moisture resistance and capillary water drainage properties ensure stability even under conventional materials subject to destruction or significant reduction in isolating properties.

Soft hemp mats with a bulk density of 30-40 kg/m^3 are manufactured from purified hemp fiber by mixing with a binder, either bicomponent fiber based on polyester, or corn starch that is required in the order of 10 %. Against combustion, hemp insulation is treated with a retarder, such as edible soda. The final blend is shaped into an endless belt and fused together at 150 °C, at the end being then cut to standard dimensions.

Insulation from hemp fiber has a thermal conductivity in the range of 0.38 to 0.4 W/(m^{2}K), which is among the better products. With these parameters, hemp insulation is aligned with conventional insulation. But they also have other features that make them totally different from these products [1].

The hemp insulation is flexible enough because of the toughness of the hemp fiber, after a short time of compression, it returns to its original shape. This feature is particularly important when assembling when it is not possible to avoid squeezing the mat when inserting between the structural members. At the same time, the cannabis insulation maintains its shape for a long time, so that it does not settle and cause the cavities to develop in places where the insulation should be.

The specific heat capacity of cannabis is c = 1600 J/K.kg. That is, 1 kg of this insulation holds 1600 J at a temperature difference of 1 °C. In practical terms, this means that if the temperature increases, hemp can receive heat. On the other hand, heat is released when the temperature drops. When comparing this parameter, we find that hemp insulation absorbs almost twice the amount of heat than mineral insulation, thus doubling the thermal comfort in the interior twice more effectively. This relatively high heat-storage capacity of hemp is comparable, for example, to wood. The composition of a low-energy wooden building using hemp thermal insulation then has a phase shift of 8 to 9 hours.

In order to eliminate moisture, the widest range of diffusion-open layers should be used in the outer edges of the perimeter structure, which, thanks to their breathability, easily draws moisture away. The diffusion resistance factor of hemp insulation has a value of μ = 1.9, which corresponds to a very permeable material and therefore optimal for this application.

In addition to high diffusion throughput, cannabis insulation has yet another advantage, namely the ability to redistribute moisture. In general, it is the ability to balance and transmit moisture through its entire volume, that is to say a kind of moisture conductivity. The moisture is transferred to the entire volume, resulting in a multiplicity of surface area for easy ventilation. Thanks to this, cannabis insulation is able to accommodate and cope with a large amount of moisture. Bulk density may increase up to 20 % without reducing the efficiency of insulation. For comparison, mineral insulation results in loss of thermal insulation properties at 2% volume moisture.

Natural fibers are long, flexible and tough, so they are able to vibrate at the same frequency with sound waves. By their behavior, the intensity of the sound waves greatly weakens. Thanks to this, cannabis insulation is used as a high-quality acoustic insulation of sandwich composites for attenuation of airborne sound [1].

With all these benefits, cannabis insulation applications in the constructions and work processes are identical to conventional materials. Handling cannabis is pleasant, however, without the use of protective gloves or respirators, without the risk of skin or airway damage.

4.2 Hemp concrete

Lime-hemp building material is a revolutionary building and thermal insulation material that can be used to build the whole house, no bricks or other thermal insulation.

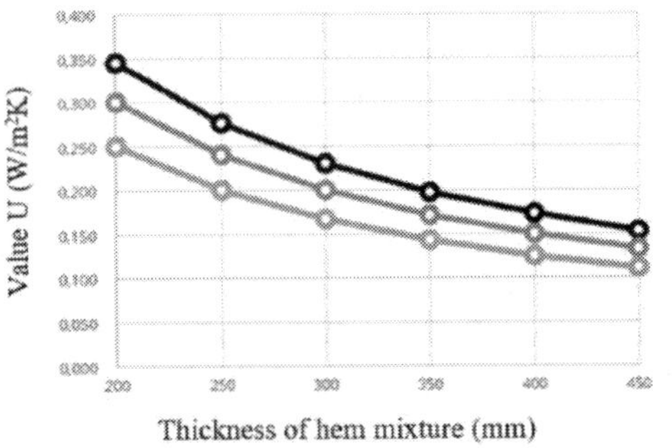

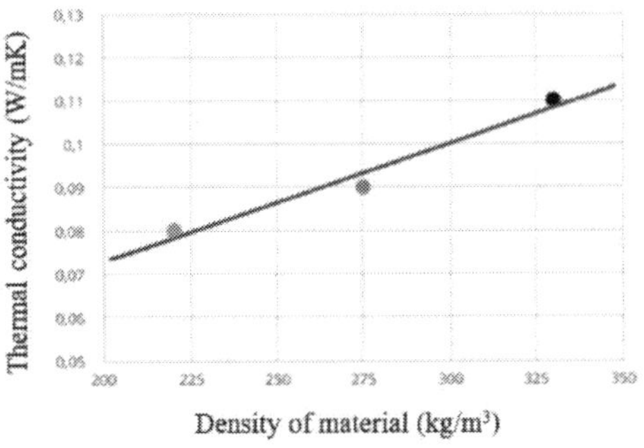

MIXTURE	200	250	300	350	400	450
Roof	0.250	0.200	0.167	0.143	0.125	0.111
Walls	0.300	0.240	0.200	0.171	0.150	0.133
Floor	0.345	0.276	0.230	0.197	0.173	0.153

	Density (kg/m³)	Thermal conductivity k (W/m.k)
Roof	220	0.08
Walls	330	0.09
Floor	375	0.11

Fig. 3. Hemp concrete - value of heat transfer coefficient, thermal conductivity vs density [5].

Cannabis masonry, sometimes also hemp concrete, form a mixture of cannabis (hemp stem inside), hydraulic lime and water.

In practice, lime is obtained by limiting the calcined limestone and burning at temperatures above 825 °C. This process releases CO2 from the process. We are obtaining lime. By addition of water, the slaked (hydrated) lime becomes the heat. However, the classic Hydrated lime is not used, but the so-called Hydraulic lime, which is the one whose components react with water and thus promote a rapid increase in strength and partial water resistance at the time of the hemp-lime building substrate aging [4]. Hydraulic lime contains pozzolans, which include silicates. Lime is again transformed into limestone, respective "stone", due to absorption of atmospheric carbon dioxide. Materials covered by the lime are also transformed. The shive contains a large amount of silica, so contact with lime results in petrification - hardening to fossilization, making the material ideal for building. During the mineralization, the material drains from air the CO2 and purifies it. This material is ideal for construction. The material absorbs carbon dioxide out of the air during mineralization and thus cleans it [5].

The ratio of raw materials varies depending on whether it is a mixture on a wall, roof or floor. The resultant hemp lime concrete composite is lightweight (up to 7 times lighter than conventional concrete) and is therefore easy to work without even heavy technology. It is a recyclable material that offers high thermal and sound insulation.

Table 1. Hemp concrete mixtures according to use [5].

Amount of material per 1 m^3 of hemp concrete according to use			
	Shive (kg)	**Binder (kg)**	**Density (kg/m^3)**
Roof	110	110	220
Walls	110	220	330
Floor	100	275	375
Plaster	200 - 250	500 - 700	700 - 950

Working with the building mix does not require special skill. When building the walls, the hemp concrete is most often squeezed into the formwork. When insulating the roof, it is rather looser, when the floor is finally wiped away, tightening. When plastering, it is classically cast, using a spray gun. For precise plasters it is worthwhile to sift and use only finer particles. Gloves and eyeglasses are good at work, as lime is slightly aggressive to earth.

Apart from the fact that hemp concrete is a universally functional and affordable building material, it is also environmentally friendly, its environmental balance is A+.

Fig. 4. Hemp concrete [4].

Fig. 5. Hemp oil and varnish [1].

4.3 Hemp oils and varnishes

Most natural constructions need more or less wood for their construction to protect them from the effects of mold and pests, outdoor elements also from the effects of weather. Very effective protective in these cases showed hemp coatings.

Hemp oil, like the whole plant, is very resistant to mold and repels the pest. It is a completely natural, biodegradable product, characterized by an increased resistance to soaking the wood with water. This oil also retains a water vapor permeable surface and thus can the wood breathe. As a general rule, the interior construction is sufficient to repaint pure cannabis oil, which, due to its pure natural composition, is also suitable for coatings of furniture, toys etc.

However, since this oil is not completely UV-resistant, it is necessary to treat the exterior elements with other means, such as a hemp lacqueric lacquer, which also contains mineral pigments to protect them from UV rays, and the wood is thus perfectly protected for many years [1].

4.4 Honeycomb plaster

Mixing cannabis, lime and sand creates a unique insulating plaster that can be applied up to a ten centimeter layer. In addition to working well and shaping it, it has the biggest advantage of breathing and regulating air humidity. The clay-hemp plaster looks natural, does not contain pollutants. The hemp plaster absorbs a lot of water and releases it gradually. Slower blasting does not cause it to crack.

4.5 Hemp funiculs, ropes and strands

The rope is made of 100% hemp without artificial impurities. Ropes are manufactured in different diameters and have high strength. They are used in joints for installing windows and frames, sealing fillers in frames, sealing expansion joints, sealing dilatation joints, building and repairing houses, bridges, tunnels, wooden cottages, etc.

4.6 Other building materials of cannabis

In ground and water construction, both classical and minerally enriched cannabis geotextiles can be used. Cannabis plastics are environmentally friendly. For example, they do not emit harshly harmful phthalates and do not cause allergic reactions. Recent research has shown that these plastics exhibit greater flexibility and resistance to pressure than synthetic plastics. It is possible to produce the installation materials needed for building a house, which are otherwise usually made of PVC, such as pipes and knees.

Other applications include unburnt bricks molded from cannabis and clay. Because of the insulating properties of cannabis, clay-hemp masonry does not require any additional insulation layer. However, bricks can not be used to build load-bearing walls (three-quarters of bricks make clay, 25% cannabis).

Thanks to the use of cannabis seals, for example, it is possible to avoid frequent damage caused by condensed water on roof windows in the area of the frame. Hemp felt impregnated with 3-5 % sodium carbonate can, for example, bind moisture out of the surrounding air and then gradually evaporate it. This protects the interior from moisture fluctuations [1].

5 Experience in the implementation and during using of the building

When constructing hemp concrete, it must be remembered that it is a natural material, ie. without different chemical accelerators or other additives. The investor has to expect a longer drying time, approximately 6 - 8 weeks depending on the weather.

According to the designer, the object can be based on spread footing foundation or on slab-on-grade foundation. The supporting structure usually consists of a wooden structure. All the technical equipment of the buildings (water, waste, electricity, etc.) are prepared before the application of hemp concrete. The application of cannabis is then applied by spraying or more often into the formwork. A great mover for mixing a hemp concrete mix is a panmixer without which construction would take much longer, because its volume is more than three times larger than a conventional mixer. The mixture poured into the formwork hardens within a few hours, so that after approximately twelve hours the formwork can be removed and moved further, and the mixture is automatically resistant to pests, molds and fire. In the realization of the masonry, one has to do in one step not only the masonry, but also the insulation and it can be said, for the most part, also the surface treatment, which is a great advantage.

When using hemp concrete for thermal insulation of the roof, the material isolates not only in winter but also against overheating during the summer. In addition, no vapor barrier or vapor barrier films are usually used in the application. Hemp concrete will handle moisture [4].

Classical lime plasters, hemp lime plasters, clay plaster, wooden cladding and others can be used as a surface finish. It is important to ensure good water vapor diffusion. Hemp concrete should not be used in the construction below ground level.

The use of hemp insulation usually requires the same procedures as for mineral insulation.

The ready-made construction is a reward for the investor because the plants, the animals and he feels good in it, do not breathe the spores of molds and other products resulting from excessively moisture-free. Natural materials additionally support the preservation of so much for our health of important negative ions in the air.

Fig. 6. Hemp concrete walls and floors [4].

6 Conclusion

Many building materials with excellent properties are used in the building industry. However, their production has direct or indirect impacts on the environment, as well as products from these materials. For example, stone melts of 1600 °C are required to form mineral-based fibers; for comparison: hemp mats are thermally stabilized at a temperature of 150 to 160 °C. A further comparison can be the amount of production emissions from the processing of petroleum products for the production of synthetic materials, as opposed to the fragrant work of drying and "breaking" of plant raw materials.

Cannabis as a fast-growing annual herb in its growth is highly absorbed by CO2. In the processing and extraction of building materials, CO2 is produced on the contrary. This also occurs in the case of lime. However, lime alone in its reaction, when mixed with water (that is, at the time of application itself), CO2 again in the air again massively absorbs. The result is a passive CO2 balance using cannabis concrete, which derives from both binder and cannabis. For an idea: a detached house with a floor area of 100 m^2 with 500 mm hemp wall thickness, contains about 120 m^3 of hemp concrete. Captured about 13 tons of CO2 will be captured [4].

Cannabis is an excellent alternative to wood. This even surpasses in some respects. Building materials from it are several times more flexible and healthier. Cannabis isolates greatly, breathes and does not attract molds. The fact that it grows very fast for cannabis also speaks. The time for obtaining good wood material is several times longer. It takes 4 months to grow the material. We do not have to worry about allergic reactions - they do not evoke any.

If we are looking for environmentally friendly building material, our health and excellent properties, cannabis use is a possible option.

References

1. M. Urbancová, *Konopí ve stavebnictví,* Czech Republic (2016). Available on https://www.asb-portal.cz/stavebnictvi/materialy-a-vyrobky/tepelne-izolace/vyuziti-konopi-ve-stavebnictvi
2. W. Stanwix, A. Sparrow, *The Hempcrete book: Designing and building with hemp-lime* (2014)
3. S. Allin, *Building with hemp* (2012)
4. P. Majringer, Konopný-beton, Czech Republic (2004). Available on http://www.konopny-beton.cz/
5. M. Šedivý, Konopné stavby, Czech Republic, Available from : http://konopnestavby.cz

Handling construction waste of building demolition

Terezie Vondráčková[1,*], *Luboš* Podolka[1], and *Věra* Voštová[2]

[1]Institute of Technology and Business in České Budějovice, Department of Civil Engineering, 370 01 České Budějovice, Czech Republic
[2]Czech Technical University in Prague, Faculty of Transportation Sciences, Horská 3, 128 03 Prague, Czech Republic

Abstract. Some building defects lead to their demolition. What about construction and demolition waste? According to the Waste Act 185/2001 Coll. and its amendment 223/2015 Coll., which comes into force on January 1, 2017, the production of waste has to be reduced because, as already stated in the amendment to Act No. 229/2014 Coll., the ban on landfilling of waste will apply from 2024 onwards. The main goals of waste management can thus be considered: Preventing or minimizing waste; Waste handling to be used as a secondary raw material - recycling, composting, combustion and the remaining waste to be dumped. Company AZS 98 s. r. o. was established, among other activities, also for the purpose of recycling construction and demolition waste. It operates 12 recycling centers throughout the Czech Republic and therefore we have selected it for a demonstration of the handling of construction and demolition waste in addressing the defects of the buildings.

1 Introduction

The contemporary construction is trying to reflect the requirements of sustainable development. The basic principles of sustainable development include efficient use of energy, water and other resources, the protection of indoor climate of buildings and human health, reduce waste and pollution [1]. Construction and demolition waste is a result of various construction activities, construction, maintenance and reconstruction of buildings, various construction modifications of completed constructions and removal of buildings. Over half of the total production of all waste in the Czech Republic consists of construction and demolition waste. Construction and demolition waste arises:
- when setting up buildings (buildings),
- their maintenance,
- when changing buildings (buildings),
- removal of buildings (buildings).

This waste is currently almost completely used because construction and demolition waste is a significant source of secondary raw materials. The Waste Management Plan of the Czech Republic for the period 2015-2024 sets a target for construction and demolition

* Corresponding author: 13260@mail.vstecb.cz

waste and prepare at least 70% of this waste for re-use. This objective is based on Directive 2008/98 / EC of the European Parliament and of the Council on waste.

Under the term construction and demolition waste (CDW) we can imagine what remains after the modification, repair or demolition of a building. They are bricks, masonry, tiling, wooden structural elements, and anything else that remains after that or that construction. Construction waste and demolition waste are also considered to be remnants of various wiring. So electrical wiring, water and waste piping. Or the excavation rock. All of these materials are recyclable. However, separation is required for their successful recycling (or further processing). First of all, the separation of contaminated components and then the material separation. It should also be noted that such waste can be polluted by materials that legislation perceives as hazardous waste. Then they must be treated as hazardous waste or detoxified for recycling [2].

According to the Waste Catalogue, construction and demolition wastes are included in group 17 and their basic subgroups are listed in Tab. 1.

Table 1. Basic CDW Subgroups, according to the Waste Catalogue of Annex 1 [3].

Catalogue name of the waste subgroup	Waste
17 01	Concrete, bricks, ceramic products and bags
17 02	Wood, glass and plastics
17 03	Bituminous mixtures, tar and tar products
17 04	Metal (including their alloys)
17 05	Soil (including excavated soil from contaminated sites), stones and extracted debris
17 06	Insulating materials and building materials containing asbestos
17 07	-
17 08	Plasterboard construction materiál
17 09	Other construction and demolition waste

The whole CDW process has its origins in the mining of building materials, which we take as primary building materials. After mining, these raw materials and materials are processed on the product, followed by the use of a statement in land or road construction, followed by demolition and ending with landfill removal, whether CDW is used for landscaping or reclamation. An integral component is also the waste generated during earthworks - the excavated soil, which occupies the largest proportion of the volume of construction waste.

A simple breakdown of construction waste is divided into four groups - Figure 1, based on the average percentage of the main CDW components:
- (65 - 75%) Soil from excavations and debris.
- (5 - 20%) Demolition building rubble.
- (10 - 15%) Asphalt mixtures, bitumen.
- (5 - 15%) Waste from construction sites

The share of individual fractions of construction waste is highly variable within the Czech Republic. The reason is different urbanization, industrialization, different construction technologies, density of built-up area and number of demolitions in individual regions.

The production of construction waste in 2008-2015 can be seen in Table 2. Data in the table come from the CENIA - ISOH database. Table 2 lists the groups of construction wastes that are most often produced and therefore their volume is higher than other groups and the groups of the most commonly used construction waste for subsequent recycling and production of recyclates. It can be seen from Table 2 that the production of construction and demolition waste is rather constant. An exception to the amount of production is formed by group 17 05 - soil, debris and stones, which has been declining in recent years.

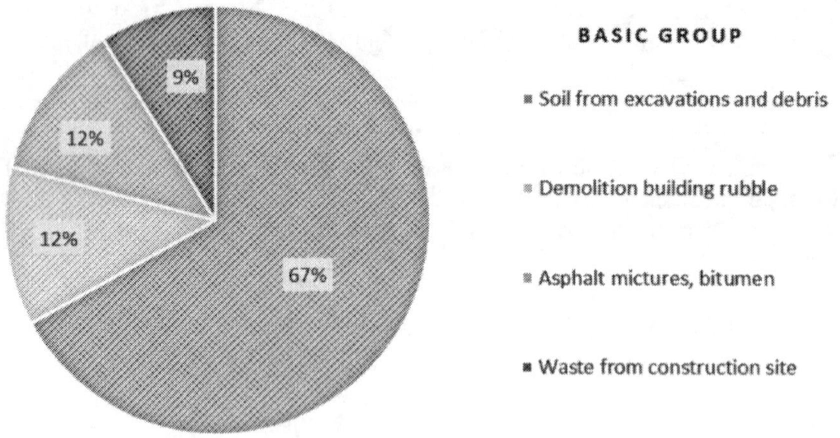

Fig. 1. Segmentation of basic CDW groups.

Table 2. CDW production in the Czech Republic [4].

Cat. n.	**Waste**	**Year**							
		2008	**2009**	**2010**	**2011**	**2012**	**2013**	**2014**	**2015**
		[10^6 kg]							
17 01	Concrete, bricks, tiles and ceramics	2934	2998	3167	3033	3444	3545	3625	3962
17 03	Bituminous mixtures, tar and tar pitch	445	516	466	443	531	560	584	575
17 05	Soil (including excavated soil from contaminated sites), stones and extracted debris	11396	10708	10845	9053	8908	10094	11632	16726
17 08	Construction material on the base of gypsum	6	7	7	8	7	7	9	8
17 09	Other construction and demolition waste	497	580	614	630	496	502	530	620
Total		15278	14809	15099	13167	13386	14808	1380	21891

2 Recycling of construction and demolition waste

As can be seen, under construction waste we can imagine what remains after the demolition of a building. They are bricks, masonry, tiling, wooden structural elements and anything else that remains after one or the other building, road, or the like. Construction waste also includes remnants of various types of building wiring. So electric wiring, water and waste piping. Or the excavation rock. Although it may sound odd, these materials are recyclable. However, separation is required for their successful recycling (or further processing). Above all, the separation of contaminated components and then separation by material. Excavating soil, for obvious reasons, is not necessary to separate. Account should also be taken of possible contamination of materials that the legislation then treats as hazardous waste.

Recycling of construction and demolition waste is largely understood as a "simple" treatment during which its physical properties change. Building wastes are a valuable source of materials that can reduce the need for extraction or production of new raw materials and materials. Based on experience from Europe, it can be said that the recycling boundary (or other recovery of construction and demolition waste) may be close to 90%. With recycling systems, building waste can return to the construction industry as a full-featured material.

The recycling of construction waste began in Czech Rep. since the 1990s. The most common way of recycling CDW is crushing. Subsequently, everything is sorted to make the fractions easy to use for further construction, replacing raw materials from natural sources. The main benefit of recycling of construction and demolition waste is the reduction of the need to extract raw materials and reduce the cost of new building materials. This can limit the opening of new quarries or sandbanks. In addition, the building recyclates, which are replaced for instance natural aggregates, can be purchased at the fractional price of the extracted natural stone.

The concept of CDW recycling technology can be called the complete process of converting unsorted, mixed or different construction waste into sorted recyclates of differently selected fractions of 4 mm to 125 mm. The procedure is such that materials that can be re-offered for sale are first separated by hand. These are mainly brick elements, roof tiles. Other materials are also separated - glass, metals, paper, plastics, wood (with fuel and building resolution) and finally unusable waste accumulated in the appropriate containers, which is taken to a municipal waste landfill. The next step is shredding and sorting.

2.1 Technology of recycling

There are two ways of recycling technology:

Recycling process carried out directly on the site where the CDW is being built - A CDW storage facility is provided at the construction site, its processing and use is directly conditional on the site. The benefits of using this recycling method are to minimize costs and charge for CDW removal and landfill. An important advantage is the use of recycled material for further construction. The disadvantages of using this recycling option include obtaining the necessary recycling permit at the site of construction and providing for the recycling process, ie, the place where the recycling will take place and the need to use the recycled product [5,6].

The recycling process carried out at the place of the permanent establishment. Establishment provide the complex CDW disposal. This activity involves the reception of waste, its processing and, as a result, its sale. The removal and sorting of SDOs into recycling plants is only possible in accordance with the rules of the relevant facility. The benefits of using this recycling process are: minimum space requirements, recycling of CDW types and continuous waste disposal. Among the disadvantages are financial factors such as landfill and transport.

3 Recycling technology at AGS 98, l.l.c.

AZS 98's first recycling center was established in 2004 in Pilsen, Borská pole. The company has built and operated recycling centers in some districts at its own expense. To date, a total of 14 recycling centers have been operated. In some centers, besides recycling of construction waste, they also deal with the crushing of natural stone (Barbora - Jáchymov), reclamation of the area after mining (Uhlířské Janovice). The company operates recycling in both ways, ie at both the waste site and recycling centers.

Thanks to the quality of services and other activities, the company has been certified by Czech Building Standart (CBS) EN ISO 9001: 2001, CBS EN ISO 14001: 2005, CBS OHSAS 18001: 2008 and is a member of the Association for the Development of Building Materials Recycling.

Mobile recycling technology at AZS 98 l.l.c is carried out by bringing a recycling line containing a mobile crusher, screener, loader and possibly an excavator to the collection site of the building waste. The next step is the crushing and sorting according to the required fraction, and then recyclate is used at the construction site. The quality of the resulting recycled material also results from the processes before direct crushing. The maximum quality of the recyclate is ensured by a magnetic separator and a lining strip and a calibrated weight for accurate pricing. Mobile recycling is only possible if the volume of construction materials exceeds 500 tons.

The recycling of CDW at the building waste recycling center takes place similarly, with the difference that construction waste is imported by both the company itself and other suppliers. The reception of construction waste is in accordance with the valid operating rules, which are approved by the relevant Regional Authority. The CDW is then accepted at the recycling center and classified according to the Waste Catalog. Crushing and sorting according to the fractions takes place after the accumulation of quantities economically beneficial for recycling. Whether recyclates are suitable for further use is verified by regular testing by an accredited testing center. It assesses not only the environmental impact, but also technological analyzes in terms of their use. All of these processes take place at fourteen recycling centers of AZS 98 l.l.c., located in four regions of the Czech Republic.

3.1 Material flow

The material flow is considered to be CDW's movement from its production, utilization or further disposal. Material flows at AZS 98 are listed in Table 3 [7].

Table 3 compares the volume of intake of construction and demolition waste with expenditure of the recyclate. This is the entire CDW income that was processed and issued by AZS 98 l.l.c. The difference in values may indicate either a larger quantity of the material being issued or, on the contrary, a larger quantity of material received. If the delivery value is greater than the acceptance value, it means that the demand for recycled material in the monitored center was higher and therefore the material was taken out of stock or out of stock from the previous year. This process is similar in most recycling centers. Graphic processing of Table 3 is in Figure 2.

Table 3. Material Flows Survey for 2014 at AZS 98 l.l.c.

Amount of CDW receipt and issue of recyclate in 2014					
Center	**District (NUTS4)**	**Region (NUTS3)**	**Receipt [t]**	**Issue [t]**	**Difference [t]**
Valcha	Plzeň	Plzeňský	74 585.49	76 200.43	1 614.94
Jáchymov	Karlovy Vary	Karlovarský	0.00	2 077.74	2 077.74
Stříbro	Tachov	Plzeňský	6 583.11	5 907.30	675.81
Blovice	Plzeň – jih	Plzeňský	2 446.59	2 872.44	425.85
Podbořany	Louny	Ústecký	5 621.65	3 988.10	1 633.55
Sadov	Karlovy Vary	Karlovarský	11 315.12	12 741.49	1 426.37
Tachov	Tachov	Plzeňský	2 103.21	4 589.99	2 486.78
Újezd	Domažlice	Plzeňský	2 623.55	445.73	2 177.82
Cheb	Cheb	Karlovarský	8 043.86	664.00	7 379.86
Ostředek	Benešov	Středočeský	4 623.22	1 291.50	3 331.72
Rokycany	Rokycany	Plzeňský	21 612.17	14 941.93	6 670.24
Zavlekov	Klatovy	Plzeňský	8 064.34	6 655.94	1 408.40
Chmeliště	Kutná Hora	Středočeský	1 228.32	0.00	1 228.32
Libočany	Louny	Ústecký	3 200.00	25,00	3 175.00
Total in tonnes			**152 050.63**	**132 401.59**	**19 649.04**

Although the direct waste recycling process does not run evenly every day, AZS 98's concept is to use recycling centers as "bottomless landfills". The principle is to recycle the received SDO volume at the recycling center and sell the same volume of CDW already recycled. Which, according to the volumes, meets this company.

It is also important to mention the use of recycled material by the company itself, not only for backfills and the like, but also for the production of brick concrete, from which the concrete cubes for dry reinforcement are also produced. The company is still trying to improve the quality of the recyclate and its further use, therefore it has long been cooperating with CTU in Prague, the Faculty of Civil Engineering. There are several projects underway to research possible recycling of recycled material.

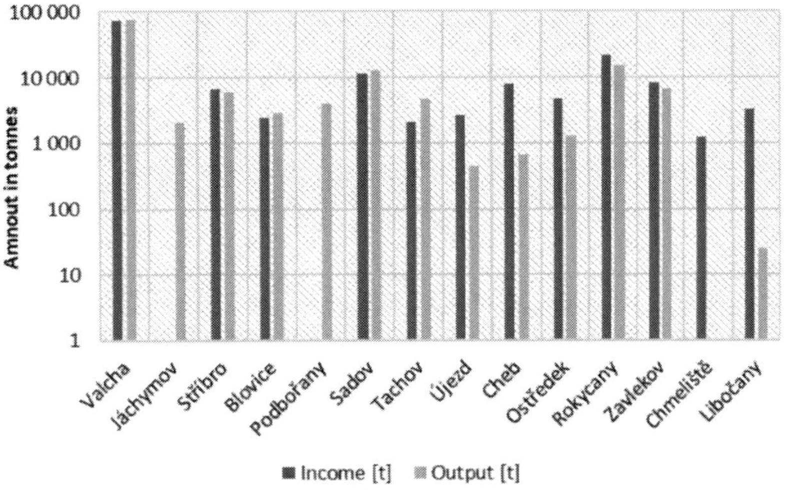

Fig. 2. Comparison of Material Income and Output at AZS 98 l.l.c.

4 Conclusion

Company AZS 98 l.l.c. Successfully solves the situation with the handling of construction and demolition wastes, as evidenced by the high weight of CDW used [8]. The system of disposal of construction and demolition waste in this company may be an example of good recycling and general waste management for other companies. The contribution came from the preparation of the grant plan.

References

1. M. Kraus, D. Kubečková, AMR **1041**, 3-6 (2014)
2. M. Škopán, Systems for the recycling of construction and demolition waste. Available on http://stavebni-technika.cz/clanky/systemy-recyklace-stavebnich-a-demolicnich-odpadu
3. CENIA. Available on http://isoh.cenia.cz/groupisoh/
4. Czech Statistical Office. Available on https://vdb.czso.cz/vdbvo/uvod.jsp
5. V. Voštová, V. Vlastimil, *Logistics of waste management (CTU*, 2009).
6. S. Szabo, Journal of Applied Economic Sciences **10**, 1153-1159 (2015)
7. The official website of the company AZS 98 s.r.o. Available on http://www.azs98.cz/
8. T. Vondráčková, V. Voštová, V. Nývlt, Matec Web of Conferences **93**, 03005 (2017)

Facade renovation - replacement and restoration of the panels in a monument protected object

Michal Novotný [1,*], and *Barbora* Nečasová [1]

[1] Brno University of technology, Faculty of civil engineering, Institute of Technology, mechanization and construction management, Veveří 331/95, Brno, 60200, Czech republic

Abstract. The article deals with problems of reconstruction of the facade and the associated problem of replacement or repair of the panels. In conventional buildings it is a smooth operation, but it is problematic in monument-protected objects. In the case of a common building, it is possible to choose any modern panels and simply replace them, but for historical objects we have to follow the claims and the rules of monument protection. In practice, it usually means the impossibility of use of modern panels, but at least a combination of old and modern technologies. Another possible solution to the problem is renovation, or repairs to the original state of the existing panels, of course with respect to the functionality of panels. The implementation of such repairs must always be based on the technical and historical survey of the condition of the object and the repairs must be professionally designed. Subsequently, corrections are made, during which it is necessary to pay particular attention to observance of the technological procedures, rules and instructions particularly in terms of monument protection. However, the functionality of the works or elements made with regard to the quality of the environment within the building is not negligible. A common problem is the lack of control of technical requirements and functional requirements. Underestimation of the problems then leads to difficult repairs. The article points to the mistakes and problems of one such construction project on a historically protected chateau building.

1 Introduction – Purposes and problems of the construction

At the turn of March and April, our institute was approached by the director of the Jevišovice Retirement Home, located in the new chateau in Jevišovice. At the time this object was undergoing the first phase of the facade restoration and we had to make a professional assessment of the works that had been carried out with regard to their quality, timeframe and functionality. According to the director, the quality of the works was not sufficient, the technical supervision of the construction was not performed well, and some parts of the implement work differed from the project documentation. The construction was also delayed from the original plans. Authors were entrusted to conduct local surveys and to make the assessment. This assessment can be divided into the following key points:
- Finding and exploring problems and input parameters,
- Local investigation and identified defects,
- Recommended solution.

* Corresponding author: novotny.m1@seznam.cz

2 Finding and exploring problems and input parameters

Our investigation related to the new chateau in Jevišovice, which, together with the castle park, is included in the list of cultural monuments. Due to this fact, the repairs and renovations are handled in cooperation with the Monument Office. Therefore, one part of the project documentation was a restoration project, but also granulometric measurements, a record of the facade elements, a static assessment, and also a technical report describing the demolition and reconstruction works. This documentation was provided to us for review, and some of its necessary parts were also provided to us in electronic form. At the same time, we were provided with records of the inspections, and additions to the contract for work between the ordering party and the contractor (the contract itself is confidential until the end of the project). The overall intention was to restore the facade of the new tower, repair the worst damaged parts of the adjoining facades, and to replace or refurbish the doors and windows on these facades. The director of the retirement home, as the representative of the client, raised the following questions during the opening meeting:

- Is the scope of the work already implemented in accordance with the documentation and the contract for work?
- Is the quality of work in accordance with the applicable laws, standards, documentation and contract for work?
- Were any defects and failures found during the inspections and the local survey, which are not in accordance with the required state of implemented work according to the documentation and the contract for work?
- Were any defects been found, which strictly prevent the handover of the work? (This was linked to the completion deadline and the handover of the work.)

Based on these questions and communicated facts a local investigation was carried out, which focused mainly on functional and fundamental defects of the work. The professional commission's objective was to answer the above mentioned questions. In the first phase of the background and documentation survey, a time schedule for corrections was found, which contained delays due to climatic conditions and consequent failure to meet deadlines. Furthermore, insufficient definition of the conditions to make a decision whether to replace or repair the panels was found – even though in this situation the decision should be made based on the documentation with a list of joinery elements, in which it was determined whether there should be an exchange or a repair. On the basis of the inspections and records, a settlement agreement was found, which read that several elements were to be repaired instead of replaced. Another finding was the inadequate definition of the work of the technical supervisor, whose duties were determined only in a framework. The next step was the local investigation and record of the detected defects.

3 Local investigation

Our local investigation was conducted three times, with time gaps, so that the contractor could correct any defects and errors. The dates were set at 5/4/2017, 19/4/2017 and 24/4/2017. Based on the first investigation, basic mistakes were identified and the director, who is also the client, was recommended several steps to take as a remedy. Also, the deadlines allowed the executive company to correct its mistakes. The measuring devices used were: a measuring wedge, a tape-measure, and a camera. Laboratory tests and examinations were not required.

The basic errors found were:

- surface defects of the repaired parts – on the facade and panels - dirt, scratches, residual paints, mortars and adhesives, uneven coloring, unfinished coatings (fig. 1a), missing sealing, surface damage etc.
- incomplete or not started repairs of the elements (fig. 1b), missing elements such as railing, tiling of ledges, and roof entrance,

Fig. 1. Aesthetic defects: **1a** - Left - incomplete facade surface, **1b** - right - unfinished window frame coatings [1].

- mechanical problems of the functionality of the panels - impermanence (large gap between wings and frame - up to 5 mm joints – fig. 2b) and crossed or hinged casements, failure of the fitting or its absence (fig. 2a), crossing of the casement in the frame, missing seals,
- aesthetic defects - imperfections of paintings - different shades, splashes of paints and other materials on final surfaces, scratched glasses, paintings, etc.

Fig. 2. Functional defect – **2a** - left - inoperative window fitting, **2b** - right - large gap between the wings of the exterior door - measured by the measuring wedge - gap of approx. 6 mm, [1].

Our local survey focused mainly on the functional defects and unfinished work in order to answer the questions - connection with the handover and completion of the construction. These were summarized in the list and made available to the client. As part of the control investigations in the next two dates, some defects have been rectified and most of the missing elements have been added. The remaining problems to solve for the client, the site manager, and the contractor were mainly the defects connected to the function of the panels, especially the windows. These were mainly flaws in the panel´s functionality - i.e. tightness, mechanical fitting, opening and closing, safety and visual appearance of the surfaces.

4 Standard and law requirements

The assessment of the problems was carried out mainly with regard to the applicable Building Act and related binding decrees, laws and standards. Unfortunately, this topic was not specified in the documentation, so the customer was in a difficult position to negotiate with the contractor. Especially the aesthetic problems and problems with the functionality of the panels were not captured. Subsequently, when assessing the state of the object and forming the expert opinion a special attention was paid to the client's requirement on the functionality and usability of the repaired parts, and also the visual appearance of the panels and facades. Above all we evaluated the mechanical functionality of the elements - closure, fittings, tightness of the casements in the frames, the shape and design of the panels, the finishing, and the overall condition of the casement. At the facade we observed primarily the finishing surfaces and the extent of repairs according to the documentation. The recommended solutions were in accordance with the requirements of laws and standards.

Following the Building Act 183/2006 [2] and its amendment 350/2012 [3], the decree 268/2009 Coll. about Technical Requirements for Buildings [4], specifies in Section 8 the essential requirements for the protection of health of persons and animals, healthy living conditions and the environment, protection against noise, and energy saving and thermal protection. These requirements, however, are not compatible with the poorly functioning or leaky windows. Section 10 - General Requirements for the Protection of Health, Healthy Living and the Environment - states that the building and its components must be designed and constructed in such a way as not to endanger the people by the consequences of insufficient thermal and sound insulation properties of the structures and its components. Because this is a home of retired people with permanent and temporary residence here, the thermal and noise conditions are paramount. It is also closely related to Section 16 of the Decree, which deals with energy saving and thermal protection, which states that the construction must guarantee the thermal comfort, required thermal parameters of the structure, thermal moisture conditions (leaky windows = draft, large heat losses) and low energy consumption of the building (to its possible extend – it is a historic building). Section 26 deals with the properties of the panels, and it is important to mention that "The construction of the apertures must be of sufficient stiffness that no collapse, hanging or other deformation occurs during normal operation and must withstand the load including its own weight and wind load even when the casement is open, without damaging, shifting, distorting or worsening the function." This means that the problems with the casements crossing, unattached frames and large spaces between casements and window frames are inadmissible. However, some problems can be solved by means of seals and additional structures. Parameters of air permeability and thermal parameters are set by standards. Standards also define the shapes and tolerances of the panels during the installation as well as the quality of their fabrication.

In the assessment of the state of the structures and the panels - especially the windows – we derived mainly from the standards (Czech Technical Standard – hereinafter CTS) 746101 for Wooden Windows - basic specification [5], and CTS 746077 for Windows and External Doors - installation requirements [6].

The standard CTS 746101 for Wooden Windows states the following facts - point 8 Appearance and Finishing – "Wooden windows must be built with a complete surface finishing applied on all surfaces after assembly and before glazing, including surfaces, which become hidden after assembly. The appearance of the surface treatment is judged from a distance of 1 m and compared to a reference sample. The degree of gloss and color shade must be uniform throughout the surface and correspond to the paint used. Surface treatment of wood must cover its entire surface, scratches and similar damage must be repaired. Final surface treatment must be continuous without surface defects.

Surface defects are not considered to be:

- a slight trace of the characteristics of the technology used for the surface treatment with the use of paints, and marks caused by wood grinding with sandpaper higher than No. 80;
- visible relief of wood texture, especially with the use of water-borne paints;
- color difference of surface depending on wood texture (spring and summer wood) and color difference of surface depending on wood absorption;

- allowed are hair cracks on the coat in the corner joints of the frames and casements but they can only reach a maximum of 1 according to CTS EN ISO 4628-4 [7] – i.e. very small with a size visible only at magnification (according to the above standard, up to size of 0.2 mm);
- it is allowed to mechanically secure the components of the panels such as glazing strips, cover rails, etc.
- allowed are hair cracks in the vicinity of fasteners (for example fittings) that are not subjected to direct weathering."

Point 9 - Installation - specifies the variance tolerance, for us were important the values of the rectangular dimensions of the frames: the difference in diagonal lengths, which is 3 mm for windows up to 1800 mm, and 5 mm for windows up to 3000 mm. The gap between the contact surfaces of the casements and the frames, and the frames with respect to each other, is permitted at the maximum value of 2 mm - which can be considered difficult to repair, and therefore, an additional solution is needed, for example by sealing.

In terms of the standard CTS 746077 for Windows and External Doors - Requirements for Installation [6] was important part no. 4 Design Requirements, subpart 4.3 Construction Opening, which reads: "Built-in windows and external doors must not impair their performance or become a health and safety risk for people. The maximum permissible deviation of the rectangularity of the frame profile (deflection of the frame profile to the longitudinal axis) of the embedded product is 3 mm for length and width up to 2 000 mm (inclusive), and 5 mm for length and width over 2 000 mm. The maximum permissible deviation of the vertical and horizontal dimensions of the frame of the embedded product of size up to 3,000 mm (inclusive) is 2 mm per meter, but overall maximum of 3 mm. In the case of strip windows assembled from individual frames that are reinforced, the tolerances apply to individual frames. The maximum tolerance for the rectangular dimensions of the frames is 3 mm for windows and doors up to width of 1 500 mm and height of 2 200 mm (inclusive) and 5 mm for windows and doors of width of 1 500 mm and height of more than 2 200 mm and less than 3 000 mm".

In terms of control during the installation of the elements, it is important to check the following points:
- design and properties of the panels, vertical and horizontal installation, rectangularity;
- anchoring the product to the aperture and making the connecting joint before it's covered
- opening and closing function of the moving casements;
- appearance of the product regarding possible defects and damage.

A record in form of a transfer protocol is made, which states the progress and completion of the work. So much to the background in terms of standards and laws. Unfortunately, there are not many legal documents, which deal with the finishing or the installation of the panels. There are more standards, but they are not binding unless they are required by the contract of work.

5 Conclusion – summary and recommended solutions

After evaluating the findings and assessing the problems resulting from the local investigation, various repairs were recommended – for instance - the removal of defects and faults during the subsequent construction works. Unfortunately, some of the defects were not of a character to be repaired – it was recommended to replace the concerned panels. This, however, mainly concerned the repairs not yet implemented. On the basis of all the facts, the following conclusions can be drawn:
- the need for high-quality documentation with clearly defined responsibilities of the participants, with specific workflow plans including details, precise steps and procedures, and the specification of the working hours of the contractor or the site manager,
- the need for a detailed contract of work (hereinafter CoW) with precise obligations and requirements for the individual participants in the process, including a description of their activities – i.e. no subsequent works cannot be required, if they do not form part

of the contract – e.g. the technical supervisor was not obliged to check the aesthetic parameters of the panels or the façade,
- clearly define the parameters of repaired structures – either renovation, replacement or repair of elements - not to change these parameters during the construction based on unilateral decision and, in case of uncertainty of the state of the construction or a problem solution, consult an impartial expert,
- document immediately any delay in construction or problems arising during construction by photographs or records – e.g. non-standard temperature conditions, non-observance of working hours - influence on later assessment in case of failure to observe CoW terms,
- ensuring a high-quality site manager and writing a detailed contract, defining his duties, especially if there is a requirement for control beyond the applicable standards and laws – e.g. the aesthetic aspect of the panels,
- insist on detailed element controls - especially when it comes to their functionality and quality - swift checks may not reveal, for example, unattached window casements, inoperative fittings or hanging casements (cannot be detected without measurement),
- not to accept the work with the promise of completing it after the handover or after the deadline - strictly insist on observance of the contractual conditions – i.e. the necessity of good contract of work.

Clearly, the assessment implies the need for quality documents, contracts and, above all, for clearly defined obligations of the participants, especially the site manager. Any failure to comply with the qualitative requirements will result in problems with the usage and warranty or post-warranty repairs.

Outputs and defects detected will also be used to adapt the working methods and test procedures for new facade technologies, which are addressed in grants from the Institute of Technology, Mechanization and Building Management at the Faculty of Civil Engineering with numbers FAST-S-17-4329 and FAST-S-17-4255.

References

1. Author´s archive. On site photographs. Jevišovice (April 2017)
2. *Act no. 183/2006 Coll. on land-use planning and building regulations (Building Act).* (Collection of Law. Prague, 2006).
3. *Act no. 350/2012 Coll., amending Act no. 183/2006 Coll., on spatial planning and building regulations (Building act), as amended, and certain related acts* (Collection of Law, Prague, 2012).
4. *The decree no. 268/2009 Coll. about technical requirements for buildings* (Collection of Law, Prague, 2009)
5. *ČSN 746101 Wooden windows – basic regulations* (The office for standardization, metrology and testing, Prague, 2015)
6. *ČSN 746077 Windows and exterior doors - Installation requirements* (The office for standardization, metrology and testing, Prague, 2014)
7. *ČSN EN ISO 4628-1 Paints and varnishes – Evaluation of degradation of coatings – Designation of quantity and size of defects, and of intensity of uniform changes in appearance – Part 1: General introduction and designation system* (The office for standardization, metrology and testing , Prague, 2016)

Point thermal bridges of insulated pitched roof structures

Jiří Šál[1,*], *Daniela* Štroufová[1] , and *Petra* Bednářová[1]

[1]Institute of Technology and Business in České Budějovice, Department of Civil Engineering, 370 01 České Budějovice, Czech Republic

Abstract. The current demands on building insulation are continuously increasing. It is understood that the lower the heat transfer coefficient of a particular part of a construction is, the greater the importance of systemic thermal bridges. This article compares the individual systems of insulation of pitched roofs in terms of the heat transfer coefficient. The focus is on the size of the point thermal bridges in rafter thermal insulation systems and determines their impact on increasing the overall heat transfer coefficient. However, it should be noted that point thermal bridges are individually very small and combined only contribute to 2% of the overall heat transfer coefficient of parts of a structure.

1 Introduction

The aim of the research in this paper is to determine the size of point thermal bridges in on-rafter thermal insulation systems and to determine their influence on the increase in the total heat transfer coefficient U in relation to the number of anchors and the insulation type. An analysis was carried out on the influence of the position of thermal insulation in an insulated pitched roof structure on the heat transfer coefficient. For the analysis we distinguish between sub-rafter, on-rafter and inter-rafter thermal insulation systems. The inter-rafter system itself is no longer acceptable by today's requirements for heat transfer. The sub-rafter insulation system reduces the inner space of a building, whereby space is lost that might otherwise be used for a different purpose. This is why the inter-rafter and sub-rafter systems are combined with other systems. At present, the following three variants are the most commonly used:

- inter-rafter + sub-rafter,
- inter-rafter + on-rafter,
- on-rafter.

In the first two systems the rafters act as thermal bridges. In contrast, the in-rafter system is often considered to be thermal bridge free. The on-rafter insulation system of pitched roofs has point thermal bridges at the points of anchoring (screws), which increase the heat transfer coefficient.

* Corresponding author: sal@mail.vstecb.cz

2 Comparison of thermal insulations of pitched roofs according to their location in a structure

A comparison was made of the three thermal insulation systems: inter-rafter + sub-rafter; inter-rafter + on-rafter; and on-rafter, using four thermal insulation thicknesses: 200 mm; 240 mm; 280 mm; and 320 mm. The values necessary for the comparison were calculated using the Svoboda software – Heat. The thermal insulation was composed of basalt fibres ($\lambda = 0.043$ W/(m·K) and was applied according to the recommended systemic solution of the manufacturers.

As can been seen in Figure 1, the differences in the heat transfer coefficient for the inter-rafter + sub-rafter system and the inter-rafter + on-rafter system are relatively small. However, these systems differ substantially when compared to the purely on-rafter system. This system has the lowest heat transfer coefficient and is therefore the best from this point of view. The system thermal bridges (rafter influence) were considered in this comparison, however the point thermal bridges were not.

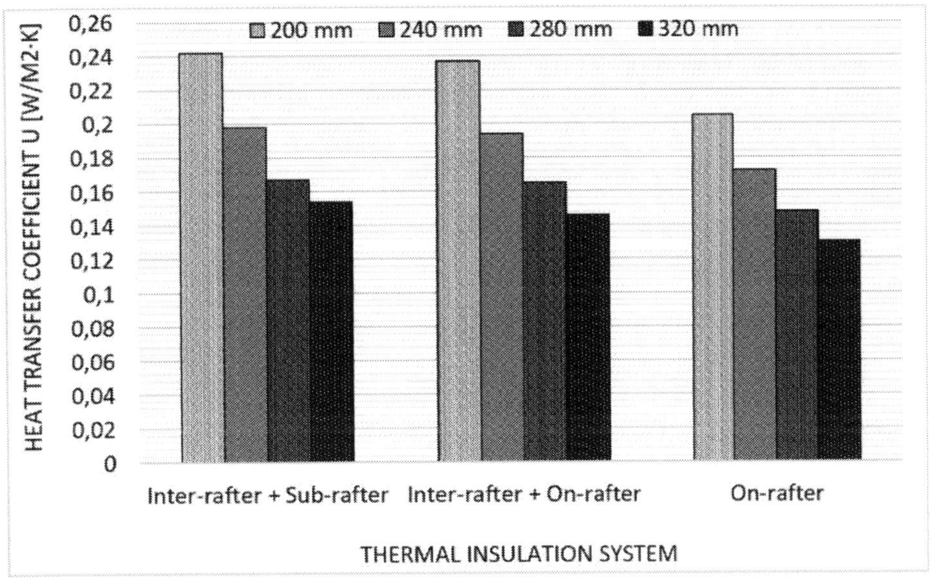

Fig. 1. Comparison of heat transfer coefficients in relation to systemic thermal bridges (wooden rafters) for pitched roof insulation systems [3].

3 Point thermal bridges

The calculations and comparison in the previous section show that the on-rafter insulation system is the most advantageous in terms of the heat transfer coefficient for all thermal insulation thicknesses. This system also has the advantage that the location of the thermal insulation layer on the rafters is compact and uninterrupted, as is not the case when the insulation is placed between the rafters. Under this system no systemic thermal bridges occur because of the poor thermal insulation properties of the wooden rafters. Nevertheless, point thermal bridges do occur where there are anchoring elements which are required to fix the thermal-insulation elements, or the supporting structure, into place. To illustrate this, three on-rafter pitched roof thermal insulation systems are described and quantified below. The systemic anchoring elements prescribed by the corresponding manufacturers were used. Different thermal insulation thicknesses were also applied for the comparison.

4 Selected thermal insulation compositions

System No. 1 - Bramac THERM TOP with insulation thicknesses of 140 mm, 160 mm, 180 mm and 200 mm. Figure 1 shows the composition of the system. The insulation material is rigid polyisocyanurate foam (PIR). Bramac THERM TOP systemic screws made of nitride carbon steel with a Duracoat coating were used as the anchoring elements. The length of the required screws was defined by the manufacturer and subject to the thickness of the thermal insulation and boarding. For the purpose of the analysis 300 mm, 330 mm and 360 mm long screws were used respectively (see Figure 2).

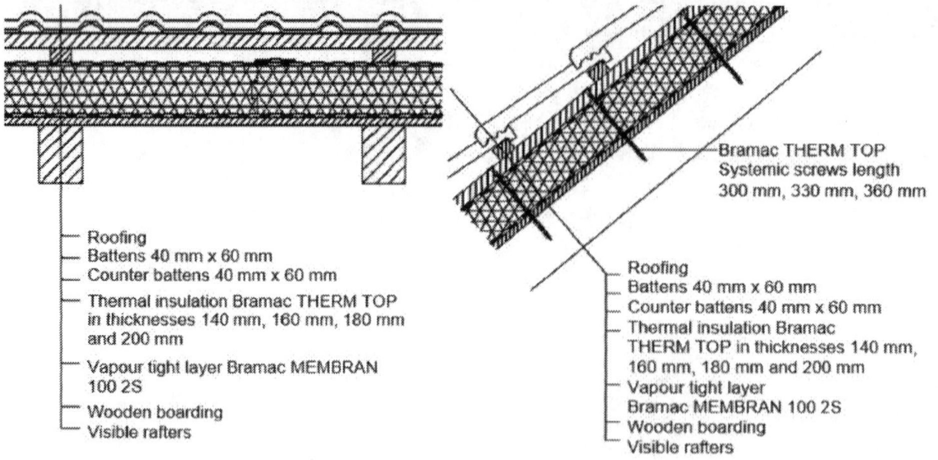

Fig. 2. Cross sections of the Bramac THERM TOP system [3].

System No. 2 - Isover UNI system with basalt fibre thermal insulation as recommended by the manufacturer for on-rafter systems. Thermal insulation thicknesses of 200 mm, 240 mm, 280 mm and 320 mm were compared. Twin UD systemic double-thread screws and Isover TRAM blocks were the bearing elements for the system. The TRAM blocks are manufactured in two variants - mineral wool blocks and foam polystyrene blocks. Both types were used for the calculations. The length of the screws used, from 400 mm to 520 mm, depended on the thickness of the blocks. Figure 3 shows cross sections of the system (perpendicular to the rafter and in the plane of the anchoring screws).

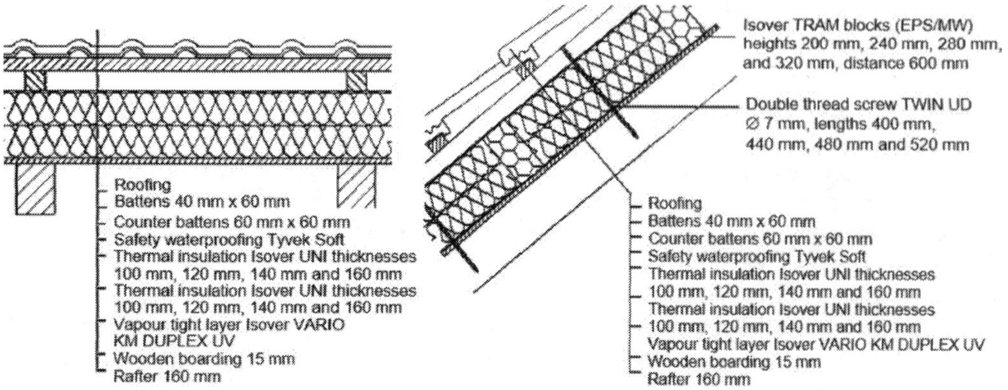

Fig. 3. Cross sections of the Isover UNI system [3].

System No. 3 – ISOTEC using panels made of rigid polyurethane self-extinguishing foam (PUR) with closed cells. The panels are covered with a layer of aluminium foil and have integrated roof battens for roof laying. The panes are anchored to the bearing structures through the battens by dowels or steel screws. The length of these prefabricated panels is 3.9 m. They are manufactured in thicknesses of 60 mm, 80 mm, 100 mm and 120 mm. For the purpose of the analysis only 120 mm thick panels were used. The presence of the integrated roof battens on the panels means that the panel thickness determines the batten distance. Each panel is anchored to each rafter. This means for example, with a rafter distance of 1 meter and a panel thickness of approx. 34 mm the number of anchors is 3 per sq. m. Figure 4 shows cross sections of the ISOTEC system.

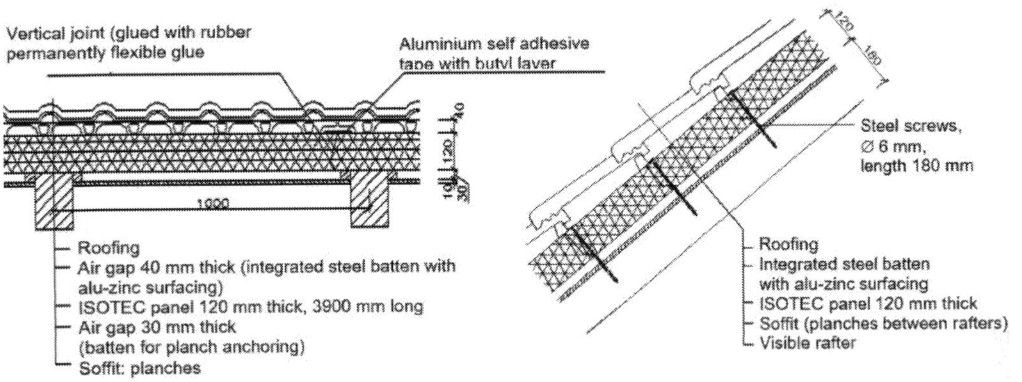

Fig. 4. Cross sections of the ISOTEC system [3].

In all the above systems the distance and the number of anchoring elements depends on the axial distance of the rafters, roofing type (light, heavy), roof inclination, and on the climatic area. Two to three anchoring elements per square meter of roof is considered to be the norm. For the analysis the heat transfer coefficient is calculated for 2 and 3 anchoring elements and for the value in between i.e. 2.5 anchoring elements.

5 Calculation procedure

The calculations were performed by means of the QuickField 5.10 software. The software enables us to quantify the point thermal bridges caused by the anchoring elements in an on-rafter thermal insulation system. It also enables us to create a two-dimensional thermal field in the x; z coordinate system, as well as solve problems with rotational axes and the cylindrical arrangement in the r; z coordinate system. Small inaccuracies occur in the calculations due to the bar elements in the composition e.g. rafters and battens, with which the anchoring elements directly intersect. These elements have to be adapted in the shape. The anisotropy of these elements can also not be considered in the calculation. The thermal conductivity of these elements was therefore entered as being perpendicular to the fibres, whereby its value is substantially lower than that of the thermal conductivity along the fibres. This also eliminated the imperfect shape input. These small inaccuracies can be disregarded as their values are in hundredths (m.W)/K, which is not significant for the evaluation of thermal loss from low energy houses.

6 Point thermal bridges and their influence on heat transfer coefficient

The values of the point thermal bridges χ [W/K] for the individual thermal insulation thicknesses and the heat transfer coefficient U [W/m2·K] not including these thermal bridges were determined for all the above described systems. The value U, including the point thermal bridges for various numbers of screws per square meter, was subsequently calculated, as was the percentage increase in U in relation to the number of anchors and the type of thermal insulation. The calculated values are summarized in Table 1 and graphically presented in Figure 5.

Table 1. Increase in heat transfer coefficient [3].

Identification	Thermal insulation thickness [mm]	Heat transfer coefficient *U* not considering TBs [W/m²· K]	Point thermal bridge size χ [W/K]	Heat transfer coefficient *U* including the point thermal bridges for various numbers of screws [W/(m²· K)]			Percentage increase in heat transfer coefficient [%]		
				2	2.5	3	2	2.5	3
Bramac THERM TOP									
A	140	0.156	0.0066	0.169	0.173	0.176	108.3	110.9	112.8
B	160	0.137	0.0062	0.150	0.153	0.156	109.5	111.7	113.9
C	180	0.123	0.0059	0.135	0.137	0.140	109.8	111.4	113.8
D	200	0.111	0.0056	0.122	0.125	0.128	109.9	112.6	115.3
Isover UNI + foam polystyrene blocks									
E	200	0.176	0.0057	0.187	0.190	0.193	106.3	108.0	109.7
F	240	0.148	0.0052	0.158	0.161	0.163	106.8	108.8	110.1
G	280	0.127	0.0047	0.137	0.139	0.142	107.9	109.4	111.8
H	320	0.112	0.0044	0.121	0.123	0.125	108.0	109.8	111.6
Isover UNI + mineral wool blocks									
I	200	0.185	0.0057	0.196	0.199	0.202	105.9	107.6	109.2
J	240	0.155	0.0052	0.166	0.168	0.171	107.1	108.4	110.3
K	280	0.134	0.0048	0.143	0.146	0.148	106.7	109.0	110.4
L	320	0.118	0.0044	0.127	0.129	0.131	107.6	109.3	111.0
ISOTEC									
M	120	0.194	0.0037	0.201	0.203	0.205	103.6	104.6	105.7

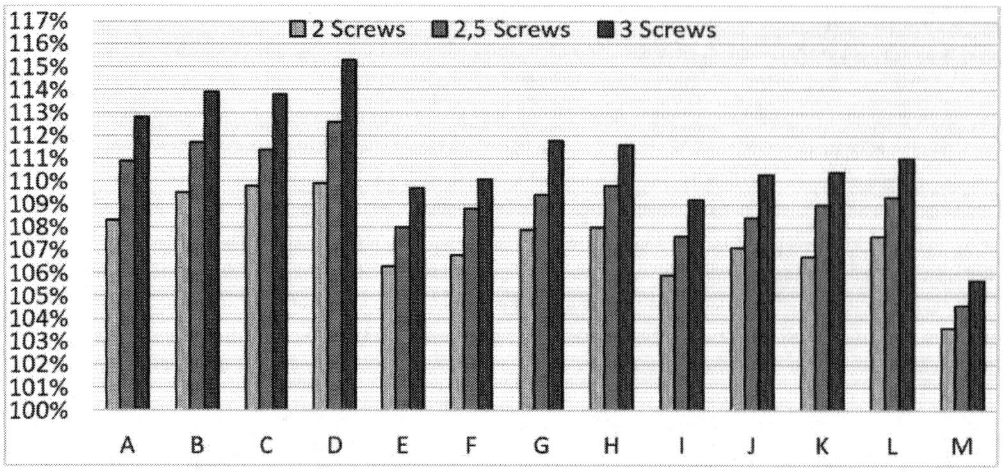

Fig. 5. Percentage increase in heat transfer coefficient U in relation to the type of thermal insulation and the number of anchors [3].

As the results show, there are considerable differences in the change in the heat transfer coefficient between the individual on-rafter systems caused by point thermal bridges even where the thickness of the thermal insulation is the same. Even though the values of the point thermal bridges are relatively low, they play an important role with regards to the low values of the heat transfer coefficient and their number per square meter.

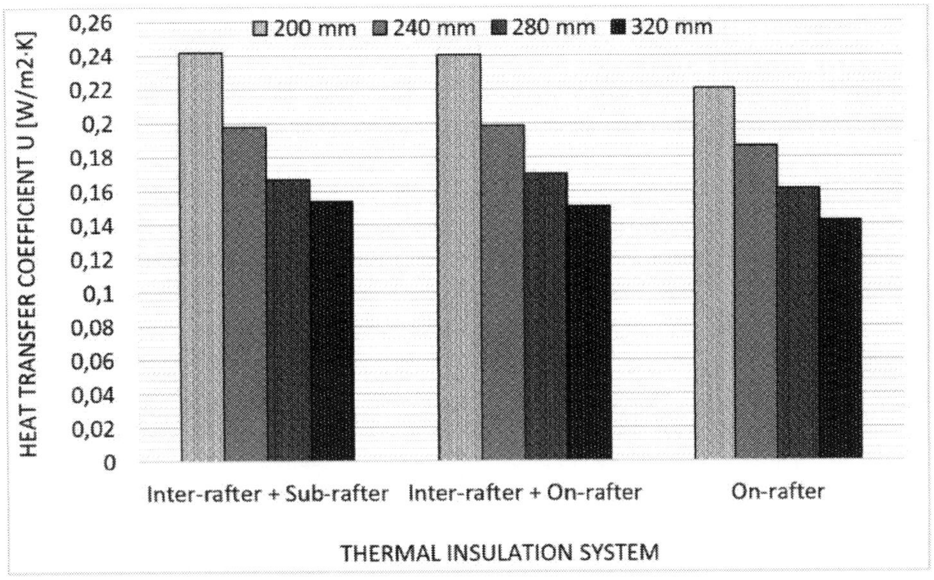

Fig. 6. Comparison of heat transfer coefficients in relation to systemic and point thermal bridges for pitched roof insulation systems [3].

7 Conclusion

The issue of systemic and point thermal bridges is of great importance in the field of thermal technology. The ever increasing requirements with regards to the thermal properties of constructions means that there is increasing pressure to resolve the issue. Even though thermal bridges contribute substantially to total thermal losses, they do not always receive the attention they should.

Having the lowest possible energy requirements should be the aim of each construction project, irrespective of whether it concerns a reconstruction or a new build. It is therefore necessary to limit energy requirements as much as possible, or at least to accurately determine the places where thermal loss occurs in a construction, or a part thereof, and to try and find a constructional solution that will resolve the problem. The lower the heat transfer coefficient of a particular part of a construction, the greater the importance of systemic thermal bridges.

References

1. R. Šubrt, P. Charvátová. *Passive Houses 2013: Collection of Contribution,* 111-116, (STU Bratislava, Košice, 2013)
2. R. Šubrt. *Thermal bridges: for low energy and passive houses: 85 Well-proven and Calculated Construction Details* (Grada, Prague, 2011)
3. P. Bednářová, D. Štroufová. *Point Thermal Bridges – Research Assignment* (ITB, České Budějovice, 2013)
4. Scudder Roofing, Available on http://www.scudderroofing.com/roofing-roofs-history.htm

Moisture problems of functional joints of wooden windows – causes analysis

Adela Palková[1,*], *Milan* Palko[1] , and *Peter* Buday[1]

[1]Department of Building Constructions, Slovak University of Technology in Bratislava, Faculty of Civil Engineering, Radlinského 11, 810 05 Bratislava, Slovakia

Abstract. The article addresses the issue of the humidity regimen of a functional wooden window joint. Current energy-saving trends in the buildings bring secondary effects associated with indoor climate change. Increase of the thermal-technical quantification, non-breathability, and diffusion resistance of the building envelope without the solution of controlled ventilation brings an extreme rise of the interior moisture. The passage of moisture through a functional joint is ensured mainly by exfiltration. The energy standard applicable after the year 2020 assumes buildings with almost zero energy need that cannot be reached without a recuperation system. The concept of the recuperation systems usually works as a pressure system, which increases the likelihood of exfiltration. Measurements done in a large climate chamber show that the diffusion of water vapor can overcome the infiltration flow of the outside air. The passage of moisture results in icing or condensation in the functional joint, which subsequently damages the surface treatment of the wood. Another serious defect is the subsequent mold growth.

1 Introduction

The problem of moisture transmission in gaps (leakages) in building envelopes is not a newly discovered issue. The moisture penetration through leakages in building envelopes has been investigated mainly in relation to slab blocks development. The leakages due to imperfect structural design and materials applied in roof envelopes formed another area for investigation. In our country, Prof. Ing. František Mrlík, DrSc. has dealt with research and moisture transmission in gaps of building envelopes [1]. Prof. Dr.Ing Gerd Hauser dealt with the similar problem area: moisture transmission through gaps, mainly for wooden houses in Germany [2]. While solving these problems the window structures were at a relatively low development level and had high leakage in terms of air infiltration and exfiltration. Due to these window characteristics as well as low requirements in energy efficiency field the relative humidity in buildings used to be approximately 30% [3]. By improving the thermal characteristics of building structures in relation to energy the relative air humidity has increased up to 50%, which is also given in STN 73 0540 [4 - 6]. Taking into consideration the change of boundary conditions and characteristics of window

[*] Corresponding author: adela.palkova@stuba.sk

structures the problem of water vapour condensation in functional gap between casement and window frame has occurred.

2 Moisture formation and transport

Indoor relative humidity is dependent on a number of factors, such as ventilation, vapour barrier in constructions, the ability to accept and deliver moisture and the source of moisture. Figure 1 shows a relative humidity graph in the living room for six model variants (processed in EnergyPlus). Variant 1 - basic alternative – gap permeability all windows is determined by the value of 0.0002 kg/(s.m.Pa), inner door 0.005 kg/(s.m.Pa) and exterior door 0.001 kg/(s.m.Pa), the evaluation sleeping zone are defined by two people daily from 23:00 to 07:00, the room is tempered to +20.0 °C. Room is oriented to the SW. Variant 2 - difference compared to the original alternative of variant 1 is in doubling the gap permeability of windows to 0.0002 kg/(s.m.Pa). Variant 3 - difference versus the original alternative of variant 1 is to remove, exclude humidity producers - people, model without counting people. Variant 4 - difference versus the original alternative variant 1 is the addition of morning and evening intensive ventilation of the room at the level of air exchange n = 3.0 l/h in the duration of 15 minutes, both in the morning and in the evening. Variant 5 - difference versus the original alternative of variant 1 is to complement the morning and evening intensive ventilation of the room at the level of the air exchange n=3.0 1/h in a duration of 30 minutes, both morning and evening. Variant 6 - differentiation from the original alternative of variants 1 is the addition of morning and evening intensive ventilation of the room at the level of air exchange n = 3.0 l/h in the duration of 60 minutes, both in the morning and in the evening.

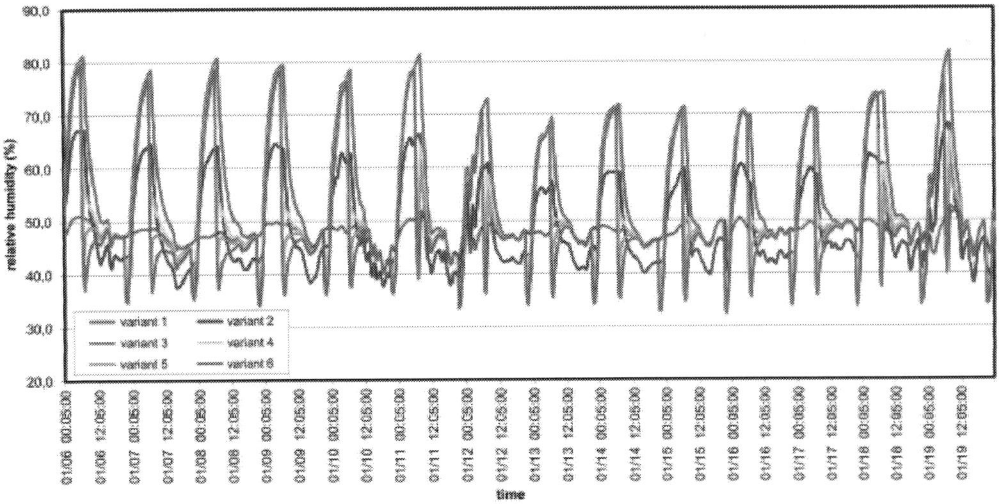

Fig. 1. Graph of relative humidity in the living room during 06.01. - 19.01.

Variants describe the real functioning of the existing buildings without the recuperation assembly. The values in the chart were determined to perform laboratory measurements on profile systems. The condensate generation in functional gap (see Figure 2 and Figure 3) is not described neither limited by the standards nor regulations. Seriousness of the problem varies depending on the material base. Considering the wooden window structures the degradation of surface finish occur and its thermal characteristics change.

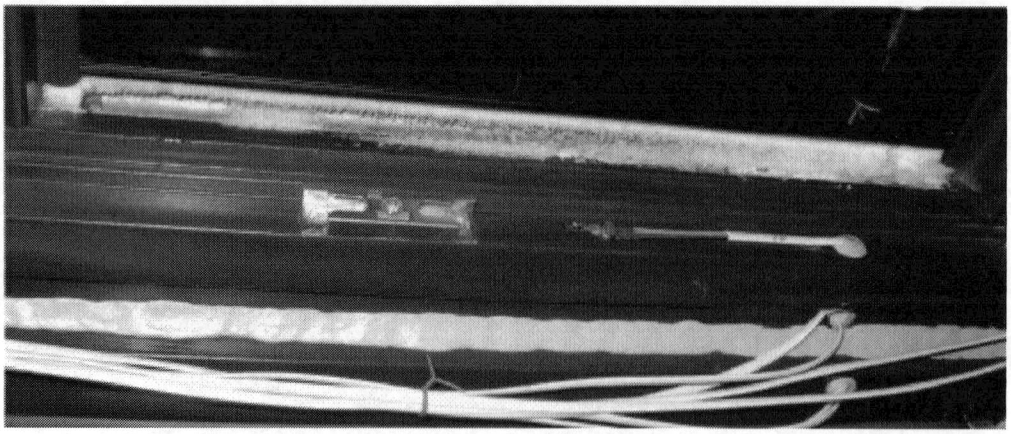

Fig. 2. Demonstration of ice coating and condensation in gap between casement and window frame at exfiltration and 200 Pa pressure difference (lower part).

Fig. 3. Demonstration of condensation in gap between casement and window frame at exfiltration and 200 Pa pressure difference (upper part).

The moisture transport through gap between casement and window frame at condensation temperature is the main cause of condensation of water vapour. The transport causes (see Figure 4):
- Air pressure differences (taking water vapour) – infiltration a exfiltration,
- Diffusion of water vapour (different saturation of water vapour in exterior and indoor air).

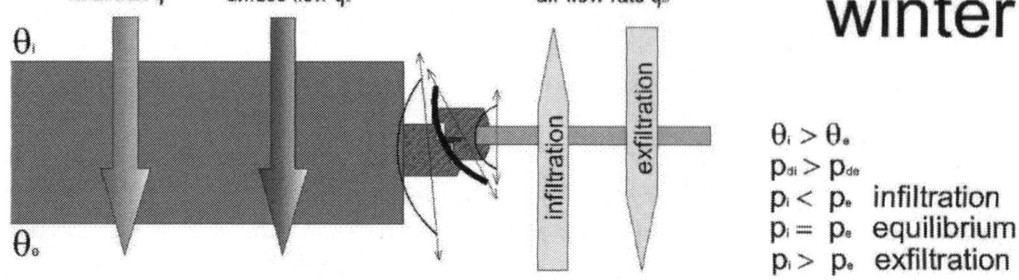

Fig. 4. Schematic representation of heat transmission, diffusion of water vapour and airflow through window structure [7, 8].

3 Laboratory measurements

The experimental measurements were carried out in heat engineering laboratories at the Faculty of Civil Engineering. Measurement models represent real window structures. The big climatic chamber illustrated in Figure 5 was used at measurements. The chamber A represents exterior climate (outdoor temperature – 11°C, pressure difference between exterior and interior from 0 to 2000 Pa, heat transfer coefficient 25 W/m²K, relative humidity 50%). The chamber B represents balancing chamber for HOT-BOX and simulate the indoor climate (indoor temperature + 20°C, relative humidity 50%). The HOT-BOX is used for measurement of heat transmission using measured element simulating the indoor conditions (indoor temperature + 20°C, relative humidity 50%, heat transfer coefficient 7.7 W/m²K). Part D is a masking panel, in which the measured window structure is imbedded. In case of our measurement the HOT-BOX was not used as the conditions with infiltration or exfiltration were to be investigated.

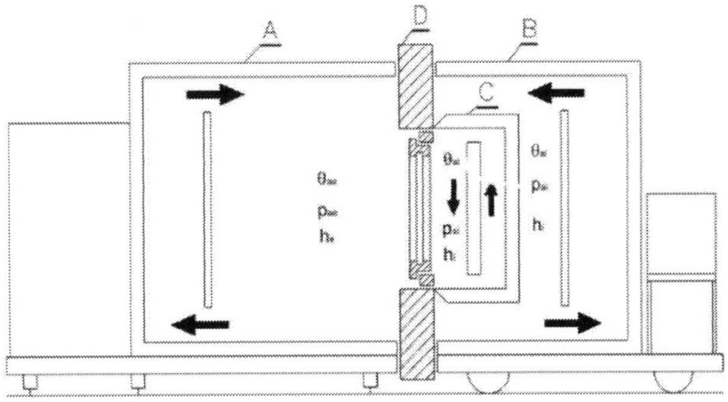

Fig. 5. Scheme of laboratory equipment of big climatic chamber [9].

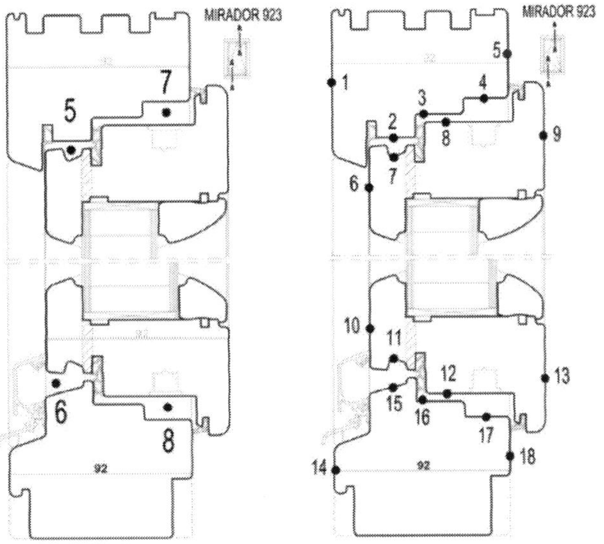

Fig. 6. Geometric parameters of window profiles and placement of measurement sensors.

Measurement sensors are divided into two groups. The first group measures the surface temperatures (PT 100). The second one measures the temperatures and relative air humidity (SHT 75). The sensors distribution is presented in Figure 6.

4 Results and discussion

At experimental measurements in big climatic chamber the occurrence and amount of water vapour and ice coating condensate was found out after each completed measurement phase. The internal surface temperatures and air temperature including relative air humidity in gap measurement points (see Figure 6) belonged into the second valuation data group. The measured values are presented in Figure 7.

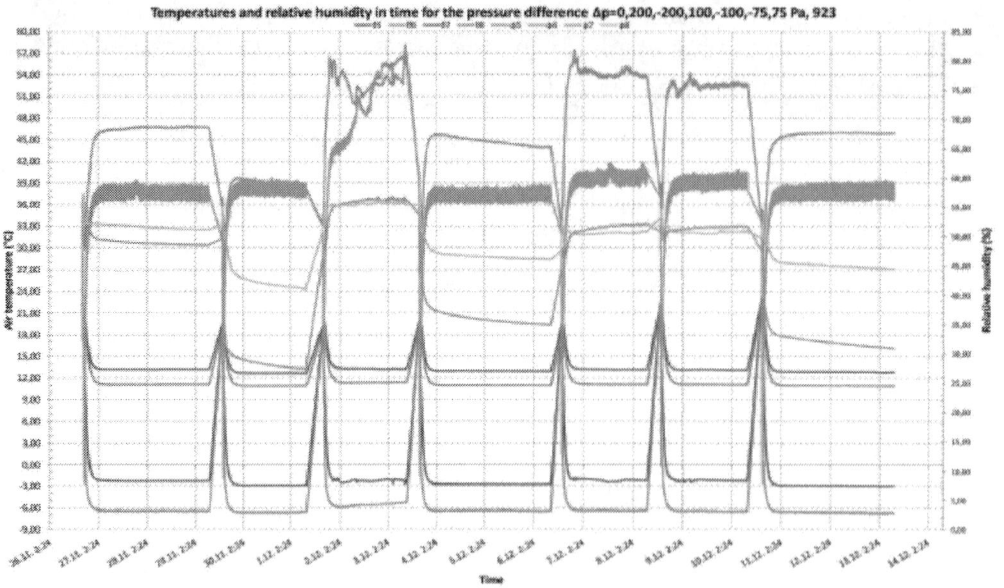

Fig. 7. Graph of temperatures and relative humidity in time for the pressure difference.

In research works carried out by Prof. Ing. František Mrlík, DrSc and Prof. Dr. Ing. Gerda Hauser on building envelope gap and joint characteristics as well as research works on windows (Dipl. Ing. Konrad Huber) it has been proofed that water vapour transported through building envelope causes problems [1, 2]. The results obtained in our research broaden and supplement the above given investigation works providing more comprehensive and more detailed knowledge of environmental physics bringing about water vapour condensation in functional gap between casement and window frame.

5 Conclusions

At an initial investigation stage it was known that condensation originated only in equilibrium state and at different pressure impact causing the exfiltration. However, it was found out during the measurements that the condensation also originates due to the infiltration. The condensation boundary for water vapour was being finding after the correct set up of circuit forging and, thus, also the sealing compression. After the measurements the value of differences of pressures between cool and warm chamber at which the condensate formation stops in the range 60 to 75 Pa. Interestingly, the condensation of water vapor in a functional gap improves the acoustic properties [10, 11].

This research was supported by Scientific Grant Agency MŠVVŠ SR and SAV under VEGA 1/0685/16. This research was also supported by Apol, s.r.o.

References

1. F. Mrlík, *Moisture problems of building materials and structures* (Alfa press, Bratislava, 1985)
2. G. Hauser, Ch. Kempkes, *Der Einfluss von windinduzierten Druckschwankungen auf das thermisch-hygrische Verhalten von durchstromten Leckagen* (Kassel, 2005)
3. I. Chmúrny, *Thermal protection of buildings* (Jaga group, Bratislava, 2003)
4. P. Oravec, AMR **899**, 184-187 (2014)
5. P. Oravec, Procedia Manufacturing **2**, 348-352 (2015).
6. *STN 73 0540-2 Thermal performance of buildings and components. Thermal protection of buildings. Part 2: Functional requirements* (2012)
7. *EN ISO 10077-2 Thermal performance of windows, doors and shurtters – Calculation of thermal transmittance – Part 2: Numerical method for frames* (ISO/DIS 10077-2: 1998)
8. A. Palková, M. Palko, Condensation of water vapor in the window structure. *Proceedings of the Conference ANSYS 2010* (Frymburk, 2010)
9. A. Palková, M. Palko, A. Puškár, *Proceedings of the 5. International conference. Buildings and environment 2010*, 204-209 (Brno, 2010)
10. D. Dlhý, AMR **855**, 233-236 (2014)
11. D. Dlhý, *ATF 2013:2nd Conference on Acoustics, Light and Thermal Physics in Architecture and Building Structures*, 21-24 (2013)

Comparison of informations from the brownfield catalog and the data needed to evaluate real estate

Zuzana Kramářová[1,*]

[1]Faculty of Technology, The Institute of Technology and Business, Okružní 517/10, 370 01 České Budějovice, Czech Republic

Abstract. At present, the topic of sustainable development of the area with brownfield revitalization is one of the current topics. The revitalization of each brownfield requires a completely individual approach, but the whole process can be generalized into several successive points: the identification and cataloging of the site, the search for the investor and the new suitable functions for the brownfield, the design phase of the project, the authorization processes for the realization of the project, realization of the project, new use of the site and realistic return on investment. One of the major challenges for revitalization is the search for an investor. This is due to many factors, among which the acquisition cost of the property, which is based on units of measurement, excels. The tool used to encourage potential investors can be, for example, databases - catalogs of existing brownfields. These contain clearly arranged basic information about individual locations. In view of the easier pricing of the brownfield price (its value), it would be advisable to align this information with the necessary valuation data. The paper therefore deals with comparing the information contained in the brownfield databases and the data needed to determine the brownfield price as a real estate.

1 Introduction to brownfields

At present, the topic of sustainable development of brownfield revitalization is one of the current topics. In the Czech Republic, this is unfortunately solved unusually and non systematically. It is affected by the current legislation, or rather its absence, in which there is almost no instrument to encourage investors to revive brownfields before building on a green meadow. In the Czech Republic brownfields are not defined or mentioned in any law or decree. The only such tool is the National Brownfields Regeneration Strategy of 2008. Unfortunately, this was not accepted by the Czech government, but only taken into account. [4]. At present, Czech Invest has been preparing a new strategy for several years, which could be submitted to the government again for approval. Without this document, the Czech Republic does not have any background that would clearly state the direction by which the whole issue and its solution should be managed.

* Corresponding author: 19511@mail.vstecb.cz

The first prerequisite for stimulating developers is to offer own brownfield sites. This is mainly done using brownfield databases / catalogs. Nor does there exist a single legislative background or methodology to implement these databases. The only legislation underpinned point is the inclusion of brownfields between so-called Recovery and Reclamation Areas, which are phenomenon 4 contained in the Subsidiary for the Analysis of Sustainable Development of the Territory, which is part of the Territorial Analytical Data for Individual Municipalities (hereinafter referred to as the TAD) [4, 7]. With this the area of legislative support in the Czech Republic is exhausted.

Because of the lack of legislative support, there is no uniform definition of brownfield. Since the author has dealt with defining the term several times [4-6] and many different authors with different definitions [7, 8], etc., the own definition will be used in following text.

"Brownfield is an unused or ineffective location. This site may consist of a part, one or more buildings or land, or a combination thereof. These properties are interconnected operationally, economically or territorially so that they form a whole. Brownfield or part of it may be contaminated."

The issue of brownfield revitalization is a multidisciplinary field combining many industries - from land-use planning, architecture and construction to environmental industries dealing with remediation of contaminated sites and nature conservation. However, revitalization of each brownfield requires a completely individual approach. This depends not only on the actual condition of the brownfield and its previous uses, but also on the local conditions (transport and technical infrastructure, morphology, limits and regulations in the territory, ...) and current needs and potential of the territory, including the socio-demographic structure in the area. It is just because of the environment that identical approaches to different localities can not be successfully applied.

Despite all the differences, the process of successful revitalization can be generalized into several successive stages - phases. The first one is the identification of the site and its subsequent cataloging, and the process continues with the search of the investor and the selection of a new suitable brownfield function. It is followed by the elaboration of the project documentation for the project completed by the permitting processes for the realization of the project. The next point is the actual implementation of the plan and the last stage is the period of new use of the site and the real appreciation of the return on investments.

One of the major challenges for revitalization is the search for an investor. This is due to many factors, including the acquisition price. In the Czech Republic, this is still less advantageous than investments on green meadow, so called greenfield. At the purchase price of the brownfield, the investor always pays not only the complete state of the buildings, but also the land is classified as building. Most brownfields in the Czech Republic are assessed in terms of the construction and technical condition as 3 (usually on a five-point scale) [5]. This means that significant investments in repairs are necessary. Another factor is the existing layout solution, which may not always meet the current needs and requirements of the current legislation, and therefore building modifications must be made to increase the initial investment. The last item is the price for your own land. This is almost always the price of a building land, which is, despite all efforts to increase the fees for withdrawal from the agricultural land fund from 2015 (Act No. 41/2015 Coll., Amending Act No. 344/1992 Coll.), still much more than the purchase of loose land, including withdrawal fees.

2 Brownfield database

In order to determine the state of the number and character of brownfields, it is necessary to elaborate their overview in the area. This is most often done in the form of an identification or localization study (hereinafter referred to as an identification study). Their output is usually a clear map of the processed area with a plot of the location of the identified localities, which is supplemented with an information database/catalog on individual localities. It is this catalog that can be used to facilitate the search for an investor, as it serves as a tender list in this case.

In the Czech Republic, as already mentioned above, there is no uniform methodology for the processing of identification studies nor the uniform form or content of brownfield databases. At the same time, there is no obligation for the administrative authorities active in the spatial planning of the brownfield site to record at all, except for the already mentioned phenomenon no. 4 in the UAP [4, 7], but it identifies the sites together with others and does not collect any additional information. Thus, if the identification studies in the territory were processed, they are based entirely on the requirements of the study commissioner and his needs - it is therefore impossible to compare them.

As mentioned above, the database collects specific information about individual locations. In general, the catalog/database is listed by content, followed by a header with summary information on individual municipalities in the area being solved, or the territorial administrative structure of the territory. In addition, the stated scales for individual criteria (for example, assessment of building technical status, designation of site types, type of ownership, etc.) are given in the summary header. Behind the headline is the list of brownfields. These are usually for the sake of clarity in the form of a table, supplemented by map drawings and photo documentation. The vast majority of catalogs are processed in the Czech Republic as one-stage, ie only with basic information slightly extended by some specifics requested by the client. These criteria are roughly 50-60 of the surveyed areas, such as: location, type of site, information on land, rough characteristics of the site as a whole, history of use, information from land-use planning documentation, ownership relations, transport and technical infrastructure, Conservation of nature and other limits, maps, photo documentation [6, 7, 9, 10]. On average, there are roughly four items on the information area, which is very low.

From the point of view of the amount of information suited to more detailed site characterization and possible selection, it is now appropriate to process these databases in two stages, ie basic information in the first stage and complementary or expanded in the second stage [6]. This will not lead to too much information being overloaded when you first get acquainted with the sites, but for the other needs in the shortlist, the data from the second level will not be missing.

For these reasons, the comparison of the data contained in the brownfield databases and the information needed for valuation of assets will be based on the recommendation for two-step databases that the author handled in a series of freely related contributions [5, 6, 9, 10].

3 Valuation issues

In the Czech Republic, many computational mechanisms are used for the pricing of movable and immovable property. The official methods used in legal assessments are laid down in Act No. 151/1997 Coll., On Property Valuation and on Amendments to Certain Acts (Act on Property Valuation) and Decree No. 441/2013 Coll., on Implementation of the Act on the Valuation of Assets (Valuation Decree). Because the pricing approaches are many and different, the next text will only be based on the procedures used by the law and the decree.

The law distinguishes three basic types of prices - the usual, extraordinary and detected price. The usual price is simply the sale price of a similar asset, including influences such as location, condition, etc., but without exceptional market circumstances, such as the seller's personal circumstances, natural disasters, etc. This is the value of the thing to be compared. The extraordinary price is, of course, the usual price including extraordinary influences. The price detected is then the price determined by law different from the usual and extraordinary prices [1]. The price detected under the Act on Asset Valuation is determined in several ways: cost, yield or benchmarking way and valuation according to the nominal value, according to the book value or according to the exchange rate or as the agreed price [1].

Furthermore, the law divides property valuation into property valuation (buildings, units, land, permanent crops, property rights to real estate), valuation of property rights, securities, other assets and business establishments. For the purpose of this paper, only the real estate sector, namely land, especially construction and agricultural, land-based buildings, including units in buildings, will be specified. Also, for the purposes of the comparison in the following text, only the price determined according to the valuation decree will be considered, and only the cost method will be considered for the choice of methods.

Interestingly, the Act on Assets Valuation in definitions and breakdown of terms such as construction and land does not rely on the definition of the Building Act and its implementing decrees, but defines them separately. Likewise, the division of buildings and land does not correspond to a common construction or urban structure.

Buildings are classified in the property valuation law as: land, civil engineering and special land, water reservoirs including ponds and other structures. As land developments are further divided into buildings, units and outdoors. Buildings are given the same status as "buildings" which are spatially concentrated and outwardly predominantly enclosed by perimeter walls and roof structures, with one or more bounded utility spaces" [1].

In contrast, the Building Law defines the construction as follows: "Building" shall mean all construction works arising from building or assembly technology, without regard to their construction, construction products, materials and structures, for use and duration ... The building that serves advertising purposes is a building for advertising." [2].

Also the concept of land is solved slightly differently. According to the Valuation Act, the division of land is as follows: "building land, agricultural land registered in the Land Registry (hereafter LR) as arable land, hop fields, vineyards, garden, orchards and permanent grasslands, forest land, which are forest land registered in LR and wooded wooded land, land registered in the LR as water bodies and other land" [1]. Among the building land according to the Act on Assets Valuation belongs: on unbuilt Land listed in the LR as built-up areas and courtyards, areas intended for building according to a valid territorial decision or other corresponding document, areas of the area registered in the LR as gardens or other areas forming a common unit with the building site, areas registered in the LR with building law, According to the LR, built-up land with real buildings not registered in the LR [1].

According to the Building Act, however, the building plot is only a "piece of land, its part or a set of land, defined and intended for the location of a building by a territorial decision or a regulatory plan" [2], ie land with valid permitting documentation, not so-called "Black building".

4 Hypothesis and Data Comparison

The underlying hypothesis is that most of the data needed for real estate valuation is already contained in brownfield databases. The research question then is whether other information can be added to a meaningful extent so that there is no enormous burden on the processor and there is no requirement for a more frequent update rate than the two-year intervals based on the UAP data update.

In order to assess the consistency or difference of the data recorded in the brownfields databases and the information needed for the calculation of the real estate price by the cost method according to the law and valuation of properties, the direct comparative method - comparison will best be used. For the sake of clarity, the element needed for valuation will always be taken into account and will be confronted with the data contained in the databases.

According to the valuation decree, there is a different procedure for determining the price of the land and determining the price of the building. However, since brownfields are in the vast majority of objects with land at least under their own ground plan, these criteria will be judged without distinction as to whether the criteria for calculating the price of the land or the building.

Data from brownfield databases (hereinafter referred to as databases) will act as data recommended for or data used in multi-stage databases [5, 6, 9, 10]. As their location within the overall database is given the division: data from the total header of the catalog and the data for the individual sites divided into the basic (primary, first degree, basic ...) and extended (additional, secondary, from the secondary stage ...).

According to the Valuation Decree, the price is always determined as the base price adjusted to the unit of measurement, multiplied by the number of units.

4.1 Comparable data:

- basic identification data (land number / number, municipality, street and descriptive number, GPS coordinates, ...) - in databases are given in almost identical content between the basic information, usually in the header of the card used to identify the brownfield
- area of land - in databases is given in m2 directly in the basic data, ie the first information step
- length (eg to determine the price for the fencing of the plot) - the data does not appear in the databases; It can be included in secondary information about land or location
- built-up area of the building - in some databases is given in m2, but not always; it is usually found between secondary information; the data can be included in the databases
- flooring - in databases are usually located in two sources: in the primary information in the description of the site, in the secondary information it is directly enumerated in the description of the individual objects as well as between the information concerning the fire safety of buildings
- floor height - in the databases are reported as the floor height between the extended information on individual objects
- floor height - in the databases are reported as the floor height between the extended information on individual objects
- built-up floor area, floor area - not mentioned in databases, only the total floor area, which is given in some databases for individual objects, in some as the total floor area for all the buildings together, is rather coarse and despite the name rather corresponds built-up floor area than a clean floor area; the data could be partly reconciled with changing the name of the item in the databases, but the precise calculation of the floor areas would be too demanding

- the building envelope (upper structure, under construction, roofing) - in databases it is usually given between secondary information to individual objects, but is not more closely divided, it is always the sum of m3 of the building as a whole, eventually only the sum of the enormous space of the superstructure and the roofing; the indication is inappropriate to align, it would be too demanding
- the built-up area of other building objects - in the databases it is given the same as the above mentioned for individual objects
- basic price of the listed municipality - not mentioned in the databases; it is not appropriate to supplement the amendment to the valuation decree, but it is possible to add it to the listed municipalities between the secondary information and the location of the site
- the size of the municipality (by population) - not mentioned in the databases; the data can be added to the overall header of the catalog as it will be the same for several brownfields in the territory and may change over time according to demographic developments
- economical and administrative significance of the municipality - the database is in some cases seemingly included in the characteristics of individual municipalities in the overall heading of the catalog, but its content is completely unresponsive; the harmonization of the content would be misleading due to the focus of studies and databases, since, in terms of valuation decree, it is not a division from a territorial point of view but a characteristic of the economic orientation of the municipality (spa tourism, tourism ...)
- the location of the municipality - in the databases, the data is included in the primary information in the form of a map drawing with a wider environment; however, the content does not correspond to the requirements of the valuation decree, which understands it as distance from economically significant settlements; it could be added to the overall header of the catalog
- technical infrastructure in the municipality - data in databases is not included directly but can be found between primary and secondary information about the brownfield technical infrastructure; data can be added to the overall header of the catalog
- civic amenities of the municipality - data in databases not included; it is appropriate to add it to the overall heading of the catalog, however, the structure required by the valuation decree does not correspond to the existing urbanistic terminology, so it is necessary to consider the content
- inclusion in the city area - not included directly in the database; the data can be added to the listed municipalities
- situation on the sub-property market - data in the database is not included; it is not appropriate to supplement the data because its characteristics require more specialized economic knowledge and knowledge of the current state of the market - the figure may change over time
- ownership relationships - the data are included in the database among the primary information but are fragmented into parts of the site categorization and ownership relations, as in the valuation decree there is also the basic characteristic of the land with the building, unbuilt land, co-ownership, ...; the data is not appropriate to directly harmonize; in the case of the introduction of the data area for the valuation it is possible to supplement it
- changes in the environment with impact on marketability - not included in databases; the information may be added
- the impact of legal relationships on marketability - in databases is implicitly included in the basic information on ownership of the site and the limits
- other non-restrictive effects - in databases is rarely included in the basic information in the notes; the data can be harmonized
- flood risk - the data is indirectly contained in the databases in the primary data between the limits; data can be refined and required in databases

- the geometric shape of the plot and the size of the plot - the data are indirectly included among the basic information in the area of the size of the plot and the drawing above the cadastral map, which clearly shows the shape of the land; data need not be harmonized
- sloping land and exposure - data in databases not included; they can be added to secondary information about land
- difficult baseline conditions - data in databases are not included; they can be added to secondary information about land
- protected areas and protection zones - in databases is included in the basic data between limits within the territory; it is not necessary to harmonize, it would be at the expense of the accuracy of the other limits described
- type and purpose of the construction - in the databases is contained in the primary information about the current or original use of the site
- the prevailing area surrounding the land and the environment - in the databases is only partially contained in the basic information, this can be read from the brownfield position drawings in the situational map; It is appropriate to supplement the primary information
- the location of the land in the municipality - in the databases indirectly included, can be read from the brownfield position drawing in the situational map; data can be added
- the possibility of linking land to engineering networks - in the databases included in primary and secondary information in the field of technical infrastructure
- civic amenities in the vicinity of the land - data in the databases is not included; it can be added to advanced location data
- transport accessibility to land - in the databases included in primary and secondary information in the field of transport infrastructure
- public transport - in databases included in secondary information in the field of transport infrastructure
- position from the point of view of commercial usability - not in the databases directly; The figure can not be added as it is linked to the use of brownfield, which may change
- population in the area - data in databases are not included; they can be added to secondary site information
- unemployment - data in databases are not included; they can be added to the overall header of the catalog
- unlisted effects - not listed in databases; can be included in notes in primary or secondary data
- parking options - listed in secondary information about transport infrastructure
- BPEJ - in the databases listed partly in the primary information when it is I. or II. class of protection; the appropriate addition, depending on the primary or secondary data on land
- categorization of the building type - in the databases partly contained in the primary information about the original purpose of the building; the information can be added between the primary information
- construction and equipment - in the databases included in the secondary information between the assessment of the construction technical condition; there is no need to harmonize
- the type of construction - in the databases included in the primary data in the description of the site as well as in the secondary data in the assessment of the construction technical state; there is no need to harmonize
- structure allowing the use of the attic - data in the databases is not included; it can be added between secondary information in the area of valuation or assessment of the construction and technical condition
- hydromelioration - partially listed in databases, the figure may appear in the primary information in a note or limits (invested agricultural land); it is appropriate to add between secondary information about land

- age of construction - not included in databases; it is appropriate to add information between the secondary information on the construction technical status or the extended site information
- estimated lifetime - data in databases not listed; it is not appropriate to replenish because it depends on the amount of investment invested in revitalization

The above comparison confirms the basic hypothesis, that in databases are or can be most of the necessary data for real estate valuation reported without the increase of personnel capacities - this is a total of 42 items contained data or data appropriate to supplement or modify / harmonize, making it the total volume of 87.5%. Unsupported and inappropriate for database insertion in any form are 6 items out of a total of 48, ie 12.5%. Reasons for not including these data in databases are always listed for each type above. This also resulted in answering the research question of possibilities of supplementation missing data into databases.

5 Concluding

Comparison of the data recorded in the brownfield databases and the information necessary for the cost method of determining the price of real estate pursuant to Act No. 151/1997 Coll. Shows one of the possible approaches how to generate or verify the utility of the collected data on brownfields. Of course, opinions on the amount of information in the database can vary, and it is possible to discuss the possibilities of inclusion or non-inclusion of the specific information, but the result of the comparison does not affect it fundamentally because most of the data is already included in the sub-blocks of both the primary and secondary levels.

Furthermore, comparison is one of the examples of how these databases can serve practical purposes. The real estate developer will have the vast majority of pre-prepared information, thus shortening the time needed to collect the input data and self-pricing.

And last but not least, this information can be used to improve initial information in the decision-making process of a potential developer when selecting a site for an investment plan - it is always true that more information helps to eliminate later complications that are common in building reconstructions.

References

1. *Law no. 151/1997 Coll., on Property Valuation and on Amendments to Certain Acts (Act on Property Valuation)* (Collection of Law, Prague, 1997)
2. *Law no. 183/2006 Coll., on Planning and Building Code* (Collection of Law, Prague, 2006)
3. *Decree No. 441/2013 Coll., on Implementation of the Act on the Valuation of Assets (Valuation Decree)* (Collection of Law, Prague, 2013)
4. Z. Kramářová. Procedia Engineering **161**, 290-293 (2016)
5. Z. Kramářová. *Brownfield?! Proč?! Problém nebo příležitost*, **65** (2014)
6. Z. Kramářová. *Ecological, functional, urban and architectural problems of brownfields in small towns*, **213** (Doctoral thesis, 2006)
7. J. B. Jackson, et al. *Metodika inventarizace brownfieldů v úrovni ORP*, **32** (2010)
8. V. Jankových *Workshop W1-410 Proceedings of workshop*, **28-30** (2004)
9. Atelier T-plan, et al. *Vyhledávací studie pro lokalizaci brownfields na území Královéhradeckého kraje* (The Main Project, 2006)
10. Atelier T-plan, et al. *Vyhledávací studie pro lokalizaci brownfields na území Jihočeského kraje* (The Main Project, 2006)

AUTHOR INDEX

EDP Sciences – Web of Conferences
17, Avenue du Hoggar
Parc d'Activité de Courtabouf
BP 112
F-91944 Les Ulis Cedex A
France

ISBN 978-1-5108-5613-4

12th International Conference on Axiomatic Design (ICAD 2018)

MATEC Web of Conferences Volume 223 (2018)

Reykjavik, Iceland
9 - 12 October 2018

edp sciences